To Lily
Enjoy and
best wishes,
Julie

Women
and the Machine

WOMEN *and the*

MACHINE

Representations
from the Spinning Wheel
to the Electronic Age

Julie Wosk

THE JOHNS HOPKINS UNIVERSITY PRESS BALTIMORE & LONDON

This book has been brought to publication with the generous assistance of the Alfred P. Sloan Foundation.

The Johns Hopkins University Press
2715 North Charles Street
Baltimore, Maryland 21218-4363
www.press.jhu.edu

Sections of chapter 3 previously appeared in "The 'Electric Eve': Galvanizing Women in Nineteenth- and Twentieth-Century Art and Technology," *Research in Philosophy and Technology* 13 (1993): 43–56.

Library of Congress Cataloging-in-Publication Data

Wosk, Julie
    Women and the machine : representations from the spinning wheel to the electronic age / Julie Wosk.
        p.   cm.
    Includes bibliographical references and index.
    ISBN 0-8018-6607-3
    1. Women — Effect of technological innovations on.   2. Women in art.
3. Technology in art.   4. Women in popular culture.   5. Technology — Social aspects.
6. Technological innovations — Social aspects.   7. Sex role.   I. Title.
HQ1233 .W67 2001
306.4'6'082 — dc21

00-011530

A catalog record for this book is available from the British Library.

Contents

Acknowledgments vii
Introduction ix

1 *Framing Images of Women and Machines* 1

2 *Wired for Fashion:*
Images of Bustles, Corsets, and Crinolines
in the Mechanical Age 45

3 *The Electric Eve* 68

4 *Women and the Bicycle* 89

5 *Women and the Automobile* 115

6 *Women and Aviation* 149

7 *Women in Wartime:*
From Rosie the Riveter to Rosie the Housewife 182

CODA *The Electronic Eve*
and Late-Twentieth-Century Art 231

Notes 239
Index 281

Color plates appear after pages 74 and 170

Acknowledgments

DURING THE COURSE of researching this book I was helped by the thoughtful comments of many people. I wish to thank current and emeritus members of the curatorial staff of the National Museum of American History in Washington D.C., including Bernard Finn, Peter Liebhold, Steven Lubar, Harry Rubenstein, Robert M. Vogel, Roger White, Helena Wright, and especially Robert C. Post for his gracious and kind words about my work. I am also grateful to the curatorial staff at the Science Museum in London for generously sharing their expertise, including Brian Bowers, Anthony Hayward, and in particular Wendy Sheridan and Michael Wright for thoughtfully answering my many questions.

I particularly wish to thank Robert J. Brugger, my editor at the Johns Hopkins University Press, for his enthusiasm and expert guidance in the production of this book, his assistant, Melody Herr, for her fine help, my editor at the Press, Joanne Allen, for her astute comments and great care in preparing the book manuscript, and Julie McCarthy, who so skillfully guided me through the production phase.

My warm thanks to my husband, family, friends, and colleagues, who offered their support throughout. I thank my husband, Averill (Bill) Williams, for his wonderful patience, love, computer expertise, help with illustration photography, keen eye for needed corrections, and good cheer. I thank my parents, Goldie and Joseph Wosk, for their enduring love and their inspiring ways. I am especially thankful to my father—who until 1996, after working for more than forty-eight years, was one of AT&T's longest-serving engineers—for his truly admirable ability to fix practically anything, for being a model of mechanical expertise, and for being a model, too, of energy, persistence, warmth, and caring, and to my mother, a superb writer of published essays and articles, for so skillfully teaching me how to edit, for being my writing mentor all these years, for seeing to it that I went to art classes at a very early age, for sharing our triumphs and pains, and for sharing with us, also, her own remarkable perceptions and gifts of observation. I thank my sister Toby Costas for her enduring generosity of heart, her empathy, and her witty rejoinders to my life's tales.

I thank my dear, late sister Paula Bhimani for her spunkiness and bright spirit, which will always be with me, as well as for stories of her brave ride alone in her Volkswagen automobile through the Khyber Pass in the 1970s, which inspired my chapter on women drivers who embarked on their own pioneering journeys.

I give my thanks to my friends, especially Sigmund and Elinor Balka, Philip K. Cohen, Robert Mark, Marlene Schwarz Phillips, and Sandra Stern, for their encouraging words.

I heartily thank the many museum and library staff members who took the time to help me with my work, including Lenore Symons, at the International Electrical Engineering Society in London, Jane Carmichael, at the Imperial War Museum in London, the staff of the Hagley Museum and Library in Wilmington, Delaware, Marie-Hélène Gold, photo archivist at the Schlesinger Library, Radcliffe College, and Ellen Kuhfeld and Elizabeth Ihrig, of the Bakken Museum in Minneapolis. I also thank Amy Beal, Alison Carter, Deborah G. Douglas, and Claudia M. Oakes, who read and commented on passages from the text. I especially wish to thank the library staff at the State University of New York, Maritime College, including Richard Corson and John Lee, for their expert help.

Finally, I am profoundly grateful to the Alfred P. Sloan Foundation and program officer Doron Weber for the generous grant that helped make possible the many illustrations in this book. I also wish to thank the Maritime College Foundation and the State University of New York chapter of United University Professions for their research and travel grants, and the Virginia Center for the Creative Arts, whose fellowship program provided me with a peaceful spot for summer writing in the Virginia countryside.

Introduction

THIS IS A BOOK with several stories to tell. One of them centers on the role of machines in helping women reconfigure and transform their lives—women living in a world in which cultural ideas about their abilities and appropriate behavior were being challenged and recast. It is also the story of images that capture this sense of transformation and change, often brilliantly bringing to the surface deep-seated attitudes about women themselves. And finally, it is the story of women using machines to help them redefine their own identities—women challenging cultural stereotypes that pictured them as timid and fearful, childlike and frail.

The idea of women being transformed by a machine appeared in an early satirical print by the British artist Robert Seymour, though here the change is simply superficial. In Seymour's 1830 print *A Reform*, Queen Adelaide, the wife of Britain's new king, William IV, turns the handle of a mill as a group of extravagantly dressed young housemaids descend into the machine and exit wearing plain aprons, "no longer dressed in silken sheen, no longer dec'd with jewels rare."[1] Earlier that year the queen had ordered all housemaids at Windsor Castle to tone down their extravagant hair styles and wear aprons, not silk gowns, and in Seymour's print the machine strips them of their finery and fashion frivolity.

Seymour's satire spoofed not only women but also royal edicts and nineteenth-century England's infatuation with new machines; however, although the women exiting the mill in Seymour's print look different, they are not very changed. But nineteenth-century artists and writers were also charting much more fundamental changes, including women leaving home to work with machines in the new factories and mills that were dotting the European and American landscapes. Later in the century, artists and photographers, as well as advertisers and writers, also began picturing the altered look and lives of female bicyclists and auto-mobilists, women whose change of clothes signaled a challenge to contemporary ideas about appropriate women's behavior, dress, and proper place in the home.

By the 1890s women wearing trouserlike bloomers and knickerbockers were pictured enjoying their new sense of freedom and mobility as they whirled away on their safety bicycles. In the early years of the twentieth century women dressed in motoring hoods and veils startled onlookers as they drove their new steam, electric, and gas-engine automobiles, and female aviators demonstrated their own skill in handling complex machines. These images made it clear that women were not only traveling far but also traveling fast as they countered cultural conventions and tested the limits of their circumscribed roles in the home.

Female artists during these years created their own images of their reconfigured sense of self. In her *AutoPortrait* the Polish-Russian artist Tamara de Lempicka presented herself wearing a driving helmet and sitting coolly at the wheel as she boldly eyes us through the framed window of a green Bugatti. Clearly this is a woman who is confident, a woman who defines herself — and her image — through the parameters of her modern machine.

During World War I and World War II women working with machines in defense industries demonstrated their ability to master jobs long considered the province of men, and again their change of clothes signaled their change of roles. Wearing factory overalls and welders' hoods, driving gloves and aviation gear, these women confounded stereotypes

Robert Seymour, *A Reform*, lithograph in the *Looking Glass*, 1 October 1830. England's Queen Adelaide turns the hand crank of a mill as fashionably dressed palace housemaids descend into the machine and exit wearing plain aprons. © The British Museum.

that envisioned them as machine-phobic and childlike without any technical skill.

Artists and photographers not only documented these changes but also promoted them as well, offering dramatic new images of women's technical capabilities during wartime. In J. Howard Miller's 1943 poster "We Can Do It!" a determined-looking aircraft-factory worker dressed in blue overalls and a red-and-white polka-dot bandanna flexes her muscles, while her pretty face with bright red lipstick is a reminder that this tough-looking female has not lost her femininity.

Miller's image, like Norman Rockwell's famed version of Rosie the Riveter, which appeared on the cover of the *Saturday Evening Post* in 1943, as well as a huge number of documentary photographs taken by the news media and American and European government photographers during two world wars, often presented women as strong and determined figures, Rosies hell-bent on demonstrating that they were capable of impressive mechanical expertise. These women, like so many female defense workers of their generation in America and Europe, showed a willingness to get dirty and master new technical skills.

But these heroic images of women taking on new roles and lives were only part of the story. The persistent doubts about women's capacities to understand, much less operate, machines, were another part. In nineteenth-century images women were often portrayed as a daffy bunch, fixated on cleanliness and baffled by machines. In the days when experimental steam carriages were first appearing on London roads, the British artist Henry Alken, in his satirical 1828 print *The Progress of Steam*, showed two well-dressed women turning up their noses in disgust at the smoke coming from a horse-shaped steam carriage being driven by a man (plate 1). Alken slyly suggests that these technologically challenged women do not even recognize that the carriage is a machine but think that it is indeed a horse, complaining that "the filthey fellow . . . feeds his horse with common coal."

British prints from the 1820s and 1830s often satirized the Age of Steam, and Alken's satire suggests the sense of disorientation that nineteenth-century men as well as women felt as they experienced a world of new machines. But Alken's print also reflects some persistent cultural convictions about women themselves. The two women are stereotypical females, aghast at the carriage's pollution and smoke, while the man driving the carriage is blithely unconcerned about his smoke-spewing machine. This and other satires often seemed to be saying that women were pathetically disconnected and alienated from the technological and that new, messy mechanical inventions were fundamentally foreign to their fastidious ways.[2]

Even when nineteenth-century women succeeded in showing these stereotypes to be wrong, their skill in bicycling and driving automobiles often met with deeply divided cultural attitudes and widespread signs of cultural ambivalence (shared by some women as well as men) about the very idea of women using the new machines.

The sight of women bicyclists and automobilists became a source of both wonder and deep dismay. From the 1890s through the early twentieth century, magazine and newspaper articles celebrating women at the wheel were countered by satirists' comic images of strapping "New Women" who carted their tiny husbands around in their bicycle baskets and maniacal female drivers, a menace on the road.

Even in women's magazines women were sometimes stereotyped in traditionally gendered terms, as childlike creatures hopelessly baffled by machines. The American illustrator Coles Phillips's cover illustration for *Good Housekeeping* magazine in 1913 shows a woeful-looking woman with a wrench in her hand sitting on the ground next to her automobile's flat tire, her pretty face picturesquely smudged with grease (plate 2). This charmingly helpless woman is clearly flummoxed by her flat tire and mystified by her machine.

There were other signs of cultural ambivalence. Advertisers for new machines tailored their texts and images to female consumers, luring women into buying machines like automobiles as a sign of their sophisticated skill and modernity, yet also continuing to cast women in an age-old guise, for example, using glamour girls in advertisements aimed at selling machines to men. Advertising their machines, Philadelphia's Pedrick Tool and Machine Company during the 1950s and 1960s featured photographs of attractive, leggy women, such as in an ad captioned "Pedrick Benders for Reinforcing Bars. So Simple That a Girl Can Run Them."

Sometimes the sexy images of women seemed almost accidental, or were they? In an intriguing photograph from the Cold War era a man working at General Electric in 1956 discusses a prototype of a G.E. atomic-powered jet engine while two pretty women listen nearby, including one suggestively straddling the phallic-shaped engine—an image suggesting that consciously or not, the photographer could not help but frame the woman as an erotic figure in the world of men and their machines.[3]

During the 1990s the alluring woman appeared in yet another guise, evoking a toughened look of mechanical chic. In 1992 a Michael Kors fashion advertisement in the *New York Times* featured a blond-haired model clad in a black leather jacket and curving tight skirt posed in front of a huge gear wheel, becoming a hard-edged industrial moll for the postmodern age (it is unclear whether this image was meant to appeal to men or to resonate with women themselves).[4]

These and other images often raise intriguing questions. Why were women so often portrayed as frivolous, silly, and mystified by machines? Why, when women were clearly undermining social stereotypes, did they continue to be ridiculed and visually grounded in humbling ways? Did these images reflect a more fundamental uneasiness about technological change and mechanization, an uneasiness felt by both women and men?

Were the images signs that men (and even some women) had mixed feelings about women's challenging conventions and mastering complex machines? Did some of the satirical images of women as huge Amazons on bicycles and wayward automobile drivers reflect men's unease and ambivalence about women themselves?

We might wonder whether the images of female drivers and bicyclists as silly and even dangerous were a reaction to the social threat women seemed to be posing to cultural conventions that assumed their proper place was in the home. To many men (and women too) these women may have seemed to be abandoning their proper maternal and caretaking

Models lend sex appeal to machine advertising, as in this advertisement from the 1950s for Pedrick Bending Machines, Pedrick Tool & Machine Company, Philadelphia, Pennsylvania. Smithsonian Institution Libraries. Used with permission of Pedrick Tool & Machine Company.

roles. The satirical images may also have reflected men's fears that mechanically skilled women were invading their turf.

Perhaps it seemed safer for men to portray women in stereotyped and culturally familiar ways than to accept women making strides in a new technological age. Portraying women toppling off their bicycles or sitting forlornly on the ground may well have been a way to keep women in their place.

These images also seem to reveal more profound concerns, including fears that women's quintessential feminine identity, as culturally defined, would become irrevocably transformed. These fears may have helped shape images of women as transformed into automatons or seductive simulacra through the agency of machines.

Sometimes the images of mechanistic, robotic women were actually more funny than frightening. In the 1860s, American and European women began reshaping their bodies by wearing the latest products of an industrial age: mammoth steel-cage crinolines and wired bustles that often required them to walk as awkwardly and stiffly as the century's newly manufactured dolls. In satirical caricatures, the women sometimes appeared robotic and ridiculous, when they were simply attempting to be fashionable and to fulfill their culture's maternal and sexual ideals.

In the work of early-twentieth-century dada and surrealist artists, with their spoofs of machine idolatry, images of women as machines were also more often sardonic than scary. The artists Francis Picabia and Marcel Duchamp limned images of mechanowomen as sparkplug and carburetor and, the quintessential emblem of eros and the machine, Duchamp's image of the mechanical bride, his *Bride Stripped Bare by Her Bachelors, Even (The Large Glass)* (1915–23).[5]

Recapitulating surrealist and dada themes in the 1970s, the American artist James Rosenquist in his large-scale painting *Gears* (1977) conflated images of shiny red lipstick, a string of women's pearls, and the gleaming metal surfaces of gears and a roller bearing in a sardonic view of a society entranced with glamour, technology, and sexual fantasies about woman as smooth-functioning machine (plate 3).

But other versions of mechanistic women presented much more troubling views, such as the women working in a New England paper mill in Herman Melville's nineteenth-century story *The Paradise of Bachelors and The Tartarus of Maids,* who have become slaves to industry and the Sultan machine. In part, these women in Melville's story reflect an ambivalence about the impact of mechanization on the lives of both women and men. During the nineteenth century, rapid industrialization and the introduction of new machines in Europe and America were often met with feelings of both elation and anxiety, including fears that mechanization and industrialized labor would profoundly alter human identity in both funny and frightening ways. One of the central metaphors to emerge in nineteenth-century critiques of industrialization was that of the industrial worker as both slave and automaton, men and women whose very identities and actions had become as rote and mechanistic as the machines they used.

But these images of mechanized women and female simulacra were often infused with an erotic charge that seemed to embody men's fears and sexual fantasies about women themselves. In Fritz Lang's 1926 film *Metropolis* the saintly Maria is replicated through an electrical machine, becoming a diabolical creature who is both seductive and dangerous. The new version of Maria suggests an important theme in works most often created by men: the anxious concern that women would become so transformed through new technologies that their very identity as women would become horrifyingly changed.

At this point it should probably be acknowledged that when thinking about the impact of machines on women's lives, we must be wary of the very idea of transformation itself. Discussions about women's changes in identity and identity transformation are a tricky business. The social impact produced by new technologies, women's identity transformation, and changing cultural definitions of gender identities and roles are actually the product of a complex interplay of social and cultural factors, not simply the result of the introduction of new machines.

As has long been recognized, the very idea of women being transformed by machines sometimes needs rethinking and qualification. For example, at first new electrical appliances like vacuum cleaners and washing machines were often envisioned as transforming women's lives by turning harried housewives into much happier beings. In her 1914 manual for women the British writer Maud Lancaster urged the use of new electrical cleaning and cooking appliances in the home and presented a utopian scene of a "hard-working husband" who is delighted when he returns home, for instead of finding "a 'neurotic' wife, worn out with the worries of housekeeping and domestic troubles, he will be welcomed by a loving woman, bubbling over with mirth and joy."[6] But later observers would seriously challenge this idea, arguing that rather than easing the drudgery of household chores, these new machines often only entrenched women more deeply into the domestic sphere and added to the tasks they were expected to perform.

During both world wars the ambiguities—and paradoxes—of women's transformation were often evoked through images of mirrors, masks, and camouflage. In some images created by government and industrial photographers, women were encouraged to picture themselves in new jobs and new roles, and their bodies were often mirrored in the shiny metallic surfaces of the machines they were helping to build, as in Alfred Palmer's photograph taken for America's Office of War Information in which an African American woman sees her own reflection as she uses a hand drill on a dive bomber at an aircraft factory in Nashville, Tennessee, during World War II (plate 4).

But women's transformation into machine-savvy workers was also considered somewhat deceptive. Even when artists and writers in wartime created images celebrating women's achievements, the subtext often made it clear that these women had not been permanently transformed. Though their change of clothes suggested that they had refashioned their identities, their new gear was often pictured as masking their truer, feminine self. In playful war stories and cartoons women camouflaged by welders' hoods and factory clothes were sometimes mistaken for men, but they soon revealed their true feminine identities, to men's surprise and delight.

Women were often reminded—and men reassured—that these women in war clothing were just taking on temporary roles. After the armistice ending World War I, the comic American magazine *Life* (not to be confused with the later photojournalistic magazine) featured an artist's drawing of female war workers, including an automobile mechanic in overalls and a motorbicyclist in pants, one of whom complains ruefully, "Girls, they're going to demobilize us, and we've got to go back to wearing petticoats and trying to be effeminate."

In World War II and after, magazine advertisers also suggested that

"Girls, they're going to demobilize us, and we've got to go back to wearing petticoats and trying to be effeminate," complains one of the women dressed for wartime work in this cartoon appearing in the comic magazine *Life* in 1919.

many women saw their war work as just a daytime role. In a 1942 advertisement for Camel cigarettes an American war worker, Betty Rice, wears a smock as she paints camouflage on miniature models of airfields. But the advertisement states that Betty's wartime role actually camouflaged her truer, more authentic self. In a nighttime view, "Camofleur Rice" appears glamorously attired as she entertains her uniformed male admirer, and the advertisement intones, "Morale experts say it's a good idea for women to be 'just women' every once in a while."[7]

During the postwar years, starting in 1945, demobilized female defense workers were asked to make another transition: to resume their primary role as housewives and to see their identities mirrored in new ways. Instead of envisioning herself as a machine maker or an operator in wartime, the woman in advertisements for Sunbeam electric appliances in 1950 was now pictured as a housewife and a consumer, seeing her own face, or that of her husband, reflected in the shiny chrome surface of her new electric coffeemaker.[8]

Rather than having their images mirrored by others, though, twentieth-century women artists would create their own versions of female identity and technology's impact on their lives. In Ovid's classical version of the Pygmalion myth the sculptor falls in love with his own creation, a beautiful image of a woman, and Venus grants him his wish, making the sculpted figure come alive. But female artists in the 1980s and 1990s seized for themselves the roles of both creator and transformer, using digital technologies to reanimate the archetypal image of Venus in the electronic age.

In the postmodern era these artists presented images of women that told another story, that of the paradoxes of being female in an era that blurred the boundaries between the simulated and the real. Their images of women as simulacra, mannequins, and dolls smile at us behind masks of glamour and beauty, masks that camouflage their secret, more authentic selves.

THIS BOOK IS not intended to trace the entire range of technological history or the full spectrum of socioeconomic forces that helped bring about technological change and social transformations. Instead, the chapters ahead look at the many ways that women and machines have been represented in written texts and visual images.

The focus is on images, from the beginning of the industrial revolution in the eighteenth century to the years after World War II, of women living in a world being transformed by the tremors of technological change, women working in early industries, women whose lives—and bodies—were shaped by new technological inventions, women using sewing machines and typewriters, bicycles, automobiles, and airplanes, women using new electrical appliances and being created by electricity themselves.

It also includes advertisers' images of alluring and seductive women, images that were used to help sell new machines to men as well as images that mirrored the hopes and fantasies of female consumers themselves. It considers female archetypes—women as goddesses of industry and femmes fatales, women as nurturing wives and mothers, women as child-like waifs. It looks at the ways these dramatically differing images of women and machines celebrate women's achievements yet also picture women in conventional and stereotyped ways, tellingly revealing the many complex, contradictory, even conflicted cultural attitudes about women and their abilities that remain to this day.

The representations are mainly of women in America and Britain, but there are also illustrations from France, Russia, and other countries as well. The wide array of examples includes paintings, satirical prints, cartoons, magazine illustrations, photographs, historical accounts, early treatises on electricity, a few chosen examples from fiction and poetry, machine manuals written by women, journalists' stories, and advertising illustrations and texts.

Some topics have been left for other writers, such as a concentrated study of images of women and machines in fiction and poetry. Also left to other writers are focused explorations of machine imagery in the work of Russian female artists like Liubov Popova and Valentina N. Kulagina and in the work of American, Russian constructivist, and Bauhaus female designers who helped shape the products of industry in the modern machine age.[9]

Ultimately, though, the wide range of ideas and the many representations of women and machines included here help to illuminate many of the central issues and questions raised when thinking about women and machines, and they may lead to other questions as well, such as whether new technologies have indeed contributed to dramatic changes in women's lives and whether the longstanding skeptical views about women's technical abilities have been finally and fundamentally changed.

In a cultural climate in which women have been saluted for their technical achievements yet also cast as technologically naive, do many women themselves still remain machine-shy and ambivalent about their own abilities? Do women who expertly drive automobiles and use their computers still turn to men to help them set up their computers, to chivalrously fix a flat tire, and to repair their errant machines (even though in this age of ever more sophisticated electronic systems it is usually difficult for both men and women to fathom much less fix complex machines)?

Finally, we may want to keep thinking about the role of representations in perpetuating social stereotypes about women and machines and in helping to refute these stereotypes as well. Have images of women as mechanically expert, as sexy lures in machine advertising, as terrible drivers, and as hopelessly unmechanical not only mirrored but also helped shape and even change public attitudes toward women and their abilities?

The pages ahead, with their sometimes funny and sometimes disturbing images, will offer a framework to help us explore these questions and controversies swirling around women and machines. Considered carefully, these images may not only reveal deeply divided cultural attitudes but also show us something about our own views and ourselves.

Women
and the Machine

FRAMING IMAGES

of Women and Machines

1 IN 1891 AMERICA's *Frank Leslie's Illustrated Newspaper* entertained readers with a revealing photograph representing a woman from the country's colonial history — a woman in a white ruffled cap sleeping quietly at her spinning wheel. With her head gently resting on the rim of the wheel, the woman looks weary but at one with her machine, content with her domestic role and her duties in the home.

But this romanticized and restful view gave absolutely no hint of a world already feeling the impact of tumultuous industrial and technological change or the changes that were already occurring in women's lives. By the 1890s women were operating machines in factories and mills, using typewriters in offices, and trying out electrical appliances in the home. During that decade women's imaginations were also being cap-

Tired, a sentimental view of a sleepy colonial-era woman at her spinning wheel. From *Frank Leslie's Illustrated Newspaper*, 15 August 1891.

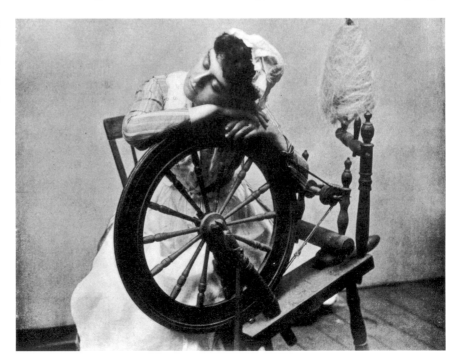

tured by new safety bicycles as well as newly invented steam and electric automobiles, vehicles that offered the possibility of increased mobility and a chance to test the limits of the pull toward home.

Women were clearly breaking new ground. Not content to just operate their machines, a few pioneering women would soon be teaching other women to service and repair their own automobiles. In a telling illustration from her early instruction manual for female automobilists, *The Woman and the Car*, the English motorist Dorothy Levitt lifts the lid on mechanical mysteries as she peers under the hood of her car to drain the crankcase to change the oil.[1]

American and European artists and writers, advertisers, and photographers skillfully charted the impact of these new technologies on women's lives, and their images often revealed conflicting attitudes about the dramatic social changes that were taking place, as well as deeply embedded cultural attitudes about women themselves.

Though stories and images made it clear that women could competently handle machines, there were also signs of an ongoing cultural ambivalence about women engaged in this new world of technological change. Images championing women's achievements often coexisted with

The British automobilist Dorothy Levitt provides instruction on draining the crankcase to change the oil. Photograph by Horace W. Nicholls, from Levitt, *The Woman and the Car* (1909).

biting satires mocking women's new skills. As contemporary magazines and newspapers celebrated the sight of women bicyclists and automobilists, other images, usually created by men, demonstrated that the unfamiliar sight of women riding bicycles and driving automobiles could be profoundly unsettling. Women bicyclists and early women drivers, like later women pilots, seemed to threaten established cultural conventions that considered technology primarily the province of men and fundamentally foreign to women's ways. To many, women who operated these new machines challenged deeply felt convictions that women's proper place was in the home.

Women driving their new automobiles were clearly on the move, and artists and magazine illustrators countered with images of women grounded — women sitting forlornly next to flat tires, women dispirited and made miserable by all things mechanical. In a world of fast-paced technological changes, in which women seemed to be moving too far too fast, these views of grounded women, again created by men, suggest a cultural longing to represent women as safely stationary and grounded by the familiar, widespread cultural beliefs about their paltry abilities and appropriately circumscribed roles.

The conflicting views of women in a fast-changing technological world and the persistent ways of viewing women themselves were often particularly vivid in three types of images: images of women and machines that incorporated popular cultural stereotypes and archetypes, including grounded women, goddesses, sirens, and runaway maids; images of women as consumers of new household machines; and images of women industrial workers up to World War I.

Grounded Women and Runaway Maids

■ Early on, women's expected place in the technological order was reflected in images like the nineteenth-century lithograph *Elopement Extraordinary*, in which a husband travels with his new bride on a fanciful steam-powered rocket to the moon. In this conventional view of proper gender roles the man steers while the woman rides behind as a passenger. (Many years later, this same idea reappeared in a 1950 automobile advertisement showing a happy housewife riding behind her husband on a rocket, a promotion for the "Rocket" Oldsmobile.)

In the 1860s a new type of image was beginning to appear; in it, women ride off on their own transportation machines and leave their husbands and their household duties behind. In 1869 the American printmakers Currier & Ives published their satirical lithograph *The Age of Iron: Man As He Expects to Be*, which prophesied a future in which women would ride away from their proper roles in the home. Several fashionably dressed women prepare to depart in their horse-drawn, iron-wheeled

Elopement Extraordinary, or Jack and his Lassie on a Matrimonial Excursion to the Moon, on the New Aerial Machine, undated lithograph. In comic nineteenth-century views women invariably sit in back while men do the steering. Science Museum / Science & Society Picture Library.

Currier & Ives, *The Age of Iron: Man As He Expects to Be,* lithograph, 1869. An early satiric prophecy of role reversal in which women do the driving and leave their harried husbands behind to do the housekeeping. Library of Congress, Washington, D.C.

carriage, leaving their harried husbands behind to sew, do the washing, and mind the children; a woman sits on the carriage, holding the horse's reins.

This comic image of dislocation in which men and women reverse their social roles, so lighthearted on the surface, reflects the anxieties that often lurked beneath the nineteenth century's pride in technological progress, the gnawing fear that new technologies would lead to explosive changes and disruptive times. In the Age of Steam fast-paced changes in transportation and invention were marked by a sense of speed and dislocation. Railroad systems were altering the travel landscape in America and Europe, and trains were often greeted with both excitement and fears about railway mishaps, boiler explosions, and the dangers of traveling at what were then considered excessive speeds, as shown in British artist Hugh Hughes's satirical etching *The Pleasures of the Rail-Road*, subtitled *Shewing the Inconvenience of a Blow up* (1831), which depicts the chaotic scene after a train's steam boiler explodes. Currier & Ives's lithographs later in the century celebrated new steamships but also presented fearsome images of frightening boiler explosions on board.

The Age of Iron also revealed the not-so-funny fear that dramatic social and technological changes would lead to equally dramatic disruptions in the domestic sphere, taking women away from their socially sanctioned roles in the home. Images of role reversal were a source of humor but also revealed signs of anxiety at the prospect that women might take on a new identity. In these views husbands often appear as helpless observers, no longer gazing at a woman as an object of love and

Hugh Hughes, *The Pleasures of the Rail-Road: Shewing the Inconvenience of a Blow up*, etching, 1831. In this nineteenth-century British print satirizing mayhem caused by an exploding steam boiler a woman rudely pushes her way out of the wrecked train. Elton Collection, Ironbridge Gorge Museum Trust.

desire but seeing her instead as a baffling figure readying herself for a new role.

Men's sense of women's shifting roles in the second half of the nineteenth century was also heightened by women's active rallying for their political and social rights in America and Europe. Centuries earlier, the artists Titian and Velázquez had painted images of a voluptuous Venus who not only looks at her reflection in a mirror but also becomes an object of the admiring artist's gaze.[2] But in nineteenth-century photographs and advertising women began to see their image reflected in new ways. In a stereographic image of role reversal titled *Women's Rights: The Rehearsal*, Venus has become a woman in a flounced dress who looks at her reflection in a mirror as she rehearses, perhaps, a suffrage speech. Her husband, with sleeves rolled up and arms immersed in a washtub, looks up in wonder at his soon-to-be-enfranchised wife.[3]

Like the Currier & Ives print and even *Women's Rights*, stereographic images during the bicycling craze of the 1890s suggested men's fears that women would soon abandon their household tasks and leave home. Spoofing the idea of women's newfound mobility and independence, comic stereographs presented role-reversal scenes in which formidable "New Women" rode off on their new bicycles, leaving their beleaguered husbands behind to do the family chores. Stereographs were two virtually identical photographs placed side by side on a piece of heavy card stock; this produced a three-dimensional image when seen through a special stereoscopic viewer. The comic images, like the one in *Women's*

Diego Velázquez, *The Toilet of Venus (The Rokeby Venus)*, oil on canvas, 1648. Venus, attended by Cupid, reflects on her own image. National Gallery, London.

Rights, were often staged scenes created by photographic companies, with models posing in theatrical roles. In one of these comic scenarios spoofing women cyclists a woman dressed in plaid knickers stands with her bicycle in the living room while her husband washes clothes in a tub. "Don't get the clothes too blue!" she instructs him audaciously as she gets ready for a ride.[4]

During the late nineteenth century female writers urged women to take solitary bicycle rides, and though American etiquette books advised proper middle-class women not to go on these rides unchaperoned lest they run into mechanical trouble or stray too far from social conventions, women did indeed go riding alone. Photographers' images of strapping women ready to ride off and leave their husbands behind suggested not only men's fears that women might become too independent but also the anxiety that women would abandon any sense of ladylike decorum and propriety.

At the end of the century, when American men and women went on unchaperoned bicycle rides in town and country, social critics fretted about opportunities for unsupervised romance.[5] They worried, too, that women and men would take sexual liberties, a theme that occurred often

Don't Get the Clothes Too Blue! stereograph, 1897. The "New Woman" takes her bicycle and leaves the laundering and childcare to her husband. B. W. Kilburn, Littleton, New Hampshire, courtesy of T. K. Treadwell.

in tales about technologies like the telephone and the telegraph, which offered new opportunities for courtship and flirtation.[6]

Artists also portrayed railroad cars as places for intimacy where men and women could feel free to engage in flirtation, courtship, and kissing, though these artists were usually amused rather than horrified at the idea. The British artist Abraham Solomon's painting *First Class: The Meeting . . . and at First Meeting Loved* (1854), depicting a young man and woman sitting in a first-class train carriage, quietly flirting in adjoining seats while the young woman's father sleeps in the corner, caused a controversy. The painting was criticized in the *Art Journal* and *Punch* for being too suggestive, and the artist appeased the critics by creating another version of his painting the following year. In *The Return*, the young female passenger smiles at her young male admirer, but her father, now fully awake, sits between the couple and talks to the earnest young man.[7]

Concerns about women's flaunting social codes also appeared in nineteenth-century stories about the railroad: male observers saw trains as a place for virtuous women as well as for women who abandoned any sense of propriety. During his visit to America in 1842, Charles Dickens traveled on a railroad to Lowell, Massachusetts, and noted approvingly that American trains had separate cars for women. He observed that "any lady may travel alone from one end of the United States to the other, and be certain of the most courteous and considerate treatment everywhere."[8] Yet Samuel Breck, in a diary of his railway travels published in America during the 1830s, complained that trains bred a loss of ladylike behavior and that women were apt to abandon their sense of decorum and gentility when they elbowed their way through crowds.[9] In a century

Abraham Solomon, *First Class: The Meeting . . . and at First Meeting Loved,* oil on canvas, 1854. A young woman in a first-class railroad carriage flirts with a male passenger while her father sleeps. This painting was controversial when it was first shown. National Gallery of Canada, Ottawa.

in which male satirists enjoyed twitting the carefully crafted facades of fastidious, fashionable women, Hugh Hughes's *Pleasures of the Rail-Road* lampooned unladylike behavior after an explosion, showing a woman with her skirt raised moving roughshod over the body of a fallen male passenger as she scrambles to exit the door.

Other artists portrayed women as reckless or inept, as in the Baltimore artist William G. Stewart's illustration *When Woman Drives*, of 1915, which shows a woman happily at the wheel, blithely unaware that her driving is terrifying her four male passengers.

Artists, advertisers, and writers also framed their images of women and machines in conventional ways, falling back on old assumptions about women as naive and cowering creatures who were baffled by the workings of both science and technology or only capable of handling simple machines. A Rauch & Lang automobile advertisement in 1910 insisted that "the most delicate woman—a 12-year-old child" could handle its machine with total ease.

Women themselves were ambivalent about their own abilities and worried that by being mechanically expert they might compromise their femininity. It was better, they sometimes felt, not to give up, or at least not to *seem* to give up, their reliance on men. Helen Bullitt Lowry, writing in the American magazine *Motor* in 1923, encouraged women to be independent drivers but also to feign helplessness if their automobiles needed repair: "We don't improve our popularity one sou by waxing efficient," she told her woman readers, adding pragmatically, "When trouble befalls you, assume an ignorance even if you haven't it. That's technique." Through mock helplessness the shrewd woman could easily receive aid from nearby men. As Lowry added, with what was probably unintended sexual suggestiveness, "A dozen male counselors have raised the lid and poked in the vitals of my engine."[10]

These dueling images of women as competent and incompetent, shy as well as assertive about their own technical abilities, independent as well as dependent on men, often suggested the difficulty of shaking off a long history of deep-rooted social attitudes about men's and women's inherent technical abilities. Men had long been portrayed as strong and technically able, women as frail and technically incompetent, or at least unsuited to engaging in complex technical operations. In eighteenth- and nineteenth-century industrial settings women were considered machine tenders, while men did the repairs. In a world of social conventions that relegated women to domestic settings, women were frequently denied training in mechanical skills, and their jobs were defined as unskilled and unmechanical. Not surprisingly, women often absorbed a conception of themselves as mechanically incompetent as part of their gender identity.[11]

Historically, the rare woman who demonstrated and relished her special technical ability sometimes felt defensive about her mechanical skills. In her remarkable *Handbook on Turning* (1842), on the use of a lathe for decorative work, first published in London and reprinted in America, the anonymous author, whom scholars identify as Mrs. Gascoigne, revealed some of the tensions experienced by the unconventional nineteenth-century woman who countered cultural expectations by demonstrating her mechanical expertise.

Before World War I, women did not often use machine tools. The small number of women who did use turning lathes were often wealthy women, such as England's Lady Gertrude Crawford and Russia's Catherine the Great, who used these machine tools in the service of art to create decorative objects such as snuffboxes and vases. Mrs. Gascoigne wished to encourage other women to use the lathe to imitate beautiful designs in wood and ivory, but she felt that it was necessary to reassure them that they could be competent without losing their femininity. "Why should not our fair countrywomen participate in this amusement? Do they fear it is too masculine and laborious for a female hand?" she asked her readers. She encouraged them to try using a lathe, asking, "What occupation can be more interesting or elegant than ornamenting wood or ivory in delicate and intricate patterns."[12]

Mrs. Gascoigne's book reveals some of the contradictions found in books written by women about machines: she draws on gender stereotypes about women's delicate touch while affirming women's special abilities and sense of self. She writes, for example, that "the taper fingers of the fair sex are far better suited than a man's heavier hand, to produce the requisite lightness and clearness of effect." Yet transcending the stereotypical image of female delicacy, the book's frontispiece includes an oval drawing labeled "The Author" showing a determined-looking Mrs. Gascoigne, who eyes the reader through her spectacles, an uncommon woman engaged in the world of machines.[13] The dual view presented in Mrs. Gascoigne's book—invoking cultural stereotypes about women's delicacy even as she offers them encouraging words—reflects the mixed cultural signals women both imparted and received about their own technical abilities.

From the very beginnings of the industrial revolution in the late eighteenth century artists often presented women as half interested, half afraid in the presence of science and technology. In the paintings of the eighteenth-century British artist Joseph Wright of Derby, men are engaged in mastering the mysteries of scientific experimentation, while women look on, fascinated yet fearful. Wright sometimes presented women as squeamish about new developments in science and technology, reflecting and reinforcing popular assumptions that they were timid about all things scientific and mechanical. In his 1768 painting *An Experiment on a Bird in the Air Pump* (plate 5), a family witnesses a scientific demonstration in which air is pumped from a glass receiver, causing a white cockatoo to begin dying. The moonlit night is tense with risk as air is about to be reintroduced so that the bird will live. The young women present at the experiment react with stereotypical female sensitivity. Anxious about the plight of the bird, they need reassurance from a kindly gentleman.

In paintings of iron-forge scenes — a theme given a mythic cast in the sixteenth- and seventeenth-century European paintings of Venus and Vulcan, with Vulcan working at the forge — muscular men work, while women are often simply attractive observers.[14] In Joseph Wright's painting *The Iron Forge*, of 1772, women watch warily or turn away. Reflecting Wright's own interest in science and technology, the forge is equipped with a water-powered tilt hammer, modern machinery for its day. Yet in the darkened space of the forge none of the females look with interest at the scene: the wife stands with her back turned to the heat, and one of her daughters clings to her waist, twisting away as she gazes outward.[15]

In art, as in actuality, forge scenes were largely a masculine domain, though women did engage in nail-making. In a very rare image, a medieval French Book of Hours dating from 1450–60 includes a small painting of a woman standing before a forge and using an anvil to create nails for Jesus's cross.[16] Before the early decades of the twentieth century the very rare woman artist who even tackled the subject of forge scenes or technology itself was considered an anomaly. In 1900, perhaps signaling a new era, the French artist Angèle Delasalle created her own painting of an iron forge.

Delasalle's work had been largely confined to conventional academic subjects — landscapes, portraits, and historical scenes — but her painting *A Forge*, of four workmen laboring in a smoky forge, was accepted for exhibition in the French Salon of 1900, where it was greeted with surprise.[17] In his catalog for the exhibit, Henri Frantz wrote that in her forge painting Delasalle was a maverick in her choice of subject and style. Frantz, who clearly had gendered notions about men's and women's art, praised her for her "vigorous" work: "Mademoiselle Angèle Delasalle's work is anything but feminine," he wrote, and he commended her for seeking "new fields for her studies while so many women-painters are contented with insipidly weak and finikin subjects."[18]

In eighteenth- and nineteenth-century Britain and France women who showed an interest in science and technology were sometimes viewed skeptically by men — reflecting once common, conventional notions of women as scientific naïfs, inherently unsuited to scientific occupations and inquiry because they lacked the necessary intellectual rigor and clarity of mind.[19] The French writer Boudier de Villemert, in *L'ami des femmes (The Ladies' Friend)* (1759), a book that was reprinted in English in London and America, insisted that although women could profitably study natural philosophy, they were "particularly not to meddle with the abstract sciences and knotty investigations, as such intrusions might cloud their minds." The idea that women might acquire any learning at all seems to have troubled men like Villemert, who insisted that "an increase of *literati* in petticoats, with their monades and scraps of Greek, is not at all to be desired."[20]

In men's satires, women interested in science often were portrayed as either silly or sexy rather than as serious students of the natural world. Two of the most important astronomical events of the eighteenth century were the transits of Venus in 1761 and 1769. Ridiculing women's scientific curiosity, the British artist Robert Sayer in his satirically titled engraving *Viewing the Transit of Venus* (1793) presents a voluptuous female sitting in a garden who eyes the stars through a portable telescope as a male admirer leers at her lasciviously and a sculpted satyr chortles in the shrubbery nearby.

These satires existed even though, as Londa Schiebinger has so deftly shown, scientific pursuits were accessible to wealthy women in seventeenth- and eighteenth-century Europe. Though excluded from universities, these women were encouraged to study natural philosophy and were renowned for their achievements in astronomy, mathematics, and physics.[21] Amid cultural conventions that limited women's access to equipment and universities and denied them scientific training, there were women whose intellectual curiosity and rigor enabled them to transcend these limitations.[22]

In artists' illustrations in eighteenth-century European treatises on electricity, women appear as eager participants in electrical experiments, like the women in Abbé Jean-Antoine Nollet's *Essai sur l'électricité des corps* (1746), who create friction by touching their hands to the spinning globe of an electrostatic generator.[23] In England, Mary Shelley was fascinated with scientific developments and galvanic experiments. As she wrote in the introduction to the 1831 edition of her novel *Frankenstein*, during the summer of 1816 she, her husband, Percy Shelley, and Lord Byron engaged in a ghost-story competition, and she listened while they discussed galvanism and the possibility of reanimating corpses, topics that may have helped shape her description, however vaguely stated, of the monster's creation in the novel.[24]

But though there were women who were intrigued by the changing world of science and technology, women were more apt to be portrayed in the more familiar guise of mythic or allegorical figures representing science and mechanics.[25] The print *Urania Coeli* (The muse Urania), engraved by Joseph Zucchi after a work by Angelica Kauffmann, presents a striking version of Urania, the muse who presides over astronomy. Wearing a wreath of laurel leaves, Urania holds dividers in her hand as she measures a constellation on an astronomical globe.[26] A century later, the American engineering magazine *Cassier's* regularly featured a classical muse of the mechanical as a decorative motif.

While women were elevated as muses of science and technology, they were also ridiculed in images that pictured them not only above but also below the fray. In nineteenth-century satires women were often portrayed as charmingly regressive and resistant when it came to scientific and technological change, empty-headed creatures whose romantic temperament was maladapted to the modern age. A cartoon published in *Punch* in 1892, captioned "Abominations of Modern Science," spoofs a woman clinging to outmoded ways. In a wry reference to Tennyson's medieval heroine in his poem "Mariana," the cartoon depicts Mariana, with her long, flowing hair, gazing in dismay at an incandescent light

Urania Coeli, motus scratatur et astra, engraving by Joseph Zucchi after Angelica Kauffmann, 1 January 1781. Urania, the muse of astronomy, uses dividers to measure a constellation on an astronomical globe. Science Museum / Science & Society Picture Library.

hanging overhead — a woman pitifully clinging to the past in the face of modern technologies. This mock Mariana returns to the moated grange and "finds to her sorrow, that her room is warmed by hot water pipes and lighted by electricity."

Reinforcing persistent cultural beliefs that science and technology were primarily a man's preserve, women in nineteenth-century art sometimes appear as uneasy witnesses to scientific and technological discoveries made by men. One of the myths surrounding James Watt's development of the separate condenser for steam engines was that he was inspired while observing steam escaping from the lid of a teakettle of boiling water.[27] In several nineteenth-century elaborations of this apocryphal story the young Watt stares intently at the boiling teakettle while a female observer stands nearby. In the 1849 engraving *Watt's First Experiment of Steam*, based on a painting by R. W. Buss, Watt's aunt, Mrs. Muirheid, stands sternly behind the table holding a timepiece. The print is captioned with a quotation from an early Watt biography, which portrays Watt's aunt as unsympathetic to his scientific curiosity, for she "upbraided him on one occasion for what seemed to her to be listless idleness — taking the lid off the kettle and putting it on again . . . watching the exit of the steam from the spoon spout, and counting the drops of water into which it became condensed."[28]

In what is probably a comic reference to the same tale, the British satirist Robert Seymour, in an 1829 etching titled *Locomotion*, presented two women happily riding in a fanciful teakettle-shaped steam carriage. Seymour's print takes comic aim at the new steam-powered vehicles of the day, including experimental steam carriages as well as steam railroads.

Woman as classical muse of the mechanical. From Current Topics, *Cassier's* magazine, 1890s.

The woeful Mariana, of Tennyson's poem, is dismayed by the power of electricity. "Abominations of Modern Science," *Punch*, 30 January 1892, 53.

Watt's First Experiment of Steam, engraving by James Scott after a painting by R. W. Buss, 1849. In an apocryphal anecdote, the young James Watt, the future inventor of steam engines, observes steam coming from a teakettle as his disapproving aunt stands nearby. Science Museum / Science & Society Picture Library.

Robert Seymour, *Locomotion*, hand-colored etching, 1829. In this British satire a man walks with his steam-powered legs while women steer their teakettle-shaped carriage powered by steam "made with a strong infusion of Gunpowder Tea." The Metropolitan Museum of Art, Gift of Paul Bird Jr., 1962 [62.696.15]. All rights reserved.

Theodore Lane, *Mechanics, Discovering Perpetual Motion (in a Wife's Tongue)*, etching, ca. 1820s. A satirical view of the inventor's wife as harpy. Science Museum / Science & Society Picture Library.

The gentleman on the left, reading a book called *New Inventions*, walks along with the aid of steam-powered legs. The man on the right flies in his steam-propelled flying machine, while the two women merrily go forward in their kettle powered by an infusion of gunpowder tea.

Women also were the butt of the joke in some satires ostensibly about politics and steam power. A hefty woman in the comic print *The Steam King*, of 1829, reaches far down to stoke the fires of a steam boiler that is topped with the head of King George IV, slyly labeled "Safety Valve." A nearby man cautions the woman, "Feed him gently or he'll burst and send us all flying," while he admires the view of her red-dressed buttocks prominently rearing into the air.

In other, more scalding satires women appear as a hindrance to mechanical invention itself. In a comic etching from the 1820s by the British artist Theodore Lane a hapless male inventor holding his head in anguish sits next to several of his contrivances, including his machine for mending, as his wife stands in the doorway scolding him while he works. Here the wife is a stereotypical sharp-tongued harpy and a maniacal mechanism herself, as suggested by the print's sly title, *Mechanics, Discovering Perpetual Motion (in a Wife's Tongue)*. Echoing the wife's vituperative outpouring, the painting on the wall in their home depicts a cannon blasting away at a group of fleeing warriors.

In nineteenth-century prints satirizing the age of mechanical invention women are turned into erotic figures, alluring yet also a dangerous sexual distraction. In his satirical engraving *Shaving by Steam* (1829) Robert Seymour takes aim at an age infatuated with new steam inventions by presenting a group of men sitting in a fanciful shaving emporium waiting their turn at a "shavograph," with mechanized brushes, lather makers, and razors. The automated machine requires that the sitter be "firm and steady," but one of the men has become so distracted by the female attendant that he allows the sharp razor of the machine to cut off his nose.[29]

The Goddess and the Siren: Women Celebrating and Selling an Industrial Age

■ In Seymour's *Shaving by Steam* the female attendant is presented as a seductive, distracting figure in a man's world of machines. Yet two familiar female archetypes—the alluring siren and the lofty goddess—played an important role in promoting new machines. These emblematic female figures were often used to celebrate and also to sell the new products of a burgeoning industrial age. The goddess archetype helped lend an aura of dignity, legitimacy, and stability to a world of rapid mechanization and technological change. Much less unsettling than women cyclists in their knickerbockers, these towering goddesses and sexy sirens cast not only technology but also women in comfortably familiar terms.

After visiting the exhibition of machinery at the Paris Exposition in

1900, Henry Adams, in his seminal chapter "The Dynamo and the Virgin," in *The Education of Henry Adams*, somewhat ruefully wrote about the role of the dynamo in America's industrial society, seeing it as supplanting the creative force once exemplified by the Virgin in a coherent medieval culture of architectural and artistic achievement. The dynamo had now become the symbol of a society devoted to new science and technology, especially in America, where "neither Venus nor Eve had ever held value as force."[30] But if neither Venus nor Eve was a reigning American deity, as Adams argued, larger-than-life mythic images of women did play a central role in helping to promote new technological developments in the modern industrial age, and advertisers' images of beautiful women often added glamour, excitement, and sex appeal to newly invented machines.

The use of mythic and erotic images of women to frame new technologies did not begin in the nineteenth century: decorative images of women had long been used to literally frame or ornament machines. An early version appeared in *Le Diverse et Artificiose Machine* (1588), a technical treatise written by the Italian military engineer Agostino Ramelli. Ramelli's book includes engraved plates illustrating fancifully decorated hydraulic and rotary water pumps.[31] In one illustration a man turns a pump crank as water pours in streams from the breasts of a mythic female figure at the spout, half woman and half serpent complete with scales and wings. Though decorative motifs were common in Italian mannerist art, here the figure of the female is more than ornamental: it also has a humanizing function, infusing an otherwise impersonal machine with maternal and erotic associations. Four hundred years later tiny paintings of mythic women would adorn the frames of nineteenth-century sewing machines.

But during the nineteenth century mythic female figures were much

Engraved illustration of a water pump fancifully ornamented as half woman, half mythical creature. From Agostino Ramelli, *Le Diverse et Artificiose Machine*, 1588, pl. 20.

more than decoration: they loomed large as goddesses of industry and the machine, becoming the apotheosis of modern industrial progress itself. Elevated rather than grounded, these towering women were larger-than-life goddesses of electricity, icons of industry. In some images the female figures are surrounded by new mechanical inventions as the women become the locus of a new economic order, a still point in a fast-changing industrial world. In 1900 the cover illustration for E. T. Paull's sheet music *Dawn of the Century* featured a classicized goddess with an electric bulb in her hair surrounded by seminal inventions of the age: a sewing machine, a camera, a railway locomotive, a telegraph, a telephone, an electric generator, and an automobile (plate 6). Huge sculptural women reigned over the Centennial Exhibition in Philadelphia in 1876 and the Columbian Exposition in Chicago in 1893.

These towering female figures have been characterized as universal abstractions, representative of familiar sculptural conventions in which women appear as personifications of civilization, virtue, and progress.[32] Serafina Bathrick, in "The Female Colossus," viewed the classicized figures as representing nineteenth-century efforts to lend legitimacy to new technologies by associating them with the prestige and traditions of classical antiquity. As "inspirational nymph" and "great mother," the sculptural figures represented an "earthbound nurturant guardian whose symbolic presence in the new industrial city or exposition was anchored in an ancient culture."[33]

But other huge sculptural figures were much more than mother figures and nymphs. Rather than being models of motherhood or em-

August Kiss's zinc sculpture of an Amazon tackling a fierce tiger, displayed at New York's Crystal Palace exhibition in 1853–54. From *Illustrated News* (New York), Crystal Palace supplement, 23 July 1853.

bodying contemporary definitions of femininity (like the dainty, decorous, wasp-waisted female so fashionable during the era), the massive
statues were often huge Amazons with features usually associated with
masculinity—broad shoulders, a square jaw, powerful arms and legs—
like August Kiss's large figure of an Amazon cast in zinc exhibited at
London's Great Exhibition in 1851, New York's Crystal Palace exhibition
in 1853–54, and, later, at America's Centennial Exhibition. They were
goddesses of war, Athenas of industry, carrying spears, staffs, and shields.

Images of goddesses also helped celebrate and sell new transportation
machines. A 1900 poster advertising Liberator bicycles featured a wide-
hipped, bare-breasted goddess with a winged headpiece and legs wrapped
with straps, a formidable figure conjuring up strength and power. From
1890 to 1910 monumental figures of women were also featured on posters
advertising early automobile exhibitions. The French artist Georges
Rochegrosse's lithographic poster for the exposition of the Automobile
Club of France in 1901 shows a goddess looming upward as she presides
over the scene in her strapless dress, her bustier made of sprocket wheels,
and her headdress ablaze with electric lights (plate 7). One year later the
club's poster figure had become a formidable art-nouveau deity sitting
with one hand on an automobile steering wheel, her neo-Grecian dress
ornamented with a sprocket wheel as a brooch (plate 8).

While images of women as goddesses helped lend stature and strength
to emerging technologies, by the early twentieth century women began to
appear in another familiar guise—as seductive lures in factory-machine
advertising. More decorative than mythic, they are shown lounging in
lavish dress or with their bodies draped alluringly around machines, early
versions of what postmodern critics in the 1990s would identify as a
conflation of woman-as-spectacle with commodity culture.[34] In a color
lithograph ca. 1901 advertising Knowlton & Beach box-cutting machinery manufactured in Rochester and London, a sensuous art-nouveau female with a curved body leans elegantly over a framed image of an end-
staying machine (plate 9).[35]

The transition in women's images from goddess to temptress can be
seen in advertisements for incandescent lights from the 1880s through the
1930s. In the earlier era of oil lamps *Godey's* magazine in 1860 had included an engraving entitled *The Light of Home*, in which a mother sitting
with her three children near a fireplace hearth and oil lamp exemplifies
moral virtue and woman's appropriate role. Articulating for its middle-
class readers the commonly held view of men's and women's proper roles,
the magazine identified women as the locus of domestic, civilizing, and
religious values, while men were assigned the freedom to venture out into
the wider world: The man's role was to "shake the senate and the field,"

while the woman was the "inspirer and the exemplar of the most heroic virtues." As "the light of home," she was "the wife and mother, the centre of the family, the magnet that draws man to the domestic altar."[36]

But in the 1880s, though gender conventions remained the same, the phrase "light of home" began to take on new meaning. Thomas Edison had developed the incandescent light, and now advertisements portrayed good mothers as those who sat with their families in homes lit by electricity.[37] To advertise their new electric lights, manufacturers at the end of the nineteenth century again used images of women as goddess and mythic figures. Drawing on visual conventions of mythic women holding a torch to signify freedom, liberty, and justice, advertisements featured artists' emblems of classically draped or nude women holding a light bulb or a bolt of electricity.[38]

Early in the twentieth century women's images in electric-light advertisements changed more dramatically: while still holding light bulbs aloft, the women were now exotic and seductive. Images of vamps—

General Electric's Mazda Lamp advertisement ca. 1910 featuring a seductive woman in neoclassical dress. The Schenectady Museum, Hall of Electrical History.

sultry, naughty, daring temptresses—became popular fare when the American actress Theda Bara appeared in the movie *A Fool There Was*, based on Rudyard Kipling's story *The Vampire*, in 1915, helping to launch a sequence of movie vamps, including the film actress Gloria Swanson. General Electric, which began marketing its new Mazda incandescent lamps in 1909, featured illustrations of vamps in its commercial calendars, in which women appeared dressed in exotic Middle Eastern costumes, reinforcing the company's oriental theme (Mazda was the Persian god of light). Continuing to invest electrical advertising with erotic overtones, a promotional General Electric photograph presented a woman in exotic dress suggestively reclining on her rug-covered chaise.[39] In another version, General Electric conflated goddess and siren in its promotional photograph showing a woman in French neoclassical dress reclining on a chaise, her head and chin slightly lowered. As she eyes the viewer in a coy, seductive pose, the light in her hand is clearly an afterthought.[40]

Seductive images of women and machines would appear in the de-

Man Ray, *Érotique voilée* (Veiled erotic), 1933. The surrealist artist Meret Oppenheim suggestively intertwines her body with a lithographic press. © 1999 Man Ray Trust / Artists Rights Society, NY / ADAGP, Paris.

cades ahead as well, not always in advertisements but as embodiments of modernity—from Georgia O'Keeffe's suggestive, long-fingered hand on an automobile tire rim in Alfred Stieglitz's 1933 photograph to Man Ray's more overtly sexual photograph *Érotique voilée*, also dated 1933, in which the nude figure of the surrealist artist Meret Oppenheim grasps the hand-wheel of a lithographic printing press, its phallic crank protruding from her body.[41] With her ink-blackened hand on her forehead, she becomes a modernist icon of woman and the machine—and an emblem of art and reproduction in the mechanical age.

Selling the Idea of Women and Machines

■ During the nineteenth century and continuing into the twentieth, mythic female figures were often used to embody—and sell—a modern industrial age to both women and men, to make new industries and machines seem enthralling and exciting. And images of erotic women helped sell machines to men. Yet recognizing women as a lucrative market and as likely users of many new home and office machines, advertisers during the late-nineteenth-century rise in product advertising included images of women in their promotions to help make new machines palatable to women or at least to make clear that these machines were intended for women to use.

These advertisements and stories—selling women on the idea of machines and selling men on the idea of women using machines—were again shaped by conventional ideas about women's inherent nature and proper social roles. Infused with stereotypes, they became a revealing reflection of pervasive cultural attitudes about women's mechanical abilities and about women themselves.

Though product advertising in nineteenth-century America was largely confined to "reason why" texts that explained technical features and product uses, early advertisements, drawing on conventional ideas about women's abilities and needs, tried to make machines seem appealing to women. Images of women also helped gender the machines, promoting the idea, for example, that early typewriters were largely intended for women's use (plates 10 and 11).[42]

The American printer Christopher Latham Sholes, with the help of Carlos Glidden and others, developed the first commercially viable version of the typewriter, and the Sholes & Glidden typewriter was marketed by the arms manufacturer E. Remington & Sons starting in 1874. At first the machines were viewed as gender neutral, though by the 1880s women were the primary typewriter operatives, largely because they worked for less pay than men. The work was considered a respectable occupation for middle-class women, and typewriters did indeed help women gain entry into what were considered respectable business jobs.[43]

The Remington company's first typewriter catalog included individual woodcuts of a man and a young woman working at the machines, but later advertisements for Remington and other companies almost always showed a woman operating the machine. Early advertising images were often based on a photograph of Sholes's daughter Lillian sitting with one of his experimental machine models in Milwaukee about 1872.[44]

Images of women at the typewriter helped not only to gender the machine but also to promote the idea that women could gain new ground by being employed in respectable jobs in business and could work comfortably in a man's world, though the lurking image of the sexy female was never far away, seen in early comic postcards showing compliant female typists and their amorous bosses. (Years later, in 1947, the fusion of glamour and dignity was the subject of the Hollywood movie musical *The Shocking Miss Pilgrim*, starring Betty Grable as the featured typist—said to be patterned after Lillian Sholes—who had the "shocking" ability to work in an office alongside men.)

Nineteenth-century stories and advertisements for new machines, including electric vacuum cleaners, promoted the machines for women's use, arguing that they were easy enough for women to handle and comfortable and safe enough for them to incorporate into their homes. Some stories also managed to make the women look, if not sexy, at least at-

Left, Lillian Sholes, daughter of the American typewriter inventor Christopher Latham Sholes, with an experimental model of her father's typewriter in Milwaukee ca. 1872. Courtesy of Hagley Museum and Library.

Right, Women in industry were often assigned tasks that required precision work. Here a woman at America's Wheeler & Wilson Manufacturing Company works at punching out sewing-machine needle eyes. *Scientific American,* 3 May 1879.

tractive to men's eyes. A *Scientific American* story introducing the new Rhyston mangle in 1879 cited the manufacturer's claim that the newly designed mangle would do "more work with less labor" and featured an engraving of an elegantly dressed woman pulling delicately on the mangle's handle with one hand while gently lifting her skirts with the other.[45]

Including the manufacturer's claims in the Rhyston story suggested the blurred line between promotion and information that was invading *Scientific American*'s stories in the nineteenth century—a blurring that helps to explain why these images often created a glamorous portrayal of women working in industry and middle-class women using the new machines at home (the magazine's editors were probably also presenting images of attractive women for its male readers).[46] In its American Industries story on the manufacture of a new sewing machine the magazine included a drawing of a factory woman dressed in an elegantly draped, frilly dress sitting poised at her machine punching out sewing-machine needles with a vase of flowers nearby. This image of a poised working-class woman in her ruffled dress promoted the somewhat incongruous idea that for women, manufacturing sewing-machine parts itself was an elegant enterprise.[47]

While Mariana in *Punch*'s satirical cartoon unhappily discovered her

This woman's casual pose is meant to suggest that using her foot-operated electric sewing machine is easy. "Sewing Machine Run By Electric Motor," *Scientific American*, 19 March 1904, 232.

home being warmed by hot water pipes and lighted by electricity, the *Scientific American* in 1911 also promoted the idea that women could comfortably—and safely—incorporate new electrical appliances into their lives, as suggested by a photograph of a woman keeping warm with an electric heater.[48] By the 1890s, with improvements in halftone illustrations, the *Scientific American* was using staged photographs of well-dressed women who happily welcomed the new machines into their homes. (The well-dressed women using electrical appliances also reflected an economic reality: before the widespread availability of home electrification in America, during the period 1880–1910, the market for new electrical appliances was only 5 percent of American homes. The electrical appliances that were available were expensive, the province of the wealthy.)[49]

The photographic tableaus also reinforced the idea that the new appliances were easy enough for even women to operate. In the story *Electricity in the Household* (1904) the magazine included a photograph of a woman in a white shirtwaist blouse sitting with studied casualness in an engaging but artificial pose. Her legs crossed and one arm casually draped around her chair, she is relaxed as she operates her new sewing machine with its foot-powered electrical motor and guides the cloth under the machine's needle with apparent ease.[50]

In another of the magazine's images a formally dressed woman demonstrates her new electric carpet sweeper in her parlor to a female friend, who sits admiringly on a piano bench while her young daughter sleeps. In this constructed view the visitor smiles, albeit self-consciously, and her sleeping daughter suggests the quietness of the machine.[51] But though these images suggested that machines were labor-saving, Ruth Schwartz Cowan and others have long argued that machines ostensibly designed to ease the drudgery of housework only served to increase women's household tasks.[52]

Women Transformed   ■ Early-twentieth-century advertisements not only promoted new appliances for women but also encouraged the idea that their use would transform women's lives. Electrification occurred slowly in Britain, but in America home electrification and domestic consumption of electricity increased quickly in urban areas after World War I. In America, household appliances became more readily available and less expensive after the war, and no appliances sold better than electric irons and vacuum cleaners.[53] Lightweight, portable domestic electric vacuum cleaners became popular after Hoover began marketing its Model O in 1908, a machine made to look glamorous with its swirling gold art-nouveau motif.[54] Advertisements for vacuum cleaners promised that the machines would transform women's lives by easing their tasks, but again, rather than making carpet cleaning easier, the new machines only added to women's household duties, especially since standards of cleanliness were also raised. Cleaning carpets was no longer an annual or biannual task done by men and boys, who took carpets outside and beat them; instead, vacuuming carpets became a weekly activity for women.[55]

Some advertisements went even further, promising to make women freer, even glamorous beings. In these ads, women were asked to identify themselves — and their fantasies about freedom — with the new machine. A 1910 advertisement for the Frantz Premier Electric Cleaner Company in Ohio featured a woman dressed in a household apron talking on the telephone. But holding out the alluring possibility of transformation, signaled by a change of dress, the woman's image is mirrored in the gleaming surface of her machine, showing her not in an apron but dressed in a coat and ready to leave. By 9:00 A.M., the advertisement promised, her work would be done, freeing her to venture out to newer climes.[56]

The power of machines to transform women's lives and identities was also promoted in early advertisements for home sewing machines. Isaac Merrit Singer intended his 1851 patent model sewing machine for industrial use, but he quickly saw the potential for the domestic market, and in 1859 he introduced his "Letter A" Family Sewing Machine, competitive with the Wheeler & Wilson family machine.[57] In the booklet *Genius Rewarded; or The Story of the Sewing Machine*, written by John Scott for the Singer Manufacturing Company in 1880, the grim lives of women who sewed by hand were contrasted with the rapturous lives of women who had switched to the sewing machine. Singer's saga opens with a mournful illustration of a woman sewing with needle and thread in the attic while her young son sits nearby, his head in his hands. Accompanying this sad view is Thomas Hood's dolorous poem, "The Song of the Shirt," telling of a woman dressed in "unwomanly rags" who works with "fingers weary and worn, / With eyelids heavy and red," as she plies her needle and labors "in poverty, hunger and dirt."[58]

Promising an alternative to this grim view, the Singer booklet tells of the transforming effects of the machine. A second illustration, envisioning "Happy Homes — Effects of the Answer," shows a now cheerful woman sitting next to her machine surrounded by her happy children. The sewing machine, as the book rhapsodizes, is clearly a boon, giving women "countless hours" of leisure for "rest and refinement" and offering them new avenues for outside employment. Using the expansive, optimistic rhetoric of the nineteenth-century belief in industrial progress, the booklet intoned that whatever "brings added comfort to the matron and the maiden; whatever saves the busy housewife's time and increases her opportunities for culture," and "whatever lifts any of the heavy household burdens" also contributes to "the highest and best progress of the world."[59]

Sewing-machine companies also held out the promise that by using their new products women could become not only more relaxed housewives but also glamorous beings, as seen in a nineteenth-century trade card for the New Home Sewing Machine Company in Massachusetts and New York City. (Trade cards were small, postcard-size retail and wholesale advertisements produced in America from the 1870s through the 1890s with color lithographed images on one side and advertising texts on the other.) On the New Home Sewing Machine card, a woman peers into her huge parlor mirror and sees herself as she would appear in the future, formally dressed with a fan in hand. The image of Venus gazing at her own image has been replaced by that of the elegantly dressed woman admiring her own reflection, now accompanied by that of her equally glamorous new sewing machine.

Nineteenth-century writers also envisioned the sewing machine as the means to transform women into more technologically knowledgeable beings. *Godey's Lady's Book* in 1860 suggested that sewing machines would broaden women's knowledge of "mechanical powers" and help "wives and daughters become enlightened upon a subject now dark to them."[60] During the next century women's ability to operate sewing machines would become a measure of their potential to learn other mechanical skills. In a *Scribner's Magazine* article in 1915, "The Woman at the Wheel," Herbert Towle argued that women who could operate sewing machines could certainly learn to operate small gasoline-powered automobiles.[61] In 1918 the *Sphere*, in Britain, linked women's sewing-machine expertise with their ability to successfully engage in war work, and during World War II magazine stories and advertisements told female defense workers that if they could operate sewing machines, they certainly had the ability to operate drills, punch presses, and rivet guns.[62]

Early medical stories portrayed sewing machines as a means for turning women into healthier beings and were infused with conventional social assumptions about women's inherently frail bodies and temperaments. In 1857 the American sewing-machine manufacturer Grover and Baker sent members of the medical profession an article that had appeared in the April issue of *American Medical Monthly*. In an era when women were often seen as suffering from nervous maladies, the sewing machine, according to the article, would give women a "new preparation of Iron," which would help remedy their ailments. Sewing by hand, the magazine added, often made women sickly; the "inexorable needle" was a "source of consumption — crooked spines, side aches, stomach derangement."

Suggesting some collusion between sewing-machine manufacturer and writer, the *Medical Monthly* author insisted, however, that there was no longer any reason "for health benefits to be broken by sewing" and waxed poetic about the medical benefits of the Grover and Baker machine: now the foot rather than the hand moved the machinery, allowing the body to be nearly erect and avoiding the "tendency to deformity and injury from curvature of the spine." Instead of prescribing iron for the "pale girl" who sewed all day, physicians were urged to give her a Grover and Baker machine. Evoking the stereotype of the nervous female and a theme of transformation, the magazine offered a glowing vision of change: "How quickly will blooming cheeks take the place of pallor and how happy should we be when we look upon our patient, a healthy, robust woman, instead of the pale, nervous, dyspeptic, hypochondriacal thing that she would otherwise become."[63] (Ironically, promoters of bicycle riding in the 1890s would tout the bicycle's healthy benefits for women in contrast to the harmful impact of a sewing-machine treadle.)

In this lithographed advertising trade card a woman looking in a mirror sees her future self with a new sewing machine. New Home Sewing Machine Company, Orange, Mass., ca. 1885. The Schlesinger Library, Radcliffe College, Cambridge, Mass.

Although sewing machines were said to make women less nervous, advertisers may also have raised women's anxieties and insecurities. Long before radio and television produced the episodic stories of soap operas, advertising trade cards presented their own sequenced narratives. The trade card tales of the New Home Sewing Machine Company were sometimes illustrated scenes presented as mini-dramas in several acts. Some of these tales raised the specter of households in chaos, wayward husbands, and illicit romance. A card in one series posed a provocative choice: "A New Home Machine or a French Sewing Girl." Here, a housewife arrives home only to find her husband on his knees kissing the hand of a French girl while she sews. But in the next three scenes the family purchases a New Home sewing machine, and the last card proclaims, "All's Well That Ends Well" as the wife now happily works at her machine while her husband reads the newspaper.[64]

Other sewing-machine company trade card vignettes presented stories about household discord erased by the introduction of the new sewing machine. In the beginning scene of a Wilcox & Gibbs Company trade card series the "noisy, hard running double thread machine" is causing a "distressing nuisance": the wife's hair is disheveled, the baby is

A rare image focusing on a woman's foot operating a sewing machine. American machines were famous for their artful design, as in this machine's fluted neoclassical pillar. "Folsom's Motion for Sewing Machines," *Scientific American,* 21 April 1866, 626.

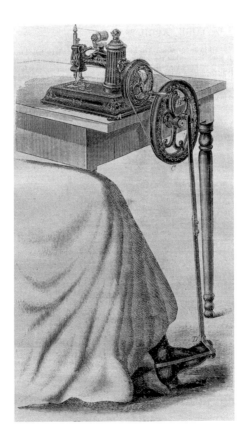

crying, the children are pulling one another's hair, and the husband, putting on his hat, is about to leave. Yet with the introduction of the new machine domestic order is restored: the children play peacefully on the floor while the baby sleeps and the husband contentedly reads.

Though they sometimes increased women's anxieties, advertisers also worked to convince women that they had the capacity to operate the machines. To encourage sales to women, Singer hired female instructors to demonstrate the machines to potential customers in the company's central office and showroom on Broadway in Manhattan, which was publicized in *Frank Leslie's Illustrated Newspaper* in 1857.[65]

The trade card illustrations and other images suggested that sewing machines could be easily integrated into women's lives and the intimacy of their homes. But some stories and images, probably aimed at male readers, went further, adding an unusually intimate view of women themselves. In its 1866 story on the improved Folsom sewing machine, said to be less tiring to operate, the *Scientific American* included a rare revealing illustration showing only a portion of a woman's skirt and leg, her dainty foot pushing down the treadle of her machine. This engraving offered an unexpectedly intimate view while promising easier work for women. In a century in which women's bodies were heavily swathed in ankle-length skirts, the mere glimpse of a female foot could be considered seductive: as one fashion historian has written, "Here we are at the heart of all male fixations, the spark that inflamed their desires, the anchor of their fantasies: the foot." (Pedal-operated sewing machines also had other sexual associations: amidst fears about the effects of sewing machines on women's health, French doctors in the 1860s also made the sensational claim that bipedal sewing machines had a harmful masturbatory effect on women as the rubbing of their thighs caused genital excitement.)[66]

The Folsom illustration was not only provocatively suggestive to male readers of the *Scientific American*; it also revealed manufacturers' efforts to design new machines that were more palatable to women. With its cast-iron pillar designed as a fluted classical column, the Folsom model was characteristic of nineteenth-century American manufacturers' tendency to ornament new domestic machines. Like the early Hoover vacuum cleaner, sewing machines, early models of the Remington typewriter, printing presses, and even machine tools were at times covered with a variety of ornate motifs, including floral decals, gilt scrolls, and even cast-iron sculptural foliage, cupids, and birds. It was probably assumed that classical imagery and decorative motifs would make the machines more appealing to women and more easily assimilated into the household, with classical columns also lending an aura of stability and dignity to the unfamiliar, fast-moving machines.[67]

Images of Women in Early Industries

■ In its promotions for home sewing machines the Singer sewing-machine company, like other early machine manufacturers, portrayed itself as a benefactor to women, easing their household tasks and, in the case of Singer, even offering them the opportunity for extra income. But while they promised an easier life, companies like Singer also validated traditional assumptions about women's role of meeting their families' domestic sewing needs. Successfully marketing machines to industry and to women doing contracted piecework at home, Singer's sanguine advertising also belied an industry in which the working conditions for both men and women were often grim.

As in so many other representations of women and machines, images of women working in the ready-made clothing industry and in other nineteenth-century industries were framed by familiar female archetypes and cultural conventions about women themselves, picturing women as pathetic saints or alluring sirens in the industrial scene. Suggesting a cultural ambivalence about women themselves, women in industry were portrayed as dignified and competent workers as well as charmingly pretty or decorative figures who put new machines, and themselves, on display. Reflecting, too, the competing claims of manufacturers, advertisers, and political polemicists, women industrial workers, like men, also appeared as either ennobled or degraded, liberated or enslaved by their work.

In a daguerreotype from about 1853, one of the earliest American photographic images of a woman and an industrial machine, a well-dressed seamstress faces the camera, seated at her 1853 Grover and Baker sewing machine, a model designed for industrial use. Though her gaze is somber, her figure is elegantly outlined, creating an aura of dignity and calm. The woman's formal pose in an industrial or studio setting was probably dictated in part by the long exposure time required for the making of daguerreotypes, which necessitated that the subject sit very still, and may also reflect her unfamiliarity with being photographed.

As one of the few surviving daguerreotypes of American working women created before the Civil War, the image has some mystery: though occupational photographs were most often taken by professionals at the request of the subject, it is not known whether the photograph was a promotional image for a manufacturing firm or an illustration for the clothing industry or whether it was a personal portrait requested by the woman shown.[68]

The quietude of this daguerreotype, whether born of photographic conventions or the woman's own reserve, does not reveal the harsh life often reflected in later images of women working with their sewing machines in industry, including American women who worked in grim conditions in centralized factories, in small shops and lofts, and in the "putting-out" system, in which they often used their own machines to work for manufacturers at home. Promotional material for the new industrial sewing machines glamorized their benefits; for example, an early Singer sewing machine advertisement pictured a woman working in a frilly dress and suggested that "a skilled female operator with one of these machines can earn $1000 a year." But the realities were different: earlier in the century women had worked long hours at home doing hand-sewing for manufacturers at subsistence pay, and even the introduction of sewing machines in the 1850s did not help them greatly increase their earnings.[69]

During the 1870s and 1880s, with wages still low and working conditions poor, America experienced mass labor protests. In this climate of unrest artists' images of industrial workers became politically charged. In illustrations exaggerated for polemical purposes working-class women at their sewing machines were presented as victims. The exhausting work of the "putting-out" system is suggested in Cornelia W. Conant's engraving *The Sewing Girl*, published in *Harper's Bazaar* in 1871. Bent over a table next to her bed, the young woman is distraught at her exhausting work.

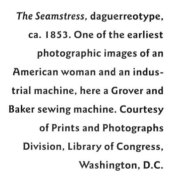

The Seamstress, daguerreotype, ca. 1853. One of the earliest photographic images of an American woman and an industrial machine, here a Grover and Baker sewing machine. Courtesy of Prints and Photographs Division, Library of Congress, Washington, D.C.

Sweatshop conditions were also dramatically portrayed in other maga-zine illustrations, such as *The Female Slaves of New York—"Sweaters" and Their Victims*, published in *Frank Leslie's Illustrated Newspaper* in 1888. Here again, the women appear harried and haggard, hunched over while being either instructed or berated by their beefy male overseer.[70]

In the early twentieth century Lewis Wickes Hine's documentary photographs and the writings of the labor reformer Elizabeth Beardsley Butler continued to publicize the poor employment conditions of women working at their sewing machines in the long, dark, dingy rooms of New York and Pittsburgh garment sweatshops. In *Women and the Trades: Pitts-burgh, 1907–1908* Butler describes women's labors as truly ensnaring, for women were at risk of getting their hair and skirts caught in the revolving wheels of their machines.[71]

Nineteenth- and early-twentieth-century polemical images of men and women working in industry often portrayed them as slaves to an industrial system that kept them compliant and oppressed. Early in the nineteenth century, Mary Shelley in her novel *Frankenstein* had presented what would become a seminal image of technology out of control. The novel's monster has grotesque origins and becomes a tormented and tor-menting creature, terrorizing Dr. Frankenstein and those close to him. Through this image of a macabre, gigantic creation, Shelley presciently illuminated early-nineteenth-century anxieties about technologies that threatened to overpower and enslave their creators. When Frankenstein ultimately rejects the idea of creating a female mate for the monster, the monster reminds him bitterly, "I can make you so wretched that the light of day will be hateful to you. You are my creator, but I am your master;—obey!" In a world often enthralled by new developments in science and technology, Shelley created what would later be seen as a central emblem of the dark side of these radical changes and hinted at the undercurrents of fear and anxiety about mechanization and techno-logical change that would increasingly lurk beneath the century's op-timistic rhetoric of progress.[72]

A version of this darker view appeared in a fictional view about women laboring in another nineteenth-century industry, a view also in-fused with an added sexual charge. Envisioning women's industrial work as both entrapping and enslaving, Herman Melville included in his two-part tale *The Paradise of Bachelors and The Tartarus of Maids* (1855) two central female paradigms, the virgin and the sexual slave. In the eyes of the story's narrator these women become both pathetic and eroticized slaves to the dominating male machine.

The Tartarus of Maids is set in a Massachusetts paper mill. During the nineteenth century women working in Berkshire paper mills like that of the Crane company sorted and cut rags, and with the increase in production of fine papers they also worked in finishing rooms, where they inspected, sorted, and stacked sheets. Like other traditionally feminine tasks, the work required dexterity, and the industry itself was considered especially suitable for women because it allowed for flexible, relatively short work hours. It was also considered safe because women worked away from heavy machines.[73]

The employment of women in this detailed work was also reflected in an early magazine story about women working in the British paper industry in 1833. They sat at paper-cutting machines, where they demonstrated their dexterity by inspecting every sheet and removing "every knot or speck in each sheet" and laying aside "those which have any rent or hole."[74]

Melville's *Tartarus of Maids* is a troubling fictional account of women at work in nineteenth-century manufacturing. In both this story and *The Bell-Tower* Melville invests the promise of technology with a sense of foreboding. Amid nineteenth-century America's heightened expectations about technological progress he presented literary parables of technological danger and destruction. In a society devoted to industrial development he steadfastly kept his eye on the troubling underside of technology's transforming role.

The women in *The Tartarus of Maids* participate in a nightmare world

*The Female Slaves of New York—
"Sweaters" and Their Victims*
(detail), in *Frank Leslie's
Illustrated Newspaper*,
3 November 1888. **Haggard
workers in an American sweatshop. Library of Congress,
Washington, D.C.**

of industry, a world of inversions in which menacing machines take on human characteristics and female factory workers take on the psychic properties of machines. They have become passive slaves, mere cogs in the diabolical mechanism of industry. Adding an edge to the tale, Melville infused the story with an uneasy subtext of female madness and sexuality that ultimately drives the bachelor narrator away in frenzy and fear.[75]

The Paradise of Bachelors recounts the narrator's evening visit to a bachelor dinner, a convivial gathering at which there is "nothing loud, nothing unmannerly, nothing turbulent." In this world of men without women there is "good living, good drinking, good feeling, and good talk."[76] But in *The Tartarus of Maids* the bachelor narrator is drawn into a murky, sinister industrial underworld that is anything but convivial: it is a marvel of efficiency yet a vision of hell.

The narrator is a "seedsman" who travels to the Devil's Dungeon Paper Mill to purchase paper for envelopes for his seed-distributing business.[77] En route to visiting a paper mill in the Woedolor Mountains, the narrator embarks on a mythic journey that will lead him through a subterranean landscape. The narrator's account is highly charged, tracing his traumatic descent not just into an industrial Tartarus with all of its associated diabolical terrors but into the underworld of Persephone—the world of women, sexuality, and madness—which has its own set of terrors for a bachelor who, we already know from *The Paradise of Bachelors*, relishes the safe, "good feeling" world of men.

The seedsman's travels toward industry become a gloomy encounter with the female psyche. Melville's description of the landscape evokes images of depression and madness, stereotypically linking the sexually unfulfilled woman with insanity: the wind-blown mountain pass through which the bachelor travels is Mad Maid's Bellows-pipe, named, we are told, after the legendary "crazy spinsters" whose hut once stood nearby.

The sexual nature of the journey is manifest: the semen-bearing seedsman travels through a phallic forest of vertical tree trunks with their "all-stiffening influence." As he moves through the "notch" of the Mad Maid's pass and the Black Notch beyond, he recalls an incident in which a runaway omnibus careened though Wren's Arch in London. In a curious merger of mechanism and sex, the narrator remembers that the runaway vehicle lurched dangerously out of control as it went through the feminine contours of the arched entryway.

The bachelor enters more explicitly into the realm of sexuality as he approaches the paper mill with its female workers. The factory building, with its "rude tower," has a forbidding masculinity, yet the nearby Blood River suggestively surrounds the scene with a menstrual flow. His is an illicit entry: "I found myself all alone, silently and privily stealing through deep-cloven passages into this sequestered spot."[78]

Entering the factory, the narrator evokes a sexually charged image of passive female workers dominated by a masculinized "great machine." The gigantic machinery that powers the paper mill autocratically demands that the women perform as servile, sacrificial figures. Perceiving a central paradox of industrialization, the narrator offers a mordant view of technology: "Machinery—that vaunted slave of humanity—here stood menially served by human beings, who served mutely and cringingly as the slave serves the Sultan." In the bachelor's view, the women are reduced to harem maids, yet their devaluation is even more drastic: "The girls did not so much seem accessory wheels to the general machinery as mere cogs to the wheels." The passive, enslaved women are dominated by a massive, implacable machine, the "inflexible iron animal" that overshadows the cringing figures of the cowering maids.[79]

Though the precision and efficiency briefly give the bachelor narrator a sense of wonder, he is filled with compassion for the enslaved workers: moving to safer conceptual ground, he now sees the women as suffering, sickly saints rather than passive sex slaves. The whiteness of the blank sheets of paper is echoed in the "sheet-white" pallor of the girls' faces. "So," he laments, "through consumptive pallors of this blank, raggedy life, go these white girls to death." In a hallucinatory vision, he sees the young women's plight in terms of sacrifice and martyrdom, their "agony dimly outlined on the imperfect paper, like the print of the tormented face on the handkerchief of Saint Veronica."[80]

The seedsman's compassion for the women increases as he learns, to his horror, of the demands of industrial pragmatism: the paper mill hires only unmarried women because they do not require time off and provide steady service twelve hours a day. Leaving this Tartarus of technology at the story's end, he pays "pained homage to their pale virginity."[81]

In Melville's story technology has invaded an Edenic world, threatening the innocence of both bachelor and maids. The seedsman's journey takes him from the paradisiacal world of bachelors to a journey through the forbidding realms of sexual knowing—from a vision of technological wonder to a nightmare of mechanized hell. But though his visit to the paper mill gives him a traumatically transformed view of industry, his traditional perceptions of women remain comfortably stable: the female workers remain safe, passive, pale virgins. As the Sultan's slaves, they are the object of his compassion, emblems of oppressive industry here safely under tow.

Like other views of women working in industry, images of women working in nineteenth- and early-twentieth-century textile mills were also infused with familiar female archetypes and cultural beliefs about women's mechanical abilities and proper social roles. Before the sixteenth century the spinning of yarn was generally done by hand by women using a distaff and drop spindle, but in the sixteenth century in England spinning was most often done by women at home using a spinning wheel, a mechanism operated by rotating a wheel using a treadle. After spinning flax or wool into yarn, women often gave the yarn to a weaver to be woven into cloth. By the eighteenth century spinning had become increasingly mechanized with the invention of the spinning jenny by James Hargreaves in 1770 and Arkwright's water-powered jenny.[82]

As mechanization began to change the textile industry in England, women were at times depicted as still happy or at least contented with their manual work. The eighteenth-century British poet John Dyer, in his long poem *The Fleece*, praised early improvements such as Lewis Paul's multiple spinning machine powered by a huge water wheel and evoked a world in which "Village nymphs, ye Matrons and ye Maids" toiled cheerfully at spinning and weaving as they worked with "health and ease." But he also assumed that female spinners would worry about being displaced by these and other new machines — a well-founded worry — and he urged them not to be alarmed: "Nor hence, ye nymphs, let anger cloud your brows," for, he added, there would always be work, and "your hands will ever find / Ample employment."[83]

At the end of the eighteenth century women in England were increasingly moving from home-based spinning and weaving into the new textile mills, which by 1800 commonly used machines powered by water and steam. In America in 1814, the Boston Manufacturing Company, which was founded by Francis Cabot Lowell, opened its first mill in Waltham, Massachusetts. It became the country's first large-scale textile mill, employing men and women to process cotton using water-powered machines. The company, joined by investors now known as the Boston Associates, soon set up mills in the town of Lowell, north of Boston, and by 1855 the town had more than fifty mills.

The rapid growth in America's cotton industry from 1800 to 1815 resulted in a large demand for women to work in mills, and during the century's first decades a large proportion of the factory workers in the cotton mills, including those at Lowell, were young women. The high numbers of women working in textile and paper mills was in part due to a shortage of labor and often reflected the assumption that the lightness of the work made it eminently suitable for female laborers. Women in these mills were usually assigned to unskilled jobs, and they were assumed to be transient workers who would work only for a short period before leaving to marry, which allowed factory owners to rationalize not giving them apprenticeship training for skilled work.[84]

The women's jobs reflected cultural attitudes about the differing abilities of women and men: mastery of heavy and fast-moving machines in the nineteenth century was considered the province of men because of their greater physical strength, while women were presumed to have the delicacy, dexterity, patience, and ability to focus on detail that made them specially suited to lighter tasks.[85] Journals regularly featured illustrations of women engaged in difficult tasks demanding precision and intricate handling, like the well-dressed woman in the *Scientific American* illustration who was pictured punching out sewing-machine needle eyes and another who was depicted grooving needles.

Men were generally employed as supervisors or overseers, mule spinners, machinists, and machine repairers, while in British and American textile mills women worked at preparing cotton and wool for spinning; at carding machines, in which slivers were fed into a drawing machine; and at roving, in which coarse strands were readied for spinning. They also ran spinning machines and minded power looms (in America in the 1830s power-loom weaving was considered primarily a woman's occupation).[86]

Early visual images of women working in the textile industry not only reflected these gendered roles but also sometimes went further, suggesting the tensions generated by men's and women's differing work roles. In the *History of the Cotton Manufacture in Great Britain* (1835), by Edward Baines Jr., engravings based on drawings by Thomas Allom present images of women and young girls working at power looms and at roving or

A female power-loom weaver crouches in an awkward pose before her male overseer. Engraving by J. Tingle after Thomas Allom in Edward Baines Jr.'s *History of the Cotton Manufacture in Great Britain*, 1835.

slubbing frames as they prepared cotton for spinning at the Swainson Birley Cotton Mill, near Preston.[87] In another, more problematic view an awkwardly crouching young girl tends to her power loom while a male overseer looks down at her. In this ambiguous view the artist seems to be presenting either a benevolent male supervisor inspecting the work of his young charge or a harsher image of domination and subservience in the industrial scene.[88]

In later images artists often framed their views of female textile workers in culturally conventional ways, presenting them in decorative, softened, and prettified ways. In his *Hand-Book on Cotton Manufacture* (1867), written for millwrights, managers, overseers, operatives, and machine builders, the American writer James Geldard included an engraving of a woman standing in front of a spinning frame. She is a pretty bit of decoration, an adjunct to the machine. Smiling, she stands facing the viewer with one hand behind her back and one lightly touching the spinning frame; like the machine, the woman is clearly on display.[89]

In William Cullen Bryant's long poem *The Song of the Sower* (1871), a wood engraving based on a drawing by Winslow Homer also presents a pretty young textile worker at her power loom. Rather than portraying her hard labors, Bryant's poem romantically fits her into the larger scheme of America's industrial progress, for she works at "The web that, from a thousand looms, / Comes forth to clothe mankind."[90]

More nuanced and telling views of women textile workers' experience appeared in the writings of America's "Lowell girls," who often skillfully portrayed the life of women textile workers in the nineteenth century. Their stories often reflected the unusual nature of the women themselves: a large number of them were from rural areas yet also educated, members of the middle class rather than the working class. These women generally stayed only a few years and then returned home, often for marriage.

In the *Lowell Offering*, edited and written by women working at the Lowell mills from 1840 to 1845, the women offered their own perspectives, sometimes confirming cultural stereotypes about a woman's point of view, sometimes viewing their work, and the machines, through a transforming eye.

Many of their stories idealize the work, reflecting, perhaps, the fact that the magazine was supported and printed by a newspaper known to be a mouthpiece for the textile corporations.[91] But their stories of mill life also reflect a complex vision: the women framed their descriptions of machinery with images of beauty and order but also lucidly recorded the mills' potential for machine mayhem.

In some of their writings, the women seemed to be validating a stereotype of women's taking an aesthetic view. The mill worker Harriet Farley, in her series "Letters from Susan," published in the *Lowell Offering* in 1844, wrote that "the girls dress so neatly, and are so pretty," and about the machinery itself she wrote, "The machinery is very handsomely made and painted, and is placed in regular rows; thus, in a large mill, presenting a beautiful and uniform appearance." Farley also took pleasure in looking at the running machines: "I have sometimes stood at one end of a row of green looms, when the girls were gone from between them, and seen the lathes moving back and forth . . . the white cloth winding over the rollers, through the perspective; and I have thought it beautiful."[92]

For Sarah G. Bagley, in her story *Pleasures of Factory Life*, the view of machinery is virtually transcendental: "Who can closely examine all the movements of the complicated, curious machinery, and not be led to the reflection, that the mind is boundless, and is destined to rise higher and still higher."[93] But an anonymous writer in the *Offering* revealed that the women were capable of recording the more somber world of working with machines. Hers is not a vision of a transcendent world but rather one of noise and monotony sometimes broken by the unexpected: "Some part of the work becomes deranged and stops; the constant friction causes a belt of leather to burst into a flame; a stranger visits the room."[94]

Also taking a chastened rather than aesthetic view, Josephine L.

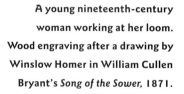

A young nineteenth-century woman working at her loom. Wood engraving after a drawing by Winslow Homer in William Cullen Bryant's *Song of the Sower,* 1871.

Two women weavers holding their shuttles in a typical nineteenth-century American photographic labor portrait. Tintype, 1860, Photographic History Collection, National Museum of American History, Smithsonian Institution.

Lewis W. Hine, *Woman at the Shelton Looms*, gelatin silver print, ca. 1910. Courtesy George Eastman House.

Baker looked at the spinning jacks and jennies and observed, "It seems but pleasure to watch their movements; but it is hard work, and requires good health and much strength." Her account is also filled with warnings of wayward machines. "Do not go too near," she admonishes, for the machines "might unceremoniously knock us over." Describing the carding room, she warns the visitor: "I beg of you to be careful as we go amongst them, or you will get caught in the machinery." The girls do not wear face masks, she adds, because it is unfashionable, and they are in danger of getting "their faces blue."[95]

Early photographs of women in the textile industry often conveyed a sense of dignity to both the women and their work.[96] During the nineteenth century both men and women industrial laborers commissioned their photographic portraits by hiring local or itinerant photographers, who produced daguerreotypes, ambrotypes, and tintypes of laborers from the 1840s through the 1870s. In these labor portraits taken in America, individual workers solemnly posed for their portraits with their work tools.[97]

In an ambrotype from 1860 two women appear dressed up for the occasion in their aprons and brightly patterned dresses. Each holds a shuttle and smiles gently while posing against a painted backdrop of an industrial town. Typical of these photographic labor portraits of the period, the ambrotype presents the women linking their identity to their work as they proudly display the tools of their trade.

In his early-twentieth-century documentary photographs of female textile workers, Lewis W. Hine also endowed the women with dignity while framing their images with an aesthetic eye. In his photograph *Woman at the Shelton Looms* (ca. 1910) the face of an attractive woman is literally framed by threads as she gazes at us intently, her image lending coherence and elegance to the factory scene.

Only a few years later the Russian cubo-futurist painter Natalia Goncharova would infuse her painting *The Weaver (Loom + Woman)* (1913) with the jagged rhythms and tremors of aesthetic and technological change, the scene charged with energy as an electric light vibrates overhead. Neither prettified nor victim, siren nor saint, this woman shakes us with her presence, surrounded but not submerged by the industrial machines (plate 12).[98]

As so many of these contrasting images demonstrate, representations of women and machines from the very beginning of the industrial revolution were often shaped by conventional notions about women's abilities and roles, yet they also revealed how women were challenging these conventionalized views. In a period when America was experiencing a wave of nostalgia about its colonial past, *Harper's* magazine in 1892 published its own version of a woman at her spinning wheel. In the engraving *This is No Courting Night*, an illustration for a drama about life in seventeenth-century Salem, Massachusetts, a young woman works intently at spinning while her handsome suitor looks on in rapt attention as the couple momentarily sacrifices romantic courtship for the woman's duties at home.[99] But while this artist has situated the young woman in a domestic framework, other image-makers, as will be seen in the pages ahead, would offer competing views, picturing women in a new romance with the machine, women who turned to new technologies and the products of industry to refashion their bodies, their identities, and their lives.

WIRED *for* FASHION

*Images of Bustles, Corsets, and Crinolines
in the Mechanical Age*

2

DURING THE NINETEENTH CENTURY'S great burst of industrial inventiveness women in America and Europe started to reshape their bodies using the latest in fashion technology. In a century that introduced steel-cage frameworks to reduce the need for heavy masonry in architectural design, manufacturers began promoting steel-cage construction to lighten the load in women's dress. Starting in 1856, women widened their skirts with the new "cage crinolines," frameworks made of a series of horizontally stacked spring-steel hoops connected by vertical tapes or steel ribs or with cloth petticoats ringed with a series of flexible steel hoops.

Spring-steel cage crinolines like this one by Thomson ca. 1860 added a new framework to women's fashions. Photograph courtesy of Irving Solero, The Fashion Institute of Technology, New York.

Women also began to wear corsets fully boned with steel and steel bustles in the form of spirals or small cages underneath their dresses. Suspended at the rear near the small of the back, these steel bustles, such as the "camel's hump" model, were designed to exaggerate the size of the buttocks. To many women who wore them, the wire bustles and cage crinolines seemed a welcome improvement over the bulky, padded bustles made of straw or cotton wool and the heavy multiple petticoats worn under voluminous skirts.[1] One fashion historian raffishly called the much lighter cage crinolines "the first great triumph of the machine age," akin to the skeletal framework of London's Crystal Palace.[2]

Wearing these new wired fashions helped women fulfill nineteenth-century ideals of womanhood and feminine beauty. Some artists of the period, like the French-born British artist James Tissot, validated the look of elegance women wished to create. In paintings like *Too Early*, showing young women at a London dress ball wearing bustles beneath their lavish silk dresses, Tissot evoked an aura of haute couture (plate 13). Elegantly clad in pastel pink and white, the women in his paintings represent the height of gentility and grace in fashionable society.

Yet in other visual and literary representations of the age women in their crinolines, corsets, and bustles looked like awkwardly moving mechanical dolls rather than fashion plates. Artists and writers presented these wired women variously as elegant or awkward, regal or ridiculous.

The idea of women as mechanical dolls or female automatons had contrasting cultural meanings in the nineteenth century. Though caricaturists delighted in spoofing the stiffly moving women, mechanized toy walking dolls invented during the nineteenth century were much admired. "Autoperipatikos," the toy patented by the American toymaker Enoch Rice Morrison in 1862, was named after the Greek words meaning "self-walking" or "self-propelled," and with its hidden clockwork mechanism that allowed the doll to walk on its little metal feet, it testified to the century's love of mechanical ingenuity. The doll was dressed in the latest fashions, for beneath its lace-trimmed dress was a small wired crinoline petticoat.[3]

But female automatons and dolls could also be associated with the macabre, as is the mechanical doll Olympia in E. T. A. Hoffmann's story *The Sandman* (1817), discussed by Freud a century later in his essay *The Uncanny*.[4] In the story, the beautiful Olympia entrances the smitten Nathanael, but he later watches in horror as her body is dismembered by the quarreling Coppelius and Spalanzani and her true artificial nature becomes known. The idea of mechanized people could also be funny. As Henri Bergson observed in *Le Rire (Laughter)* (1900), there is something inherently comical about the artificial being substituted for the natural, about the mechanical being superimposed on the living: "The attitudes, gestures, and movements of the human body are laughable in exact proportion as the body reminds us of a mere machine."[5] This humorous view was often taken by nineteenth-century artists and photographers, who satirically depicted women wearing crinolines as mechanical creatures.

Often they were depicted navigating gingerly in their stiff undergarments, unceremoniously knocking over furniture and toppling out of carriages. Nineteenth-century stereographic images of women in crinolines provided valuable visual documents of these undergarments, which were rarely photographed, but the views were mostly satirical scenes using models in posed, theatrical settings. Spoofing the hazards of wearing crinolines, a comic stereograph from the 1860s and a sheet-music cover illustration featured a woman sprawling in the street after her cage crinoline became caught in the door of an omnibus as she tried to exit (plate 14).

Unimpressed by the outer trappings of fashionable elegance, England's satirical weekly *Punch* in 1868 engaged in a form of nineteenth-century deconstruction by stripping away the elegant exterior to lay bare the framework beneath. In the magazine's cutaway drawing of a crino-

Dress and the Lady, from *Punch,* 23 August 1856. Deconstructing women's outward images, this cartoon lays bare the underpinnings of fashion.

lined woman, on the left side she is shown fully dressed, while on the right side the skeletal framework underneath her skirts is displayed. The *Punch* illustration spoofs not only fashion but also the visual conventions of technology itself and becomes a mock engineering drawing revealing the woman's underlying fashion mechanism. The illustration also exposes women themselves, demystifying their facade of gentility and revealing the unromantic underpinnings of fashion.

These idealizing and comically satirical images of women in their crinolines, corsets, and bustles — images that both elevate women's stature and ground them — became a telling gloss on the century's often ambivalent attitudes toward female fashions, new technologies, and women themselves.

In a century that celebrated the woman with a tiny waist, ample hips, and a large bosom the new corsets, crinolines, and bustles helped women fulfill contemporary cultural definitions of the ideal feminine figure. Crinolines helped create the look of broader hips and helped produce the illusion of a small waist. Bustles created the look of rounded buttocks, and corsets helped give women the appearance of broad hips and a narrow waist while enlarging the bosom.

The new underclothes also helped women fulfill the century's complex and contradictory ideals of womanhood, according to which women were envisioned as both maternal and sexual, chastely pure and erotic. The tight-laced corsets suggested chastity, strict morality, and self-restraint, while wide hips and a broad bosom gave women the look of fecundity and fertility, emphasizing their social and biological function of bearing children. Yet the hourglass figure so cherished during the nineteenth century also gave women the unmistakable look of erotic sensuality.[6] Bustles and hoops also helped fulfill dual cultural ideals: as one commentator has argued, they gave "the illusion of an enlarged pelvis and the subliminal promise of sexuality, fecundity, and easy delivery."[7]

The new undergarments also helped women who hoped to enhance their social status. In nineteenth-century America upwardly mobile women who aspired to gentility wore corsets in emulation of the garments worn by seventeenth- and eighteenth-century women of the European aristocracy. Originally associated with gentility, the tight corsets and crinolines were signifiers of the leisure class, worn by women who did not work and serving as a sign of their inactivity. As Thorstein Veblen suggested in 1894, corsets made it difficult for women to move, bend, or stoop, and so wearing one suggested that other people did a woman's work, an indicator of her husband's wealth and power.[8] At first, crinolines were given legitimacy when worn by the French empress Eugénie, and they were initially worn by other women as a sign of their own wealth and status. But crinolines' widening popularity among all classes in Europe and America, including housemaids and factory workers, may also have contributed to their waning popularity among women of rank.[9]

The uses of special undergarments to reshape women's bodies actually has a long history, and the idea of using special hoops to widen women's skirts was not at all a new one. From the mid-sixteenth century and into the seventeenth, aristocratic women in Britain, France, Italy, and Spain wore the farthingale or vertingale — originally known as the "Spanish farthingale" — which was an underskirt extended by hoops made of wood, whalebone, or wire. In France's royal court in 1580 women themselves became a type of mechanism by wearing the "wheel farthingale." As one writer described it, "It was as if the wearer were standing inside a wheel, with the skirt attached to the outer rim and falling vertically to the ground."[10] During the eighteenth century British and French women wore underneath their skirts huge, greatly exaggerated side hoops, or panniers, which extended horizontally on each side, at right angles to the waist. Whalebone hooped petticoats were again popular in eighteenth-century Europe, fashionable in France until the French Revolution and worn at the English court until 1820.[11]

The Crinoline ■ During the early nineteenth century the taste in women's fashions turned toward expansive skirts worn with many starched and sometimes flounced muslin petticoats underneath (in England the *Ladies' Companion* in 1856 reported women wearing fourteen petticoats at formal balls). These bulky undergarments spurred inventors to work at finding a way to relieve the weight of these many petticoats.[12] The introduction of the steel-cage crinolines in 1856 seemed an ingenious solution to the problem of lightening women's underclothes, and in this age of industrialization they could be mass-produced for widespread use.

In July 1856 R. C. Milliet, of Besançon in France, was granted, through his agent, the first British patent for his invention of a metal crinoline, a caged framework or skeleton made up of a series of steel hoops connected by vertical cloth tapes.[13] Sometimes the cage was worn separately, and in some versions the hoops were sewn directly into a cloth petticoat. These steel crinolines often had nine to eighteen steel hoops. In 1860 nine-hoop crinolines were worn during the day, and eighteen-hoop spring steel crinolines with ten to one hundred springs were worn under ballroom dresses.[14] (The word *crinoline* came from the French word *crin*, meaning "horsehair," which starting about 1830 and continuing during the decade was used to stiffen petticoats and stuff women's bustles. The word soon became associated with any type of petticoat using metal or whalebone hoops, even those in which horsehair was not used.)[15]

One of the leading producers of steel-cage crinolines was the American company owned by W. C. and C. H. Thomson, with factories in New York, London, France, and Saxony. The company's designs included the "Crown" crinoline and the "Zepherina" or "Winged Jupon," also known as the "Safety Crinoline," which was designed to prevent women from getting their feet entangled. In a century in which industries proudly announced their production figures *Harper's Weekly* in 1859 reported that Thomson's New York factory consumed 300,000 yards of spring steel every week, producing 3,000 to 4,000 crinolines daily, and employed an average of a thousand girls at four dollars a week, and that the Douglas & Sherwood hoop skirt factory, on White Street in Manhattan, had women working at two hundred sewing machines and consumed a ton of steel each day.[16]

Women working in the finishing room of the Douglas & Sherwood hoop skirt factory in Manhattan. *Harper's Weekly,* **29 January 1859.**

Portraying crinoline making as a glamorous business, the *Harper's* illustrations lent a look of elegance to the women working in the hoop-skirt factories. In his description of women working at Wests, Bradley & Carey's Hoop Skirt Works in New York City, John Leander Bishop, in *A History of American Manufactures* (1868), took a sanguine view, describing the rooms of "intelligent operatives at their work," where "hundreds of young women are engaged in this light and pleasant toil."[17]

During the approximately ten-year period in which crinolines were popular, starting in 1856, their shapes varied, changing from a dome to a pyramid and then becoming flattened in front without the spring steels. Turning to somewhat more streamlined designs in the late 1860s, women wore half-crinolines, including the "American cage" crinoline, introduced in 1862, which reduced the crinoline's weight to half a pound by leaving the upper half as a skeleton and encasing only the lower half in cloth. By 1866 women began to abandon the cage crinolines, turning once again to horsehair bustles and petticoats, then called tournures or jupons.

Crinolettes made of steel hoops, like this one by Thomson, ca. 1880, helped women gain the latest fashionable shape. Museum of Costume, Bath, England.

The crinolette, introduced in 1868, used steel half-hoops in the back and could include horsehair or crinoline flounces, creating a bustle. About 1880, the Thomson crinolette used wire hoops for the bustle area as well and could be worn with a chemise, a corset, and drawers underneath.[18]

From the outset, women's crinolines were often presented in contrasting views. Advertisers claimed that the petticoats, by giving women the appearance of imposing size, helped endow them with a look of stature and dignity. A contemporary British advertisement for Thomson's "Crown" crinolines insisted that the women who wore these undergarments looked "graceful, modest, ladylike and queenly," perhaps alluding to the crinoline's early associations with European royalty, including Empress Eugénie.

But while crinolines may have been seen as giving women the appearance of dignity, stature, and style, the very nature of the wired petticoats also undercut women's dignity by challenging contemporary standards of feminine modesty. The wide crinolines were meant to swing gracefully, but they had an unfortunate tendency to tilt up from behind when women walked. They also tilted up when women sat down, exposing their legs and undergarments. To cover their legs, women and young girls wore long, lace-edged linen pantaloons that sometimes reached to their ankles.[19] Edward Philpott, a British draper, insisted that his firm's "Ondina" crinoline, with its wavelike bands, allowed a lady to "ascend a steep stair, lean against a table, throw herself into an armchair" without "provoking the rude remarks of the observers, thus modifying, in an important degree all those peculiarities tending to destroy the modesty of English women."[20]

Though wearing crinolines may have helped women free themselves from wearing many heavy petticoats, the new steel-cage petticoats clearly hindered their grace and mobility and could also be dangerous: they might catch fire when women moved too near to fireplaces and stoves.[21] They also caused less serious mishaps. Skirts became so wide that their wearers had trouble sitting and getting through doors and also had to be careful not to knock over items on tables.

Given all these awkward features, it is not surprising that the steel-hooped petticoats became a source of hilarity to humorists, who satirized the plight of women in their outsized skirts as they tried to travel or walk arm in arm with their male escorts. Other observers were less amused and made testy complaints about the new crinolines. Sir Charles Eastlake, whose influential book *Hints on Household Taste* (1868) went through several editions and was widely reprinted and read in England and America, called the crinoline a "wretched invention" that had continued in popularity "in spite of the charges of indelicacy and extravagance which have been so frequently brought against it" and akin to the previous hoop or farthingale which had been "both ridiculed and seriously condemned by our ancestors." As with the earlier hoops, "neither satire nor sermons seem to have affected its use."[22]

One of the enduring stereotypes about women is that they are empty-headed creatures who happily embrace new fashions even when those fashions prove awkward, ungainly, and even, as in the case of crinolines, hazardous to human safety. The difficulties posed by the new crinolines gave ammunition to nineteenth-century satirists, who lost no time belittling women as foolish slaves to fashion's whims. England's *Punch* in 1858 parodied crinolines in the song *What a Ridiculous Fashion!* which claimed, "The more you scoff, the more you jeer, / The more the women persevere, / In wearing this apparel queer."[23]

The satires and the images of awkward women were not new. A century earlier, British writers had protested the absurdities of women's hooped skirts, describing mishaps that cast women in an inelegant, awkward light. When British women began wearing boned hooped skirts again early in the eighteenth century, the *Guardian* published satiric letters of complaint, including one by "Tom Pain" in 1713 reporting, "I saw a lady fall down the stairs the other day, and, believe me, sir, she very much resembled an overturned bell without a clapper."[24]

In a satiric poem of 1711 a British poet from Bath took his own satiric shots at the hooped petticoat as he spoofed the technology of underwear:

> The world appears all lost in wild amaze
> As on these new, these strange machines they gaze.

Contemplating the possibility of hooped women flying through the air, the poet reassures the reader that woman is earthbound:

> Let it suffice you see th' unwieldy fair
> Sail through the streets with gales of swelling air;
> Nor think (like fools) the ladies could fly. That's all romantick, for
>   these garments show
> Their thoughts are with their petticoats below.[25]

During the nineteenth century men complained that crinolines kept them at a distance, hindering them from walking closely with their female companions or sitting near them. In a series of articles on the crinoline published in 1867–68 an anonymous writer in *Harper's Bazaar* half-jokingly complained to female crinoline wearers, "You have required three chairs to make one of you comfortable; while your escort—father, or husband, or cousin—has perched near you on a fourth." Signifying that the crinolines had lost popularity in America, the writer noted in an article one month later, "We may rejoice, however, that the reign of the terror-inspiring hoop is over."[26]

Writers recognized one of the paradoxes of the crinoline: the undergarments created an effect that was alluring yet distancing, intimate yet remote. The British novelist George Meredith complained in the 1860s that the crinoline "has put an immense division between the sexes. It has obscured us, smothered us, stabbed us."[27] But a century earlier the French essayist Montaigne had seen the benefits of these petticoats, remarking wryly, "What is the use of those great bastions with which our women have just taken to arming their flanks, except to allure our appetite and to attract us to them by keeping us at a distance?"[28]

During the 1850s and 1860s *Punch* continued to target the crinolines as an obstacle to intimacy, bemusedly asking in 1857, "Now you don't suppose hoop petticoats are looked upon with favour by the masculine eye-sight? You surely can't imagine there is 'metal more attractive' to a man in half a ton of Crinoline than in nature's flesh and blood unsurrounded by steel armour?" The magazine added, "Put the question to your partner in a ball-room, and see if he approves of the fashion which makes ladies unapproachable. Whether as a waltzer or as husband, a man likes a woman he can take into his arms; and how is this possible when she is entrenched in an impregnable hoop petticoat, which when he approaches he breaks his shin against?"[29]

The crinoline as a distance-making invention was a particularly favorite target of the French artist Honoré Daumier, whose comic lithographs lampooned the discomforts caused by the new petticoats. Along the way, Daumier's lithographic drawings also poked fun at women, at the century's new technologies, and at the impact of women's fashions on men. In a lithograph from his series "La Crinolomanie," in the satiric French newspaper *Le Charivari* in 1857, Daumier satirized the plight of a gentleman who leans far over in order to walk arm in arm with a crinolined lady. The lithograph's caption warns, "If women continue to wear crinolines in steel, men will have to be invented made of flexible rubber."

Daumier's images of the crinoline brilliantly encapsulates many of the most important satiric themes associated with the new fashion, including the notion of crinoline-wearing women as rigid, ridiculous creatures daftly unaffected by the absurdities of their dress. During the nineteenth century the turnstile was a quintessential emblem of social engineering: the spoked, wheeled device allowed visitors to the century's great industrial expositions to enter the fairs in an orderly fashion. Making fun of the new turnstiles as well as women's fashions, the artist satirized the fate of two women who had gone through a turnstile at the huge Exposition Universelle in Paris in 1855. The turnstile has deformed the bustles and

Honoré Daumier, *Saprelotte! si les femmes continuent. . . ,* lithograph, *Le Charivari,* 9 September 1857. A satirical view of the crinoline's impact on the relations between women and men. The Metropolitan Museum of Art, Gift of Edwin de T. Bechtel, 1952 [52.633.1 (22)].

Honoré Daumier, *Effets du tourniquet sur les jupons en crinoline,* lithograph, *Le Charivari,* 25 June 1855. In this satirical print, the haughty women are unaware that the turnstile has made their dresses—and the wire crinolines underneath—ludicrously deformed. The Metropolitan Museum of Art, Gift of Edwin de T. Bechtel, 1952 [52.633.1 (8)].

crinolines of the haughty women, who walk with their noses in the air, ludicrously unaware that their fashionable crinoline petticoats have become deformed in a heap behind them. In a second version, published the same year, a woman has become ensnared in the turnstile's spokes, a victim of a "Machine Invented by an Enemy of Crinoline Jupons."

Airborne Women. Another comic theme in Daumier's satires of the crinoline, one that seemed to target women as well as their fashions, was that of women who had become airborne or were inflated or flighty figures in their balloonlike dresses. These were women whose fashion-mindedness gave them the look of a stereotypically frivolous female lightweight. One lithograph shows a group of women wearing crinolines walking solemnly in their flounced dresses, and the caption reads, "These are not women but balloons." In another, published the same year, women float through the air far above the earth, made light and buoyant by the outstretched skirts of their crinolines.

As seen in Daumier's prints, nineteenth-century artists and writers enjoyed envisioning women's petticoats as flying machines or parachutes. Women were pictured sailing through the air, held aloft by their umbrella-

Honoré Daumier, *Manière d'utiliser les jupons*, lithograph, *Le Charivari*, 16 April 1856. A novel use of crinolines: to help women sail through the air. The Metropolitan Museum of Art, New York, Rogers Fund, 1922 [22.61.211].

shaped skirts. Contemporary newspaper accounts told of a French girl whose crinolines had literally served as a type of parachute: a girl from Toulon was purportedly saved from injury when, after angrily hurtling herself out a window, she landed safely due to her crinoline.[30] A British newspaper fantasized that perhaps women's crinolines could be fashioned into devices that would help them "to fly by the suitable arrangement of a flighty style of dress." By adjusting her crinoline in a high wind, a woman might "get caught up aloft and transported through the air to considerable distances."[31]

Images of airborne women buoyed by their crinolines paralleled an actual, short-lived fashion that started about 1849 and continued into the 1850s: women extended their skirts with a series of hoops that were actually air-filled pneumatic tubes. The fad also extended to inflatable rubber bustles and "bust improvers."[32]

The theme of airborne woman was part of the nineteenth-century's often satirical view of fantasies about flying machines, like the aerial inventions satirized by the British artists Robert Seymour in his etching *Locomotion* (1829) and William Heath, who sometimes signed his prints "Paul Pry," in his engraving *The March of Intellect* (1829). Images of women aloft also echoed the century's fascination with ballooning, a fascination that had begun in the eighteenth century, and the balloon flights of early female aeronauts. During the early nineteenth century British and French fashions featured large, full "balloon sleeves," satirized by Heath in his print *The Flight of Fashion* (1829), which showed balloon-sleeved women comically engaged in preparing balloons for flight.

Yet underlying these satirical images of flying women sailing along in their crinolines may have been subliminal fears of women distancing themselves—taking off and transcending the domestic sphere, the carefully circumscribed life of hearth and home. The longing to ground these errant females is evident in the poet from Bath's satiric poem of 1711, in which women who sail through the air actually have their thoughts on their petticoats below.

The Gigantic Woman. Visual images of the crinolined woman as an inflated creature may also have suggested more fundamental male fantasies and fears about women as huge, enveloping, protective mother figures. The image of the hooped petticoat and cage crinoline as a type of large sheltering tent or umbrella reappeared in literature, art, and photography of the eighteenth and nineteenth centuries.

Sometimes these paradigms of the huge, enveloping woman were seen in satires that belittled women or brought them down to size. Even before the wired crinoline, the British satirist Joseph Addison in his parody of petticoats published in *The Spectator* in 1709 took a condescending view of women's fashions, ridiculing a woman wearing a wide petticoat and reducing her in stature. His satire conjured up the image of the petticoat—and by extension, the woman wearing it—as a giant, protective umbrella offering men comfort and shelter.

In Addison's satire a "beautiful young damsel" preparing for a court case being brought against her goes to Sir Roger de Coverley's house but has trouble entering because of her huge hooped petticoat. Sir Roger orders that the "criminal" "should be stripped of her encumbrances till she became little enough to enter my house."

Continuing the theme of expansion and reduction, the young woman confesses that she wore the petticoat "for no other reason, but that she had a mind to look as big and burly as other persons of her quality," but "she began to appear little in the eyes of all her acquaintance." Ultimately, Sir Roger himself comes to terms with the petticoat by transforming it through technology: he orders the making of "an engine of several legs, that could contract or open itself like the top of an umbrella, in order to place the petticoat upon it, by which means I might take a leisurely survey of it."

Spread open with the aid of the mechanical device, the petticoat ultimately becomes a large, sheltering canopy: "It formed a very splendid and ample canopy over our heads, and covered the whole court of judicature with a kind of silken rotunda." Sir Roger adds, "I entered upon the whole cause with great satisfaction, as I sat under the shadow of it." Having stripped the girl of her formidable appearance, Sir Roger can now avail himself of her in a symbolic fashion; he can take comfort in her symbolic presence, resting within the comforting, sheltering, umbrella-like maternal space of her extended petticoat.[33]

In the nineteenth-century era of crinolines, photographers continued to create satirical views of the woman as protector. In a stereographic photograph titled *Surprised*, published by the London Stereograph Company, a cuckolding husband whose wife has come home unexpectedly finds a convenient hiding place beneath the wide crinolined skirts of his lover. The mammoth size of the new crinolines prompted not only fantasies of the protective woman but also images of monstrous women, fearsome creatures of colossal proportions. In 1858 *Harper's Weekly* published an engraving titled *The Monster Lady of Crinoline at Turin*, showing a mammoth woman, her skirt extended by a flounced crinoline, presiding over a crowd fishing in the river.[34]

Probing the cultural symbolism suggested by the widened skirts and enlarged figures of women wearing crinolines, some commentators have framed their discussions in economic terms. David Kunzle, in *Fashion and Fetishism*, saw the huge skirts as signs of "economic power and the accumulation of individual wealth," while James Laver viewed the crinoline as a symbol of fertility in an age of economic well-being, noting that during historical periods of austerity and discipline, including the French Revolution, corsets and crinolines disappeared. The French fashion historian Philippe Perrot argued that while women tied themselves to the "heavy machinery" of the monumental cage crinolines to evoke the "implications of leisure, wealth, spare time, and domestic servants," the avail-

The Monster Lady of Crinoline at Turin, engraving in *Harper's Weekly,* 3 April 1858. In the age of gigantism in machines, images of monumental women wearing crinolines could be seen as fearsome.

ability of less costly versions to working-class women reflected their ability to emulate bourgeois values by acquiring "industrially produced imitation chic."[35]

But what has not been considered is that the image of the gigantic woman in her widened skirt is also suggestive of nineteenth-century attitudes toward gigantic new mechanisms. This was a century in awe of colossal technological marvels that seemed the apotheosis of social progress—machines such as the 680-ton Corliss engine, which powered the exhibits at the U.S. Centennial Exhibition in Philadelphia in 1876. Journals such as the *Scientific American* were replete with illustrations of gigantic steamship cylinders and towering Bessemer converters that loomed high over factory workers.

The crinoline, with its skeletal framework, also suggested connections with two other large-scale architectural emblems of the century's technological progress: Joseph Paxton's glass and steel Crystal Palace, specially designed for the Great Exhibition in London in 1851, and New York's own glass and metal Crystal Palace, which housed that city's exhibition in 1853–54.

The image of the gigantic woman wearing her crinoline also had a link to the huge sculptures of mythic female figures presiding over the century's industrial expositions. As Serafina Bathrick has written about the allegorical, often classicized sculptures of women at the New York Crystal Palace exhibition and Chicago's Columbian Exposition of 1893, the mammoth sculptures not only embodied the "True Woman" ideal of the nineteenth century—the woman who guarded hearth and home—but also, through their mythic associations, served to legitimate manufactured goods and industrial progress.[36]

Finally, the image also reflected some of the century's widely held assumptions and stereotypes about women themselves. The huge crinolined woman suggested subliminal wishes and anxieties about woman as a larger-than-life earth mother, a mother both longed for and feared. The mammoth woman also echoed the century's fascination with larger-than-life images of women as powerful warriors; this was, after all, a century that proudly exhibited large cast-metal sculptures, such as the one of an Amazon warrior astride her horse by August Kiss at London's 1851 exhibition, New York's Crystal Palace exhibition, and America's Centennial Exhibition, as prime examples of the industrial arts.

The figure of the gigantic crinolined woman suggested a dominating, fearsome, overpowering woman, a woman who threatened, perhaps, to push aside men, edging them out of the picture. (As if in answer to the fears of an all-enveloping woman, a comic nineteenth-century London Stereoscopic Company image features a scene in which a bearded man gleefully lowers a cage crinoline over the heads of a group of young women who sit rather anxiously on a parlor floor [plate 15].) But as the psychologist J. Carl Flügel argued in 1930, the image of the huge woman in her crinoline was also viewed with respect: "To a generation that valued pose and presence more than smartness or efficiency," the bulky crinolines "appear to confer a certain dignity upon the sex that wore them. Indeed, the crinoline has been looked upon as a symbol of feminine domination."[37]

The woman in her wired crinoline petticoat—a product of industrial progress—thus embodied some of the century's competing views of female identity: woman as both protector and monstrous threat; a laughable caricature of grotesque vanity and the height of queenly elegance; and a flighty as well as formidable creature.

Corsets   ■ Artists and writers also enjoyed spoofing nineteenth-century developments in wired corsets, which also tended to make women look like mechanical dolls or robotic figures rigidly conforming to fashion and cultural conventions.

Like the wearing of undergarments to widen women's skirts, the wearing of stiffened corsets also had a long history. Upper-class women during the European Renaissance period wore dresses with bodices made rigid and stiff with buckram or pasteboard, and during the late Renaissance the bodices were stiffened with whalebone. More extreme examples were iron corsets and the use of iron stays in the late sixteenth century. It has been suggested that these British and French iron corsets were probably worn not for fashion but for physical corrections.[38] During the eighteenth century women wore corsets made rigid with sewn-in busks of wood, whalebone, ivory, or metal.[39]

During the nineteenth century steel increasingly competed with whale boning as a stiffening for corsets. After mid-century, corsets were made of cotton or buckram, with stiff steel bands and supporting wires, like the "Coraline" corset manufactured in America by Warner. By the 1860s corsets with steel busks were being worn to cover the bosom, abdomen, and hips, and later in the century manufacturers claimed to have improved the steel busks by making them rustless and unbreakable.[40]

Other versions of corsets offered women the latest in pseudoscience and fashion technology. In 1883 an advertisement for Dr. Scott's "Electric Corset" promised not only to give women a graceful appearance and a fashionable "French" shape but also to help cure ailments such as rheumatism and kidney problems. In some advertisements, Harness' Electric Corsets in London offered to provide a cure for a weak back as well as relief from hysteria and dyspepsia, and its advertisements also promised to strengthen women's organs. Manufacturers, including the British firm of Izod, also produced steam-molded corsets that were touted as perfect for women's figures. These corsets were shaped using earthenware or metal molds, suggesting an effort to standardize women's figures to a cultural ideal. Quoting *The Queen* magazine, Izod also claimed that the molded corsets were inspired by Hogarth's theoretical Line of Beauty, discussed in his treatise *Analysis of Beauty* (1753).[41]

Women's corsets, like women's crinolines, evoked contradictory cultural meanings. By squeezing the waist, the garments created the illusion of enlarged bust and hips, promoting the cultural feminine ideal of both sexuality and fecundity. As Valerie Steele noted in *Fashion and Eroticism*, women wearing corsets were also conforming to contradictory cultural ideals that wished for women to appear both erotic and virtuous. By creating the look of a well-developed figure with a slender waist, and by emphasizing the female body, corsets became, in Steele's words, a "sexualizing device." Yet wearing a corset also helped women present them-

At a time when electricity was popularly seen as bestowing medical benefits, this Harness' Electric Corsets advertisement from 1892 promised to strengthen women's inner organs. Mary Evans Picture Library, London.

selves as puritanically moral and upright, for "the straitlaced woman was not loose."[42]

Indeed, an article on corsets in *Godey's Lady's Book and Magazine* in 1864 associated the abandonment of corsets with sexually loose behavior, citing as an example the period of the French Revolution, "when the general licentiousness of manners and morals was accompanied by a corresponding indecency of dress."[43]

The erotic aspects of wearing a corset fascinated eighteenth- and nineteenth-century artists, particularly in England and France, who slyly aimed their satire at the practice known as "tight lacing." During the 1770s, while some reformers were urging the abandonment of tight stays, status-conscious middle- and upper-class women kept struggling to make their waists and hips appear ever smaller. To create the appearance of a minuscule waist, women wearing cotton corsets often needed to have someone else pull at their corset laces to make them tighter.

Eighteenth-century artists delighted in creating images laden with erotic overtones of coquettish women having their corsets laced up by men. In Nicholas-Andre Monsiau's *An Abbé Lacing a Lady's Stays*, engraved in 1796, an elegant woman stands calmly, hand on hip, cup in hand, with her back toward the leering cleric who tugs mightily at the laces.

The lacing and unlacing of corsets continued to be associated with intimacy and sexual intercourse in nineteenth-century fiction and erotic illustrations.[44] Rather than presenting women as beautiful and seductive, British caricaturists of the period, including Thomas Rolandson and James Gillray, made both men and women appear ridiculous in their anguished struggle to tighten the lady's corset. In Rolandson's etching *A Little Tighter* (1791), the image is one of wincing and groaning as men pit their strength against the recalcitrant stays.[45]

Satirizing women's tight lacing as well as the industrial era's reliance on mechanization for problem solving, caricaturists created images of mock inventions designed to help women tighten their stays. The 1790s marked the appearance of what David Kunzle has called "the windlass theme" in caricatures. In these satiric prints women angrily grimace as their stays are tightened with the aid of a mechanical device. In a satiric French print issued by the publisher Basset titled *New Method of Lacing à l'Anglaise, for Slender Waists, by Milord Bricklinghton*, a man tightens a woman's stays by turning a windlass-type device.[46] Standing with one leg on the ground, the man lifts his other leg high, pointing at the woman's groin. The woman herself stands with her eyes aimed upward, looking ridiculous in her enormous outsized hairdo and outlandish hat.

British artists created their own versions of the windlass theme as they mocked fashion-enslaved women as well as machine mania during the early nineteenth century. In the etching *A New Machine for Winding Up the Ladies* (ca. 1830) a woman's laces are being pulled by a machine that tightens them at the waist. As a female servant turns the wheeled gadget, the stern-faced corsetted woman stands with her back to the machine, wincing and bracing herself as she grasps two poles (plate 16). (Continuing the theme into the twentieth century, the American illustrator Orson Lowell's comic drawing *Girth Control*, of 1917, presented maids operating a windlass that pulls the laces of a hefty woman in a room now lit by a newer technology, electric light.)[47]

Bustles ■ New wired versions of bustles invented in the nineteenth century provided women with another means of fulfilling cultural ideals, helping them to embody an image of femininity and fecundity and standardize their shapes. But again, women wearing the bustles were represented in competing views, as both the height of elegance and the butt of ridicule.

Again, the practice of wearing bustles was not new: women had been wearing rolled pads since the Renaissance, and early versions of bustles were made of wire or whalebone. The padded rolls, which looked like sausages, were worn around the waist and under the skirt (during the Renaissance they were sometimes derisively called "bum rolls"). In the 1770s it again became fashionable to wear rolled pads or cushion stuffing, what the British prime minister Horace Walpole in 1783 called "invisible machines."[48]

During the nineteenth century new versions of bustles were designed to support the fullness of women's skirts and also to give their posteriors a rounded look. In 1815 women wore long, sausage-shaped pads, and by the 1830s bustles were made of whalebone or rows of stiffened material. As the popularity of crinolines waned, women during the 1870s wore pannier crinolines — half-crinolines with a swelled bustle area behind — or separate crescent-shaped horsehair or down bustles that curved across the hips. By the mid-1880s women often chose the narrower "camel's hump" bustles, made of frills of stiff cotton or horsehair, or hooped steels sewn into the dress lining or half-petticoat. In other versions, introducing a more up-to-date technology, separate small wire cages were tied around the woman's waist or hung from her waistband, suspended down the center of the back between the petticoats and the skirt. By 1890, though, the fashion of wearing bustles was largely over.[49]

Like crinolines, the new wired bustles seemed to offer women a way to fulfill the cultural ideal of beauty while feeling that they were benefiting from the latest technologies. In its ad for the B.V.D. Spiral Bustle in 1870, a New York firm promised that the undergarment would impart "a graceful, rounding shape to the figure, in keeping with the Latest Fashions," while the design of its metal spiral design allowed lightness and coolness.

But even as they touted the advantages of their new "scientific" bustles, advertisers in the 1880s also revealed that these new bustles had a major drawback: they had the unfortunate tendency to compress and become deformed when the wearer sat down. The Health Braided Wire Bustles, patented in 1880–85 and advertised by the Weston & Wells Manufacturing Company in Philadelphia and the American Braided Wire Company in London, were designed to alleviate some of the problems. In 1888 the American Braided Wire Company advertised its "Myra Torsion Spring," "Comforto," and "Old Reliable" models, which would "yield to the slightest pressure, yet immediately return to their proper shape after the severest usage," and "properly sustain the heaviest drapery," so that "the wearers are never mortified by their being crushed or bent into ridiculous shapes."[50]

Bustles that were originally designed to help women avoid appearing awkward were actually Rube Goldberg–like contraptions that tended to make women look like robotic, mechanized creatures. The 1877 "Langtry" bustle, named for the famed British actress Lillie Langtry, had metal bands that rose when a woman sat down and sprang back when she stood up again. According to the advertisers, by wearing this inventive bustle the wearer could achieve the "correct Parisian shape."[51]

Though the "camel's hump" (or "horse's rump") silhouette was promoted as haute couture by advertisers, it became yet another source of hilarity to satirists—as did the women themselves. The wire bustles, like the crinolines and corsets, gave contradictory cultural signals. Though women who wore stiff corsets may have hoped to project an image of self-discipline and moral rectitude, their bustles, by emphasizing the rump, produced unintended hints of bestiality. As two clothing psychologists suggested, the bustle made the women look like they "were wearing a credible imitation of a tail."[52]

Artists presented competing views: while paintings by Renoir and Tissot lent an air of grace, charm, and dignity to women's bustled figures, the French artist Jean Béraud in his painting *La Baignoire, au théatre des Variétés* (1883) took a bemused look at the way men responded to women's exaggerated fashions. In a Parisian theater box a woman looks in a mirror as she powders her nose while her mustached male companion takes his own smiling look at her hugely protruding bustle. Meanwhile, a man in the audience, seen through the rectangular opening of the couple's box, peers at them both through his opera glasses.[53]

The French postimpressionist painter Georges Seurat in *A Sunday on*

La Grande Jatte — 1884 (1884–86) evoked a timeless tableau of Parisian men and women at leisure, with the women's stark silhouettes revealing the ungainly nature of the bustle's humped curve. In Seurat's painting, middle-class Parisians sit and stroll at leisure on La Grande Jatte, a park-like island in the Seine. Standing stiffly, the figures are stylized creations, almost robotic in their erect and rigid posture (plate 17).

A hundred years later, in the postmodern era of the 1980s and 1990s, with its analyses of industrial and media reproductions, writers such as Linda Nochlin, in *The Politics of Vision,* viewed Seurat's *Sunday on La Grande Jatte* in terms of its social and technological context. Nochlin saw it, not as a vision of classicized stasis and utopian calm, but as a critique of bourgeois, industrialized society — the painting's flattened, standardized figures with their "machine-turned profiles" a critique of factory mass production and a reference to the mechanized techniques of "endless pictorial reproductions" used for illustrations in the mass press.[54]

To Nochlin, the painting's frozen figures are an allegory of the modern social malaise, the "dehumanizing rigidity of modern urban existence."[55] Seurat's pointillist style of systematized dots of paint does indeed suggest the painter's scientific if not obsessive approach to painting technique, reflecting an age, as Meyer Shapiro wrote, that showed a "profound respect for rationalized work, scientific technique and progress through invention."[56]

But what one also sees in Seurat's painting is the way women's fashions tended to demean women by making their figures look contorted and animalistic. To the right in the painting, a man and a woman stand together as a couple, the woman's dress revealing her high, rounded bustle. Seurat emphasized the bustle by echoing its curve in the woman's parasol and the parasols of other figures in the painting. A small monkey

An advertisement for women's Health Braided Wire Bustles, including "Old Reliable," patented in 1880, which would not get crushed into "ridiculous shapes," *Young Ladies Journal,* 1888.

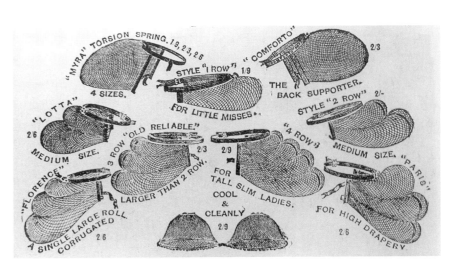

is at her feet, its arched back paralleling her own curves and undercutting her efforts at dignified fashionability. The monkey, with a running dog beside it, becomes a sign of vitality and life, vying with the world of impersonal standardization.

On some levels at least, Seurat's painting of the bustled women sums up the nineteenth century's ambivalent views of women as both elegant and silly, stylized, doll-like creatures who relied on awkward, artificial frameworks to reshape the natural contours of their bodies. In a century that valued new inventions, the mechanized woman was often seen as a technological wonder, a figure that intelligently made use of the latest techniques, however artificial, to improve on nature and achieve a fashionable look. But other views were more skeptical: writing about a new style of corset, the French *Journal des dames et des modes* in 1811 fretted about the silly look of artifice and mechanization: "It is as if a clever artist is playing with dolls that only move with springs."[57]

With the arrival of new machines for traveling, notably the safety bicycle and the automobile, at the end of the century, however, women would increasingly abandon the artificial body-shaping garments and the look of a wired doll. Women riding on their bicycles would cast off their corsets, crinolines, and bustles, leaving behind the outmoded fashions and social restrictions as they turned to other technologies and machines to help reshape their identity anew.

The ELECTRIC EVE

3 IN HIS WHIMSICAL mid-eighteenth-century poem on electricity the German professor Georg Matthias Bose presented a fanciful image of receiving a shocking kiss from a femme fatale: "I kissed Venus, standing on pitch. / It pained me to the quick. My lips trembled / My mouth quivered, my teeth almost broke." Anyone who was scandalized by the experiment, the speaker added, could throw himself into the ocean.[1]

Bose's poem reflects the century's great fascination with scientific demonstrations, and Bose himself was credited by the famed French experimenter, Abbé Jean-Antoine Nollet, with the idea of an electric kiss, produced when a spark of static electricity was generated through friction. Demonstrations of the kiss became widely popular in Europe during the century, but after the invention of the Leyden jar by Musschenbroek in 1745, the sparks produced in electrostatic demonstrations could become more lethal. In Geoffrey Sutton's words, the "charming little sparks and little shocks" became at times a "virtual weapon of terror."[2]

Bose was known for novel and attention-getting electrical experiments, including rigging a dinner table and a chair so that dinner guests jumped when sparks flew from their forks.[3] But his poem about Venus and the electric kiss is particularly revealing because it evokes a fami-

A woman operates an electrostatic generator in an experiment that sends a shock through a circle of participants. From Windler, *Tentamina de causa electricitatis,* 1747. Courtesy of the Bakken Museum and Library, Minneapolis.

liar archetype — the seductive but lethal woman — and reveals the often contingent cultural role of women in the world of scientific discovery. Women in eighteenth-century France actively participated in scientific experiments and also served in a symbolic role, as mythic figures embodying electricity's light-giving and shocking force.

During the eighteenth century women were among the audience members at electrical demonstrations, and Abbé Nollet welcomed the large number of aristocratic women who attended his lectures on electricity.[4] Eusebio Sguario's treatise *Dell'elettricismo . . .* (1747) included an illustration of a woman suspended from a swing who receives a shock as she touches a spinning glass cylinder, and in Peter Johann Windler's *Tentamina de causa electricitatis* (1747) a woman operates an electrostatic generator, helping to send a shock to a group of men holding hands in a human chain.[5]

Goddesses of Electricity and Facsimile Females

■ Women both participated in experiments and appeared in allegorical roles. In the eighteenth-century French print *L'Expérience sur l'électricité* a woman stands on an insulating platform in a demonstration of the use of a Leyden jar. In one hand she holds a long metal rod above a spinning glass globe being turned by a man standing behind her. Friction between the globe and a cushion creates static electricity, producing the light of a corona discharge at the tip of the rod. In her other hand she holds a metal rod dipped into a water-filled glass jar being held by a young black servant, who is about to touch the metal rod and receive a huge electrical shock (the woman, in whose body electricity is flowing, will receive a lesser shock).

The statuesque female figure, standing on a platform, is elevated to a goddess figure in her classical stance, her curved body gracefully draped with cloth. Rather than being shocking, she has a stabilizing role, helping to situate the experiment, with its potentially lethal effects, into a world of rationalism and order.

During the nineteenth century women again appeared in emblematic roles as dazzling goddesses celebrating the emerging electric age and as seductive sirens whose bodies and identities were wondrously — sometimes frighteningly — transformed through electricity. These contrasting images gave shape to the century's hopes and fears about electricity and embodied ambivalent views of women as well.

Women had long appeared as goddesses of light and truth: in the pre-electrical era artists had created images of classically draped nude women representing the embodiment of truth, their extended hands producing light or holding illuminated lamps (as in Jules-Joseph Lefebvre's painting *La Vérité* [1871].[6] But by the late nineteenth century, these goddesses had become emblems of the modern electrical age. Albert Robida's vision of the future in *La Vie électrique* (1890) included his drawing of an electric muse, a woman with charged light radiating from her hair. In New York in 1883, Mrs. Cornelius Vanderbilt was photographed in her emblematic role as "Electric Light" when she appeared in her masquerade dress at the Vanderbilt Ball.[7]

In an era entranced with all things electrical, William J. Hammer, Thomas Edison's assistant at his West Orange, New Jersey, plant, presented his own goddess of electricity at his "Electrical Diabolerie" New Year's party, at his home in Newark. Hammer, a consulting and supervising engineer with offices in Manhattan, was Edison's representative at the Paris Exhibition of 1889, where he supervised Edison's display, which, Hammer said, was visited by thousands of people a day who came to hear Edison's phonograph.[8]

At his party Hammer entertained his guests at "an electric supper" complete with "Electric Toast," "Wizzard Pie," "Telegraph Cake,"

In *L'Expérience sur l'électricité,* eighteenth century, the woman as allegorical figure helps demonstrate the use of the Leyden jar. National Museum of American History, Smithsonian Institution.

"Electric Cigars," and music by "Prof. Mephistophele's Electric Orchestra." Offering further entertainment, his younger sister Mary, dressed in white and mounted on a pedestal, posed as the goddess of electricity wearing tiny electric lamps in her hair and suspended as earrings, while in her hand she held a wand with an electric lamp in the shape of a star.[9]

Women as electric goddesses also became popular at the century's huge world's fairs and electrical exhibitions, like the spectacularly lit Exposition Universelle in Paris in 1900. Electricity at these exhibits was more than a theme, for as David Nye has noted, "it provided a visible correlative for the ideology of progress."[10] Here too, women whose bodies were illuminated or covered in small electric lights became the mythic embodiment of the modern electric age.[11]

In popular nineteenth-century entertainments female music-hall performers danced with their bodies illuminated by tiny electric lights, sometimes in acts designed to convey a sense of danger. New York's Koster & Bial music hall in 1892 featured the female entertainer Nada Beyval, billed as a "chanteuse electrique," singing French songs in her costume studded with rows of miniature electric lights that flashed dramatically on the darkened stage. (The journal *Electricity* dismissed the

An electrical muse in a frontispiece plate titled *L'Électricité (la grande esclave)*, in Albert Robida's *La Vie électrique*, first published in Paris in 1890.

act's danger as "probably much exaggerated" because the house circuit was of a low voltage and the only danger was the possibility of a short circuit, and it added unsympathetically that "the young lady's account of her sensations when the volts pass through her body, may perhaps be ascribed to nervousness.")[12]

Women appeared not only as goddesses of electricity but with their bodies beautified and duplicated by electricity as well. In the last decades of the nineteenth century a remarkable series of inventions generated a heightened sense of expectation and wonder. The incandescent bulb and the telephone helped enhance human vision and communication, and

Mrs. Cornelius Vanderbilt as "Electric Light" at the Vanderbilt Ball, 26 March 1883. © Collection of The New-York Historical Society.

electricity was used to create reproductions and facsimiles of the human body, the telephone and the electrically powered gramophone transmitting the human voice and early experimental motion pictures creating facsimile images of the body.

Celebrating the power of electricity, Thomas Edison in his article "Electricity Man's Slave" used the body as a metaphor to proclaim the wonders of electricity. Of the many factors that had stimulated progress during the last half of the century, he wrote, none had played so essential a role as electricity: "hardly a single nerve or fibre of that complex body we call society . . . has not thrilled and vibrated with its influence."[13]

Edison's metaphor of a vitalized social body had its analogue in cultural images of women whose bodies were enlivened or transformed through electricity. Early on, low levels of electricity were thought to provide medical benefits to women, helping to restart suppressed menstrual periods and cure ailments in their breasts and wombs.[14] Starting in the 1880s, electromechanical vibrators were used to help women achieve sexual orgasm as a cure for neurasthenia and hysteria.[15]

In the nineteenth century electricity was also used to prettify women's bodies and, as hinted in a French journal, to transform their social status as well. One of the fashion fads of the late 1870s and 1880s was battery-powered "flash jewelry" for women, first made popular in France and England. The *Scientific American* in 1879 showed three French examples that used small batteries: a hat ornament consisting of a diamond-studded bird that moved its wings; a scarf pin in the shape of a golden rabbit holding a tiny mallet in each paw, with which it beat a small gong; and a scarf pin in the shape of a skull with diamond eyes and an articulated jaw (the skull was said to roll its eyes and gnash its teeth).[16]

Women also wore imitation gemstone diadems or brooches in which

Battery-powered "flash jewelry" created by Monsieur Trouvé in France. The bird moved its wings, the rabbit beat a small gong, and the skull rolled its eyes and gnashed its teeth. From *Scientific American*, 25 October 1879.

tiny electric lights shining through colored glass created the effect of sparkling, expensive jewels. The lights were powered by tiny 2- to 4-volt batteries worn hidden in their dresses, the switch often carried in their pockets.[17] As the *Electrical World* noted in 1883, the wearing of electric jewelry had a transforming effect on women, for "a dashing *demi-mondaine* can thus make a penny worth of glass eclipse a duchess' diamonds or rubies."[18] The facsimile jewels, it added, were also becoming popular with women higher on the social scale.

These examples of electricity's uses suggest three important themes: electricity as producing enlivened, attractive-looking, transformed women; electricity as "man's slave," as Edison put it; and the nineteenth-century fascination with producing facsimile copies, the illusion of the "real thing." These themes were embodied in nineteenth- and early-twentieth-century images of simulated females created through electricity who were themselves "man's slave."

These facsimile women were present in both fiction and film, like the female android Hadaly, the character so named by a fictional version of Thomas Edison in Auguste Villiers de l'Isle-Adam's novel *L'Ève future (Tomorrow's Eve)*, published in France in 1885.[19] The Electric Eve also appeared as the diabolical double of the virginal character Marie in Fritz Lang's 1926 silent film *Metropolis*, whose script was written by Lang's wife, Thea von Harbou. These versions of the Electric Eve are a revealing reflection and recasting of gender archetypes and constructions of female identity in which women appear as saintly virgin and diabolical femme fatale.

By the period 1877–79, when *L'Ève future* was being written, Thomas Edison had developed the phonograph (for which he was famous in France) and was working on his incandescent lamp at Menlo Park, New Jersey. The fictional Edison is a central figure in the novel and seems to speak for Villiers himself. The novel centers on this fictive Edison's use of electricity to create and activate a female named Hadaly, whose name, we are told, is the Iranian word for "perfection." Hadaly is an exact copy of Alicia Cary, the mistress of Edison's visiting British friend Lord Ewald. Lord Ewald is tormented by Alicia, whom he finds to be a beautiful goddess with a mundane mind. His all-too-human lover is a twenty-year-old actress and singer, an "almost superhuman beauty," who is compared to the sculptural *Venus Victorious* and the Venus de Milo.[20] But in spite of her classic features, Alicia is also a woman whose every word is a mediocrity, who possesses a "vulgarity of mind." She is even likened to a machine in the "mechanical fidelity of her singing."[21] Lord Ewald, filled with passion, finds himself both attracted and repelled by her, which leads him to contemplate suicide.

To help his friend, Edison offers the possibility of creating a female

Plate 1. Henry Alken, *The Progress of Steam*, etching, 1828, a comic and stereotypical view of women who are cleanliness-minded and ignorant about anything mechanical. Science Museum / Science & Society Picture Library.

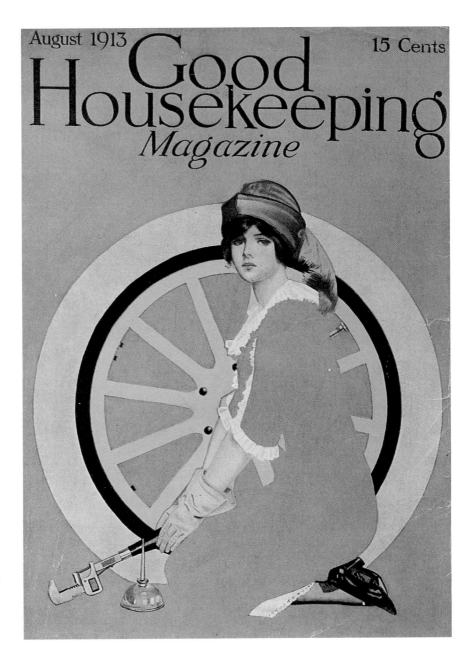

Plate 2. A familiar theme—the grounded woman forlorn over her flat tire. Coles Phillips, cover illustration, *Good Housekeeping*, August 1913.

Plate 3. A glossy view of sex and the machine in James Rosenquist's painting *Gears* (1977). © James Rosenquist / Licensed by VAGA, New York.

Plate 4. In this Office of War Information photograph by Alfred Palmer an African American woman operating a hand drill is mirrored in the surface of a Vengeance dive bomber at Vultee-Nashville in February 1943. Library of Congress, FSA-OWI Collection.

Plate 5. Young girls apprehensively witness a scientific experiment in which pumping air from a glass receiver is causing a white cockatoo to die. Joseph Wright of Derby, *An Experiment on a Bird in the Air Pump*, 1768, oil on canvas, National Gallery, London.

Plate 6. Welcoming a new era, this goddess is encircled by central inventions of the age. *Dawn of the Century,* march and two-step, sheet music by E. T. Paull, New York, 1900.

Plate 7. Lithographic poster by Georges Rochegrosse, 1901. A goddess of the machine presides over the Fourth International Exposition, sponsored by the Auto Club of France. Poster Photo Archives, Posters Please, Inc., New York.

Plate 8. A formidable female decorates this poster by the Belgian artist Privat Livement for the Fifth International Exposition of the Automobile Club of France, 1902. Poster Photo Archives, Posters Please, Inc., New York.

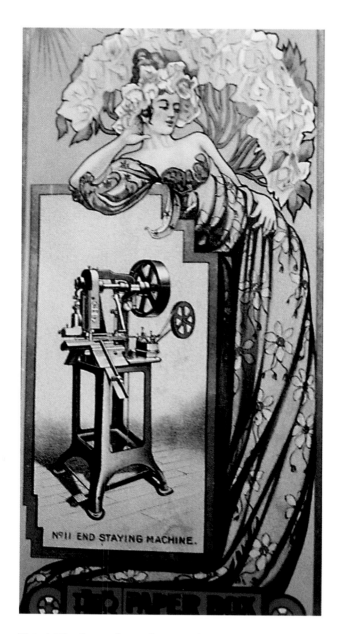

Plate 9. The sinuous figure of a woman with art-nouveau curves lends appeal to an advertisement for an end-staying machine displayed at the Pan-American Exhibition of 1901 and manufactured by Knowlton & Beach Paper Box-Making Machinery, of Rochester, New York, and London, England. Collection of The New-York Historical Society.

Plate 10. Swirling papers and money surround this typist in a poster by Leonetto Cappiello advertising Remington typewriters, 1910. Poster Photo Archives, Posters Please, Inc., New York.

Plate 11. Lithographed advertisement for a Pittsburg-Visible writing machine, published in Hamburg, Germany, n.d.

Plate 12. Natalia Goncharova, *The Weaver (Loom + Woman)*, oil on canvas, 1913. The woman weaver is energized but almost hidden amidst the textile machinery around her. National Museums and Gallery, Cardiff. © 2000 Artists Rights Society (ARS), NY / ADAGP, Paris.

Plate 13. Elegantly clad women at a London dress ball. James Jacques Joseph Tissot, *Too Early,* oil on canvas, 1883. Guildhall Art Gallery, Corporation of London, UK / Bridgeman Art Library, London / New York.

Plate 14. *Crinoline Difficulties,* stereograph. Foiled by her awkward crinoline, this woman takes a spill as she exits from a London omnibus. Courtesy of T. K. Treadwell.

Plate 15. A group of women about to be caged by a steel crinoline. Stereograph, London Stereoscopic Company, late nineteenth century. Courtesy of T. K. Treadwell.

Plate 16. *A Correct View of the New Machine for Winding Up the Ladies*, etching, ca. 1830. A laughing look at a windlass-type mechanism designed to help women with tight lacing. Science Museum / Science & Society Picture Library.

Plate 17. Georges Seurat, *A Sunday on La Grande Jatte—1884*, oil on canvas, 207.6 cm x 308 cm, 1884–86. The curved silhouettes of fashionable women at leisure on the island of the Grande Jatte near Paris. Helen Birch Bartlett Memorial Collection, 1926.224. Courtesy of The Art Institute of Chicago; all rights reserved.

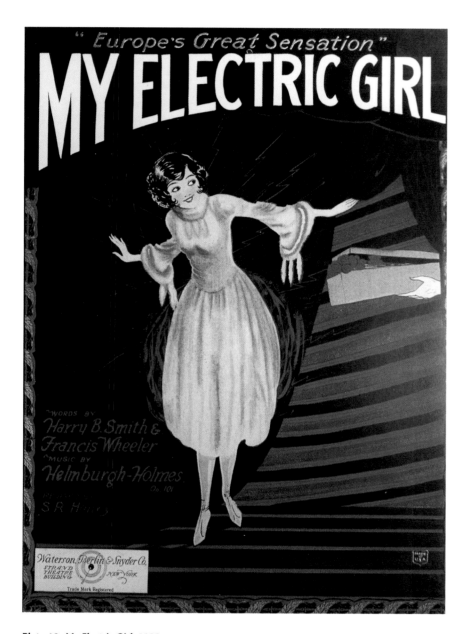

Plate 18. *My Electric Girl*, 1922.
Sheet music, DeVincent Collection, Archives Center, National
Museum of American History.

Plate 19. Holding a coaster brake aloft, this cyclist celebrates one of the improvements that made bicycling easier for women. Poster Photo Archives, Posters Please, Inc., New York.

Plate 20. A formidable woman warrior helps advertise Liberator bicycles in this copy by Emile Clouet of a poster by PAL (Jean de Paléologue), 1900. Poster Photo Archives, Posters Please, Inc., New York.

automaton that will be an exact copy of Alicia, with the same features, hair, and even body perfume but with none of her mediocrity of mind. She will be created through a mysterious photosculpting process and activated by electricity. Electricity, "this vital, surprising agent," will give the automaton "all the soft and melting qualities, all the illusion of life." Most tellingly, Hadaly will be not only an exact facsimile but a technologically created reproduction that surpasses the original. "This copy, let's say, of Nature . . . will bury the original without itself ceasing to appear alive and young," says Edison. "Oh," he adds, "it's better than real!"[22]

Lord Ewald is enthralled by Hadaly, whom he mistakes for the actual Alicia, and takes her, encased in a coffin, on a steamship voyage home to England. But as Edison learns by telegram at the end of the novel, the steamship suffers an explosion produced by turpentine and gasoline. Hadaly is destroyed in the ensuing flames, and Lord Ewald kills himself in despair.

In one sense, the story of Edison and his Electric Eve reveals the nineteenth century's absorption with new technologies and electricity. The Edison of the novel is, to a certain extent, the stereotypical mad scientist—like Dr. Frankenstein, intent on making scientific discoveries and not always mindful of the consequences. In developing methods to avoid railroad collisions, for example, the fictional Edison arranged to have two trains go at each other full force. This experiment failed, however, and many people were killed.

In his experiment to create the electric android Hadaly, Edison has a vision of both hopefulness and foreboding. He optimistically envisions the advantages of mass-producing female androids: not only will he provide life-giving solace to Lord Ewald but he will provide men everywhere with a beneficial alternative to what he considers the destructive influence of marriage-breaking mistresses. In making the android, though, Edison also has a premonition of a tragic outcome for his new technology.

The hopes and fears invested in the fictional Edison's facsimile woman echoed the larger currents of ambivalence associated with nineteenth-century electrical inventions themselves. Nineteenth-century journals wrote optimistically about the benefits of electricity but also ominously about its dangers. While the real Edison's experiments heightened expectations for a new world illuminated by electric lamps, nineteenth-century engineering journals also warned of electrocution, and Edison himself was later engaged in the debate surrounding which type of electricity to use for the electric chair, a lethal device that was considered scientifically efficient and a humane way to kill.

The novel's references to steamship explosions and a railroad wreck early in the novel also reflected actual horrifying railroad and steamship catastrophes that were reported regularly in the American and European press. For all of the rhetoric of progress in the nineteenth century, these disasters were a constant reminder of the dangers that accompanied new technological developments.[23]

In Villiers's novel Hadaly's creation is presented in pseudoscientific jargon. Edison announces that he will use "electromagnetic power and Radiant Matter" to create her and that she will be covered with artificial flesh woven with induction wires.[24] Though her appearance will be convincingly lifelike, she will be very much an artificial being: her interior will be made of metal armor, and she will give forth puffs of smoke. Alicia's facial expressions and physical gestures are recorded in a pattern on the android's central cylinder. Hadaly even seems to breathe, for she takes in air and her breath is warmed by electricity. Hadaly's words are recorded on two golden phonographic disks set in motion by electricity that contain seven hours of language taken from famous writers and recorded by performing artists.

Hadaly's phonographic core had an analogue in a phonographic doll that Edison himself patented in 1878 but that was not mass-produced until 1889–90 at his West Orange, New Jersey, factory. The doll had a German bisque head and jointed wooden limbs, and its steel torso housed a miniature motorized phonograph with wax cylinders and a small needle tracing grooves on the cylinder. The doll's facsimile voice was created by girls working at Edison's plant, who recorded the words and sang nursery rhymes, such as "Mary Had a Little Lamb." By 1893 the French doll manufacturer Jumeau was producing its own "Bébé" phonographic doll, which was advertised as being able to talk and sing and having a vocabulary of seventy-five words; it had phonographic cylinders that allowed it to speak in French, English, or Spanish.[25]

But whereas in the novel Hadaly's words are melodious to Lord Ewald's ears, the actual phonographic dolls were reportedly cacophonous. By 1890 the production of dolls by Edison's manufacturing company was suspended because of the dolls' fragile mechanisms, and there were complaints that the voices were unpleasant to hear.[26] At the 1900 Exposition Universelle in Paris young girls sat in a special room at recording benches recording words for other versions of phonographic dolls, and a contemporary description evoked a surreal scene: "Each one sits before a large apparatus, singing, reading, crying, reciting, talking with all the appearance of a lunatic! She dictates to a cylinder of wax the lesson that the little doll must obediently repeat to the day of her death with a guarantee of fidelity."[27]

Rather than being grotesque, however, Hadaly is a benevolent version of an android, not only a product of nineteenth-century fantasies about technology but also an extension of the eighteenth century's love of mechanically driven toy automatons and the Cartesian and French philosophes' fascination with the human body as machine.[28] Particularly fascinating to European audiences were automatons such as a mechanical lady musician created by the Swiss craftsmen Pierre and Henri-Louis Jacquet-Droz in 1783, which played the clavecin and rolled her eyes and whose chest rose and fell as she breathed.[29]

But in *L'Ève future* Edison mocks these early automatons as "monsters" that, with their jerky movements, are obviously artificial and adds that in his era "the techniques of reproduction . . . have been rendered more precise and perfect."[30] His electrically wired Hadaly is superior, for she moves without jerkiness and will appear completely natural and lifelike, an exact simulacrum of a human being, a double of Alicia.

Edison's insistence of Hadaly's verisimilitude in part reflects the nineteenth century's fascination with producing imitations, reproductions, and facsimile images, evident in the development of the phonograph, photography, motion pictures, and techniques like electrotyping for reproducing exact copies of original works of art. During the 1840s William Henry Fox Talbot, in England, developed calotypes, his method of photographic reproduction using negatives, and chromolithographs were also later developed to mass-produce copies of art. Starting in the 1820s and 1830s, manufacturers also began producing decorative cast-iron reproductions of original sculptures. Electricity too was used in the "imitative arts."[31]

Phonographic doll manufactured at Thomas Edison's West Orange, New Jersey, factory. Young girls recorded the dolls' voices on wax cylinders. From *Scientific American*, 26 April 1890.

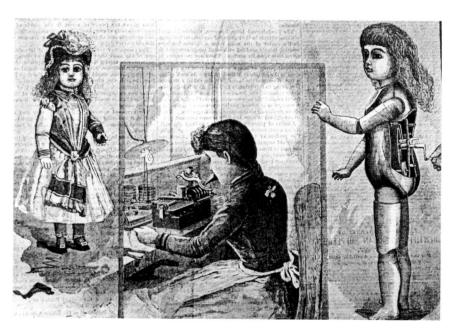

In 1838 Moritz Hermann von Jacobi, in Russia, and Thomas Spencer, in England, simultaneously developed a method of electrotyping, creating replicas of sculptural objects by using an original object or first making a mold in plaster of Paris, wax, or clay, coating the mold with a conductive material such as graphite or lead, immersing the mold in a chemical solution and subjecting it to electric currents, which left it coated with a thin layer of copper. The resulting replica was then electroplated with a thin layer of silver. Electrotypes of famous museum masterpieces, including a chalice by the Renaissance sculptor Cellini, were highly valued as works of technological ingenuity.

Electroplating was a process developed in the 1840s whereby decorative objects such as die-stamped tableware and vases made of base metal, including nickel and zinc alloys, were similarly immersed in a chemical bath and subjected to currents, which left a coating of copper or, more often, silver deposited in a thin veneer. The resulting ornately decorated tea sets and vases were a commercial success in both Europe and America.

Nineteenth-century British art critics, such as A. W. N. Pugin and John Ruskin, fretted that these imitative arts were inherently deceptive and degraded, for, as they complained of electroplating, the thin veneers

Lady musician, a clockwork automaton created by Pierre and Henri-Louis Jacquet-Droz in 1783 that rolled her eyes and played a clavecin, or harpsichord. Musée d'art et d'histoire, Neuchâtel, Switzerland.

of precious metals simply covered base metals. They also complained that the imitations and plated wares were simply crude versions of delicately wrought originals. But nineteenth-century manufacturers, the popular press, and cultural commentators often praised these reproductions, insisting that they were equal if not superior to the original models.[32] In 1876 the *Art Journal* concluded confidently that the high aesthetic, educational, and moral value of electrotyping had virtually eliminated the differences between reproductions and originals: "For all the purposes of Art—to give pleasure, to refine taste, to convey instruction—the electrotype is quite as good as the original in costly metals of gold or silver; indeed, it may be a question which would be preferred."[33]

The character Edison in Villiers's novel similarly argues that Hadaly is "better than real" because, among other reasons, she will be an idealized woman, transcending human frailties: "The Android knows neither life nor illness nor death. She is above all the imperfections and all the humiliations."[34] She is also ideal in that she is a woman whose every response is technologically controlled, whose words are beautiful, elegant, and predetermined. By phrasing his questions properly, Lord Ewald can receive one of Hadaly's artfully articulated programmed responses to match his every mood. Referring to Alicia, the original model, Edison insists, "The present gorgeous little fool will no longer be a woman, but an angel; . . . no longer reality, but the IDEAL!"[35]

The Electric Eve in Villiers's novel is a female facsimile copy, a simulacrum created through the use of electricity. She is also a revealing paradigm of nineteenth-century stereotypes about women as obedient slave, as saint or angel, and as alluring and dangerous siren. In her complexity she is a version of the type of female figures made mythic in Victorian England, where, as Nina Auerbach observed, women were often envisioned with mutating identities: they existed in a world where angels were "irrefutably female," where demonic iconography was predominantly female, and where an angel could metamorphically "modulate almost imperceptibly into demon."[36]

Drawing on the image of the biblical Eve as well as on these sentimentalized, nineteenth-century views of feminine nature, Edison in the novel sees women in polarized terms, as either innocent or evil, either self-sacrificing angels—"purified, elevated spirits" with a "dedication to duty, by self-abnegation and free devotion"—or demonic temptresses. As both primal Eve and the Virgin Mary, women are at once innocent virgins and sexual temptresses, the degraded and the sublime. Hadaly herself is conceived of as virginal and pure, and she and Ewald will have no sexual relations: she tells Ewald, "You had better not touch that deadly fruit within this garden!"[37]

The Alicia of *L'Ève future*, however, is a fallen woman. Seduced earlier by her fiancé, she can no longer be married. Like the nineteenth-century electroplated vase, she has a beautiful exterior but is essentially base. Her body fulfills the ideal of feminine perfection, but her voice is an "empty instrument." As conceived of by Villiers, women like Alicia are deceptive types who lure men to their doom. In the novel, Edison is motivated to create an idealized version of a woman by the experience of his married friend Edward Anderson. At a performance of Gounod's *Faust* Anderson is seduced by the evil ballet dancer Evelyn Habal, and he later commits suicide in despair.

Evelyn herself is a "deadly female," one of those "creatures of man's second fall, these inciters of evil desire," "sirens who destroy." (But while Edward Anderson was led to bankruptcy and suicide, his wife in the novel appears as a type of savior. Anny Sowana—a play on the name Anderson—emerges in the novel as a type of artist and mystic: it is she who sculpts a model of Alicia, and it is she who animates Hadaly with her supernatural being. And like the ideal of the self-sacrificing woman, she dies in the process of giving Hadaly a semblance of soul or life.)

Often, however, real women in Villiers's novel are portrayed as inherently artificial and mechanistic. When Ewald voices his fear that Hadaly can "never be anything but a doll without feeling or intelligence," Edison tells him to contrast a real woman with the automaton and see if "*it isn't the living woman who seems to you the doll.*" The dancer Evelyn Habal herself is a mechanical woman: she is "the Artificial giving an illusion of life." She has her own buried machinery in the form of metal corsets, wigs, and makeup. Women like Evelyn are morally degraded as well: with their superficial physical attractions affixed like nineteenth-century electroplating and veneers, their thin exteriors covering base metals, they are essentially frauds. Edison concludes that "any woman of the destructive sort is more or less an Android." It is to save the world's marriages from these duplicitous temptresses that Edison proposes to create his facsimile women. Not only will the facsimiles be benign and beautiful rather than evil but, as true products of the new technological age, they will be prototypes for mass production, created and reproduced without the need for sex.[38]

Hadaly herself has a mechanized existence, one called into being by Ewald as an "objectified projection of your [Ewald's] own soul." She is, as the real-life Edison envisioned electricity, a version of "man's slave," literally a push-button automaton, aroused or activated by touching one of the rings on her fingers, which have sensitized stones. Says Edison, "Living women too have rings one must press." While Alicia's words to Ewald often sounded a "dissonant note," Hadaly's words, those of eminent writers encoded on metal cylinders or phonograph records within her, will offer him pleasing answers to his every thought, making her a mirror rather than a source of bafflement.[39]

(Creating his own version of the mechanized, mirroring woman, the real-life Thomas Edison was photographed at his West Orange laboratory with a woman who was using one of his inventions, an "Ediphone," a proto-dictaphone in which the typist, connected by a cord to the machine, transcribed his words.)

The nineteenth century's fascination and fears about technology, as well as about women, are also evident in the novel's references to *Faust*, to Milton's *Paradise Lost*, and to the story of Noah in Genesis, where God repents for having created humanity. All of these suggest the dire consequences of flirting with evil, like the new technologies that seemed to transcend the "natural" and cross dangerous frontiers into the artificial. Revealing the century's dual views of technology, dire news of the steamship disaster comes via a telegram, itself one of the nineteenth century's new electric wonders. And like these wonders, Hadaly herself is ultimately both life-saving and life-destroying, for though she enlivens Lord Ewald, her own death leads to his suicide at the end.

In the early twentieth century the idea of an electrified woman again

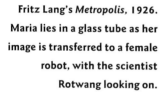

Fritz Lang's *Metropolis*, 1926. Maria lies in a glass tube as her image is transferred to a female robot, with the scientist Rotwang looking on.

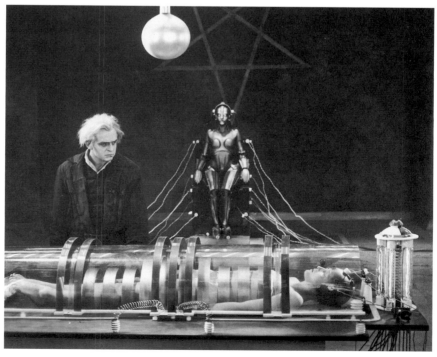

captured the popular imagination, like the light-hearted American song *My Electric Girl* (1922), about a young woman whose touch brings high-current romance rather than a shock (plate 18). But in Fritz Lang's 1926 film *Metropolis* the idea of a woman created through electricity also evoked a grimmer view. As in *L'Ève future*, technological catastrophe figures in the film, which opens with an explosion produced by excessive pressure in a steam boiler in an underground factory world peopled by robotic "machine men," depicted as slaves to the factory system and dominated by the Master of the Metropolis, J. Fredersen. Fredersen's son, Freder, is haunted by factory accidents and the oppressive, clock-work labor demanded of the workers.

In another scene, beneath the city, the saintly Maria offers comfort to the workers in a sanctuary among the catacombs and pleads with them to remain peaceful. Freder is captivated by the angelic Maria, but his father feels threatened and asks a mad scientist, the inventor Rotwang, to build a robot in Maria's likeness to sow discord among the workers and destroy their confidence in her.

The simulated version of Maria is created through electricity amidst flashing rays, floating rings of light, and bubbling tubes. Rotwang had earlier created a fabricated, stylized metallic female robot-worker, which he now connects by cables to the real Maria, who is encased in a glass box. Subjected to electricity, the robot is transformed into Maria's exact double. In contrast to the pious, nurturing, angelic Maria, however, the android Maria is a seductive dancer and *belle dame sans merci*. "The copy is perfect," says Fredersen, and he orders the facsimile Maria to stir the workers to perform criminal acts, which will allow him to destroy them in the ensuing melee. Maria becomes a version of what Joy Kasson in her study of nineteenth-century sculptural images of women sees as an enduring view of female transformation: "The persistent vision of woman's transformability suggests a nightmarish world where good and evil, safe and dangerous, domestic and demonic, prove indistinguishable."[40]

In a hallucinatory, surreal montage in the film, a vamplike version of Maria as siren rises out of a huge cauldron, the centerpiece at a party attended by male guests in black tie, who reach toward her with leering eyes. Rushing to Maria's underground sanctuary, Freder, the son, finds only images of the seven deadly sins. Meanwhile, the false Maria incites the workers to riot and to destroy everything, while the real Maria is held prisoner at Rotwang's house. The angelic Maria escapes, and her diabolical double is burnt at the stake and reverts back to the look of a mechanistic android. In a hellish Armageddon the electric dynamos explode amidst a huge flood, and the fleeing workers collide with black-tie celebrants. Lang ends the film on a pietistic note. Fredersen, Freder, and Maria stand with hands joined as the screen's text proclaims the value of the heart as mediator between the hands that build and the brain that plans.

The Electric Eve in *Metropolis* presents not only a dualistic view of women but also the lurking fear that women are not what they seem. What looks like the savior Maria is really a subversive siren. The film also presents a technological world that is both transcendentally uplifting and also degrading. As with the nineteenth-century factory-made imitative arts, the elevated, original Maria is rivaled and almost supplanted by her degraded double.

Women Consuming and Celebrating Electricity

■ Women had been portrayed as goddesses of electricity, exotic androids, and seductive advertising lures, but during the early decades of the twentieth century manufacturers also asked women to envision themselves differently: as consumers of new electrical products of industry. Rather than being passively switched on like Hadaly, they were courted as customers for electric automobiles, electrical appliances, and electric

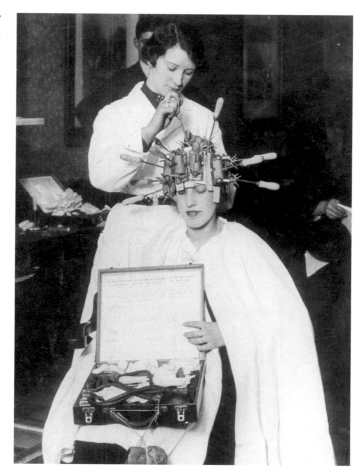

A woman demonstrating a new electric permanent-wave machine. Photograph by James Jarches in the London *Daily Herald* ca. 1937. National Museum of Photography, Film & Television / Science & Society Picture Library.

lights in their home. (Lady Randolph Churchill, whose London home was lit by electricity in 1880, purportedly had a pear-shaped electrical light switch in her bedroom that she mischievously called "Randolph.")[41]

After 1918, American appliance manufacturers and utility companies stepped up their efforts to promote the use of electricity by housewives, and by the 1930s in Britain, when most of the country was linked by the national grid, the national power-transmission system, women were similarly encouraged to use electricity in their homes. To promote the use of electricity, the Electrical Association for Women (EAW) in England published the magazine *The Electrical Age for Women* from 1934 to 1983. It also published several editions of the *Electrical Handbook for Women*, which told women how to use appliances, how to do their own electrical repairs, how to teach the uses of electricity to other women and children, and how to sell appliances to other women.[42]

In their appeals to women consumers, advertisers changed their strategies: the barely disguised erotic imagery of scantily clad "fairies" of electricity and General Electric's Mazda ads with seductive women in harem clothes were increasingly supplanted by images of apron-wearing women using electric appliances in their homes. Not only were women cast as consumers of electricity but they were photographed using new electrical technologies to prettify their hair. In 1907 in Paris, Charles Nestle demonstrated his electrical machine for giving women's hair a permanent wave, and in 1918 in England, Eugene Sutter improved the machine further. In the process call *frisée forcée*, hair was soaked in an alkaline solution and subjected to heat from an electric current, creating waves that could last six months.[43]

Though hair salons were associated with the idea of feminine beauty and elegance, newspapers and magazines featured much less glamorous images of women sitting under permanent-wave machines. The electrical muse in Albert Robida's *La Vie électrique* radiated shining rays of light, but the woman shown getting an electrical permanent wave in a photograph in London's *Daily Herald* in 1937, with her hair in metal cylinders splaying from her head, looked more like a machine-age Medusa than a goddess of the machine.

While women were imaged as housewives enthralled with appliances and as beauty-minded maenads, female writers and artists created their own conceptions of electricity as a potent force in modern life. Natalia Goncharova exhibited her paintings *Electric Lamp* (1912) and *Dynamo Machine* (1913–14) at the "No. 4" exhibition in Moscow in 1914.[44] With its electric blues and jagged flashes of yellow light, *Dynamo Machine* captures the dynamic energy and excitement of an era entranced with modern technologies.

In her painting *Electric Prisms* (1914) the French artist Sonia Delaunay presented whirling color disks as a bold image of the early-twentieth-century infatuation with electricity. In her memoirs Delaunay wrote "j'aimais l'électricité" (I loved electricity) and recalled strolling down the Boulevard St. Michel in Paris with her husband Robert in their early years and being inspired by the circles of light from the new electric lights that were replacing gas lamps: "At night, during our walks, we entered the era of light, arm in arm. . . . We would go and admire the neighborhood show. The halos made the colors and shadows swirl and vibrate around us as if unidentified objects were falling from the sky, beckoning our madness."[45] *Electric Prisms*, with its concentric, disjunctive disks of bright color, captured the dazzling sensation of radiating light and the whirling of wheels.[46]

Electric Women   ■ In her own celebration of electricity's potent force the American poet Harriet Monroe, in her poem "The Turbine" (1917), went a step further, fusing features of a turbine and electrical generator to create a formidable image of woman-as-machine. Monroe, who founded the influential magazine *Poetry* the same year, captures in her poem the speaker's worship and terror of the fearsome yet also disarming machine queen, whose "royal business" is to light the world. Drawing on familiar cultural conceptions of femininity, Monroe presents this complex queen sitting on her throne as "ladylike and quiet as a nun" yet also moody and changeable, for "she's a woman / Gets bored there on her throne" and "Tingles with power that turns to wantonness" and "truant mood."

Monroe's queen is prone to unexpected upheavals: showering sparks and light, she is beset by "Destructive furies" as she suddenly laughs and "Bedevils the frail wires with some mad caress," calling down "ten thousand lightnings / To shatter her world, and set her spirit free." But she is also capable of tenderness, for she shares with the speaker a sense of intimacy as they exchange confidences, jokes, and "little mockeries."[47]

While Monroe envisioned her woman-as-machine as an alluring but fearsome queen beset by her own "Destructive furies," dada and surrealist artists produced much more sardonic images of electrically charged women, both seductive and coy. In his sardonic drawing *Portrait d'une jeune fille américaine dans l'état de nudité*, of 1915, Francis Picabia, the French artist who was allied with the New York dada artists and the Stieglitz group, created an image of a young American female as a spark plug, mechanically both cool and hot. Giving a short burst of generative power, this female spark plug bears the word FOR-EVER, coyly holding out the promise of the eternal and enduring.[48] Two years later Picabia created an image of a young American female as an Edison light bulb on the cover of his journal *391*. (In 1920 the Berlin dada artist Hannah Höch went further, spoofing machine idolatry as well as cultural constructions of feminine beauty in her witty photocollage *Das schöne Mädchen* [Pretty girl], in which a young woman, surrounded by cut images of automobile insignia — sports a light for a head.)[49]

In his own image of the electrified woman the photographer Man Ray drew on an archetypal image, the classical Venus de Milo.[50] Born Emmanuel Rudnitsky in Philadelphia in 1890, Man Ray became a close friend of Marcel Duchamp's in New York, and in 1921 he went to Paris, where his circle of friends included Duchamp and Picabia. Using a technique first used by Talbot in his investigation of light, Ray produced what he called *rayogrammes*, in which objects were put on unexposed sheets of photographic paper in a developing tray and then exposed to a light source, usually a plain light bulb, producing a monoprint of abstracted forms. In 1931 he published *Électricité*, a portfolio of ten photogravures of rayographs created using electric light and commissioned by the Companie Parisienne de Distribution d'Électricité.[51]

The portfolio included images of an electric fan, a toaster, and an incandescent bulb, as well as two images of female torsos, *Salle de bain*

Électricité, a Venus-like embodiment of electricity, photogravure of a *rayogramme* by Man Ray. From *Électricité: Dix rayogrammes de Man Ray*, 1931. Spencer Collection, The New York Public Library, Astor, Lenox and Tilden Foundations. © 1999 Man Ray Trust / Artists Rights Society, NY / ADAGP, Paris.

and *Électricité*. In the image titled *Électricité* a photograph of a woman's torso with no head or arms, suggesting the classical Venus, overlies a second nude torso and is crossed diagonally by flat, white wavy lines signifying electrical current. This Venus embodies an erotic fantasy and is also a metaphor for electricity itself, the white wavy lines exuding a high-voltage charge.

Ray's images in his portfolio pay tribute to the wonders of modern electrical appliances, and as Pierre Bost wrote in the portfolio's text, they conjured up the magic of push-button machines: "'*Presser sur un bouton' est devenu le geste magique des contes modernes et futurs*" ("Pressing a button" has become the magic gesture of modern and future tales).[52] But Ray's female torsos also tell another important story: they are a version of Villiers's Alicia in *L'Ève future*, the sensual female who was also likened to the Venus de Milo and to the push-button Hadaly. (His actual model for his electric torsos was the American Lee Miller, who was his lover, his assistant, and his model in 1929–31 and who had her own career as a fashion photographer.)[53]

Ray's photographic torsos in *Électricité* clearly reflect his obsessions with electric light, with truncated female torsos, and with provocative Pygmalion-like images of the erotic sculptural female figure come alive. During this period he used the process of solarization to create photographs of mysteriously ethereal, truncated, and isolated female torsos, keeping his technique secret until 1932.[54] His obsession with fragmented and disembodied female figures may have been fed by Miller's appearance in Jean Cocteau's film *The Blood of a Poet*, in which casts of classical sculp-

Left, Nancy Burson, *Untitled* (girl with doll's eyes), digitally produced from a Polaroid Polacolor ER Land Film print shot from the computer screen. Copyright 1988 by Nancy Burson. Courtesy of the artist.

Right, Nancy Burson, *Untitled* (mannequin with real girl's mouth and eyes), digitally produced from a Polaroid Polacolor ER Land Film Print shot from the computer screen. Copyright 1988 by Nancy Burson. Courtesy of the artist.

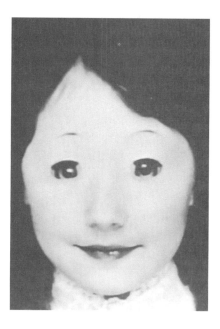
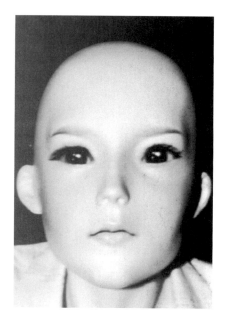

ture, including the Venus de Milo, seemed alive and had the heads of living persons. During the 1930s Ray continued to produce photographs of stone female figures coupled with living models.[55]

In 1937 the French magazine *L'Art vivant* published a story about the Electrical Exposition being held in Paris that year. The story featured photographs of two archetypal images of women and electricity: Man Ray's *Électricité* and a nineteenth-century photograph of a woman posing as a "fée," or fairy, of electricity complete with an electrically illuminated star on her wand.[56] And many years later another archetype, the Greek goddess Elektra, was the reigning muse of a major celebration of electricity and electronics held in a Paris art museum in 1984.[57]

But by the late twentieth century female photographers were using digital imaging techniques to fashion their own images of female identity. Using morphing techniques to fuse two digitally scanned photographs, the New York artist Nancy Burson created eerily penetrating composite faces of women-dolls. One of these startling photographs presents the fusion of a girl's face with doll's eyes, and a second fuses a mannequin's face with a real woman's eyes and mouth. The ambiguously artificial girl with doll's eyes invites us in, draws us closer with her demure and friendly smile, while the mannequin with woman's eyes and mouth remains impassive, impersonal, and impenetrable, keeping us at a distance.[58]

The representational images and discourse of postmodernism helped bring images of the Electric Eve full circle. In a world dominated by electronic means of reproduction, film and video images, digital photographs, and animation mediate between the viewer and a direct perception of the "real thing." Digital imaging and computer-generated virtual-reality environments created artificial worlds that reevoked the nineteenth century's hopes and misgivings about the copy displacing the original and the artificial replacing direct experience.

As Donna Haraway wrote in *Simians, Cyborgs, and Women*, we live in a cyborg world, a world of electronic communications in which the difference between the artificial and natural remain ambiguous, a world in which "our machines are disturbingly lively and we ourselves frighteningly inert." It is also an age, as she wrote, of "transgressed boundaries, potent fusions," an age in which women confront the task of retaining their feminine identity while being unafraid of embracing "partial, contradictory, permanently unclosed constructions of personal and collective selves."[59] The work of photographers such as Burson, with her images of cyborg women, continue the dialogue of Villiers's *L'Ève future*, presenting a compelling and disturbing vision, through the medium of electricity, of our own problematic "potent fusions" and fractured identities.

WOMEN *and the* BICYCLE

4 DURING THE 1890s, women in America and Europe joined in the bicycle-riding craze, delighting in the sense of mobility and freedom that it gave them. "What enjoyment to a cramped and warped women's life is the whirl of the wheel," wrote America's *Wheelwoman* magazine in 1896, adding that bicycle riding raised a woman's thoughts above "household cares," letting her "rejoice in the feeling of liberty and delight in her own strength."[1]

The boom in bicycle riding among women was prompted by the availability of new two-wheeled "safety bicycles," relatively stable vehicles with a dropped-frame version specially adapted to women's needs. Mastering these new machines helped women challenge the stereotypes that portrayed them as timid and inept. But while women were rejoicing in their new-found feelings of liberty, their riding almost immediately caused critics to fret.

Women riding bicycles had long been a tempting target for satirists, who laced their comic images with sexual innuendoes. Particularly popular were early satires of women riding the two-wheel version of the draisine, patented by Baron von Drais in Germany in 1817 and manufactured in France, England, and America.

The standard draisine, also known as a hobby horse, or "hobby," and called an "accelerator" in Britain, had two wooden wheels that were propelled by the rider, who straddled the high backbone and walked forward. Because of its tricky steering mechanism and because the rider had to balance the machine and propel it forward, operating the draisine was difficult for both men and women, particularly given the poor condition of the roads. To accommodate women's skirts, Denys Johnson, in England, produced the first women's two-wheel model with a dropped frame in 1819, and Hancock & Company in London produced a lady's hobby, also known as a pilentum or accelerator.[2]

Attempting to attract female consumers, a contemporary advertisement featuring a woman riding her hobby took pains to point out that the lady's model was equal in power to the gentleman's but was less liable to upset. Sitting on a wooden seat, women could lean forward and rest their arms on a cushioned board, while the drapery of their dresses flowed "loosely and elegantly to the ground."

But while advertisers conjured up images of handling the draisine with dignity, British satirists created comic images of fashionable male dandies astride their new machines and lampooned women riders, who were seen as risqué in their efforts to ride. In Robert Cruikshank's 1819

Robert Cruikshank, *The Ladies Accelerator,* 1819. This print satirizes the new fad of riding hobbies, or accelerators, and takes aim at women as risqué riders of the new machines. © **The British Museum.**

A Pilentum or Lady's Accelerator, ca. 1820. A delicate-looking woman holds the steering wires of her accelerator as she rides through her estate. Science Museum / Science & Society Picture Library.

engraving *The Ladies Accelerator* two buxom women riding a man's version of the vehicle suggestively straddle the hobby's high frame, their breasts supported on a special steering-mechanism bar, as one of the women comments, "I do not see why Ladies should not have a Lark as well as the Gentlemen." In another engraving of the same year with the titillating title *The Female Race! or Dandy Chargers Running into Maidenhead* Cruikshank's comic glee is aimed at two elaborately dressed women who also lean provocatively over the hobby's high frame with their legs and petticoats exposed. As they walk forward with their vehicles en route to "Maidenhead Thicket," one says happily, "When inclined to give one's legs a stretch, I don't know a more delightful exercise."

Nineteenth-century skeptics also suggested that because of women's delicate and frail temperaments, they would find bicycles difficult to ride and control. Indeed, a British engraving from ca. 1820 shows a fashionably dressed woman riding daintily on her pilentum, a three-wheeled version of the draisine for women.[3] Her feet, in tiny slippers, touch the machine's treadles, and her delicate hands hold the two wire controls extending from the front wheel. As one bicycle historian noted, "the lady might perhaps make a few circuits of the grounds of her country mansion, but her dainty arms would hardly be strong enough to take it far along a rough road."[4]

Women on Velocipedes and Tricycles

■ As later versions of the bicycle were developed, comic views of women riders continued to appear. The velocipede, an early two-wheeled bicycle with a crank and pedals attached to the front axle on each side of the front wheel, was first developed by Michaux in France in 1865 and manufactured in both Europe and America, becoming particularly popular during the years 1868–69. Like the draisine, it was not easy to ride: weighing at least sixty pounds, it lacked brakes, was difficult to mount and dismount, had two wooden wheels with iron tires, required strength and coordination for pedaling and steering with the same (front) wheel, and also had few springs, all of which made for a jolting ride and helped to earn the machine the nickname "boneshaker." For women accustomed to wearing long, full skirts, riding the velocipede was particularly hard because of the placement of the pedals and the high backbone.[5]

In his history of the bicycle Andrew Ritchie argued that few British women rode the velocipede, saying that only a rare woman "would have been willing to expose herself to the ridicule that velocipedes encouraged."[6] Yet British and American women, as well as women from other countries, did ride the machines, including female acrobats, who rode velocipedes in circus high-wire acts. Velocipedes became particularly popular in America in 1867. While some French society women rode velocipedes outdoors on the boulevards of Paris, American women were more apt to ride them in clubs and riding schools in cities like New York, Boston, and New Haven; in 1869 *Harper's Weekly* pictured them at the Hanlon Riding Hall in New York.[7]

In France and Belgium intrepid women even rode in velocipede races, though into the 1890s there were separate categories for women in major events and even separate tournaments.[8] In 1868 the French journal *Le Monde illustré* provided graphic imagery of women willing to race their velocipedes in public, seen in a lithograph of four women riding in a race at Bordeaux that November. The story and illustration also appeared in

Bare-legged French women compete in a velocipede race in Bordeaux. From *Le Monde illustré*, 14 November 1868.

In this detail of an American version of the French illustration showing French women competing in a velocipede race in Bordeaux, the women's legs have been covered. From *Harper's Weekly*, 19 December 1868.

Harper's Weekly the same year, but with a difference: in the French illustration the women's legs and even their thighs were exposed, but in the *Harper's* version their legs have been discreetly covered with lace-edged underwear. Even if tempered, the *Harper's* story gave American readers a rare image of women's ability not only to ride but to ride with daring.[9]

Still, there were only a small number of women velocipede riders, and artists' images of them tended to be highly charged, with the women's exposed legs a sign of their daring. Female velocipedists appeared as emblems of fierce determination and social progress, as well as emblems of naughty women whose behavior was scandalous and outré. The French

Studio photograph taken in Trieste of an outré woman entertainer sitting on her velocipede, 1860s. Bildarchiv / Österreichische National Bibliothek, Vienna.

magazine *Le Vélocipéde illustré* in 1869 featured a drawing of a woman wearing a helmet, her hair blowing in the wind, dressed in short pants with her legs exposed as she carries a flag bearing the word *Progress*, the lamp of her velocipede illuminating the way. But in studio photographs of the 1860s women who posed with their velocipedes were often figures of social marginality, as suggested by their risqué clothes, their provocative poses, and their exposed legs.

Photographed with their velocipedes and dressed in tights and costumes, female acrobats conveyed an air of professionalism, in contrast to female dance-hall entertainers, who sometimes posed with a look of both detachment and daring, as in a photograph taken in Trieste in the 1860s in which a cigarette-smoking woman sits astride her machine.

Some male writers were offended by women challenging conventions on their riding machines. During the height of the velocipede's popularity the British writer Joseph Firth Bottomley, in *The Velocipede*, chided American women who insisted on riding by themselves on a two-wheeled velocipede rather than riding with their husbands or on a more stable, three-wheeled version. The writer conjured a glowing vision of America's colonial past, a time of a "patriarchal method of travel" when sensible wives rode with their husbands on a horse. To keep her balance, the wife would "lean lovingly against, or even to half embrace, her husband," and "thus unconsciously her sense of dependence was strengthened." But the modern-day American velocipede rider, Bottomley wrote disapprovingly, was of a different species, a "strong-minded and independent" woman who "felt that her independence was to some degree at stake." These American women were "ladies who want a little of the risk and dash," he said, adding that women were hardly "content with a machine that cannot possibly upset or run into somebody."[10]

Bottomley preferred that women ride a tricycle, or better yet, a tandem, or double-seated bicycle, driven in front by their husband, cousin, or lover, with the woman seated sidesaddle behind. He wrote ruefully that American women "cannot rest content with such a machine, but we in Europe are still so old fashioned as to prefer propriety to sensation." He evoked what seemed to him a horrible vision of women riding about on their own velocipedes with their friends, insisting in curmudgeonly fashion that "custom and nature revolt against it." After all, he argued, the reason women and men were interested in each other at all was due to their "respective dependent and protective positions," but "when a lady velocipedes she destroys all this kind of subtle interest."[11]

Satirists too took a dim view of women riding velocipedes. Though velocipedes had an awkward steering mechanism that made them hard to handle for either men or women, caricaturists mocked female riders as barely able to control their machines. In a French lithograph two wild-eyed women velocipedists with their hair flying take a rocky ride, their bare legs frantically reaching for the pedals.

A more complex view of a woman and her velocipede appeared in a painting ca. 1869 by the French artist Maurice Betinet of Blanche d'Antigny, a French dancer and courtesan, who modeled for the painter Gustave Courbet. The unsmiling cyclist is pictured wearing knicker-bockers under her short dress as she stands on a country road with one hand holding a kerchief and the other resting casually on the machine's handlebars, her serious face and poised stance lending dignity and legiti-macy to both model and machine.[12]

After the velocipede craze waned — gradually in Britain, in America in 1869, and in France a year later — another type of bicycle, the high-

Lithograph of two velocipedists, ca. 1869. Velocipedes were notoriously hard to manage, but in this satirical French illus-tration the women are frenzied riders on their wayward machines. Warshaw Collection, Archives Center, National Museum of American History, Smithsonian Institution.

wheeler, was introduced in 1870. Later known as the "ordinary," the new machine was ridden primarily by young men and remained popular for at least twenty years. With a very high and large front wheel, the bicycle was difficult to mount and dismount, particularly for women wearing the full skirts of the period, and there were few accounts of women who were not entertainers riding them.[13] During the 1870s and 1880s, however, tricycles became popular and were considered the preferred machine for women.

The tricycle, a heavy machine designed for adults, had two large wheels and a small one up front for steering. The "sociable" was a tricycle with an extra seat to the side, permitting two people to sit side by side, though more popular were the tandem versions, which allowed one person to sit in front of the other.[14] Men were more apt to ride tricycles during the 1880s, though women also rode them since the placement of the seat could more easily accommodate their full skirts. Tricycles helped make cycling a viable possibility for women and also helped heighten their mobility.[15]

Lithographed advertising poster for a Columbia tricycle and high-wheeler ca. 1885. © Collection of The New-York Historical Society.

But though women tricyclists were photographed riding alone, artists and advertisers were more apt to legitimate their riding by picturing them riding with a companion. The British bicycle illustrator George Moore's *En route to the Lake*, of 1885, shows a man and woman riding their separate tricycles side by side, and American illustrators in *Harper's Weekly* and other magazines showed couples riding their sociables along Manhattan's Riverside Drive.[16] In 1880s posters advertising Columbia high-wheelers and tricycles, some of which were sold as works of art, women on their tricycles appear as tiny figures either trailing behind a man on a high-wheeler or riding next to him. Given the significant disparity between the heights of the two machines, in one Columbia advertisement the female tricyclist appears petite next to the man's looming machine.[17]

The New Safety-Bicycle Craze

■ The tricycle was popular only until the mid-1890s, when it was supplanted by the new "safety bicycle," which immediately prompted a cycling craze. The first version was developed in England in 1885 by John Kemp Starley as the "Rover safety," and safeties became hugely popular in America during the bicycle-riding boom of the 1890s. The machine had two wheels approximately the same size, a diamond-shaped frame, and rear wheels driven by a chain and sprocket, and by 1888 it also had pneumatic tires to help absorb shock and vibrations, all of which made for a comfortable ride. The woman's version had a drop frame with a lowered tube between the legs and no crossbar. With the addition of springs, enclosed gears, and, later, a coaster brake, women could ride with less chance of getting their skirts tangled, which added to the bicycle's popularity, though the tricycle was still regarded as more stable and easier to mount and dismount (plate 19).[18]

Almost immediately the idea of women riding the new safety bicycles was greeted with dramatically contrasting views. In its special bicycling issue in 1896 America's middle-class *Munsey's* magazine portrayed the bicycle as one of the nineteenth century's "epoch making discoveries," ranking it with what it considered the century's greatest achievements—the railroad, the steamship, the telegraph, and the telephone. Magazines like *Munsey's* portrayed traveling by bicycle as women's journey into a new realm: "To men, the bicycle in the beginning was merely a new toy, another machine added to the long list of devices they knew in their work and in their play. To women, it was a steed upon which they rode into a new world." It was, in fact, "the best gift the nineteenth century has brought her."[19]

Munsey's viewed the effect of the safety bicycle on women as truly transformational, turning woman into a "new creature" whose rides into the countryside were teaching her a "new language." Invoking cultural stereotypes of women as fragile and frivolous, *Munsey's* marveled at the change: "Hitherto a weak, helpless creature, she can at her own sweet will cover great stretches of country without appreciable fatigue, and all the delights of motion, sun, and air come without any more effort than has been given to dawdling about the streets."[20]

Though *Munsey's* alluded to the conventional view of woman as a "weak, helpless creature" in its bicycle issue of 1896, in a later article in the magazine Anne O'Hagen portrayed this stereotypical view as a cultural construct, a fictive formulation all too readily adopted by misguided women themselves. O'Hagen envisioned bicycles helping women to displace their romanticized self-conceptions and dramatically change their temperament: "The general adoption of athletic sports by women meant the gradual disappearance of the swooning damsel of the old romance, and of that very creature, the lady who delighted, a decade or so ago, to describe herself as 'highstrung,' which . . . meant uncontrolled and difficult to live with." Through sports and exercise, the author insisted, women could develop not only firmer muscles but also "a more equable temper, and a dethronement of the 'nervous headache' from its high place in feminine regard."[21]

Though O'Hagen made wry reference to the "nervous headache," medical writings in the 1870s and 1880s were often filled with discussions of neurasthenia, originally viewed as a condition experienced by both men and women, a disorder marked by insomnia, depression, and dyspepsia, among other symptoms detailed by writers such as George Beard in *American Nervousness: Its Causes and Consequences* (1881). In such works, as one writer has suggested, neurasthenia was portrayed as a modern disorder, "as much ideology as malady," experienced by both women and men. Its supposed symptoms in women were wide-ranging, including weeping, irritability, depression, stomach upsets, heart palpitations, and morbid fears.[22]

In the eyes of late-nineteenth-century observers, one of the bicycle's biggest advantages for women was that it helped relieve them of some of these ailments and let them venture out of their constricted lives into the much healthier countryside. The *Handbook for Lady Cyclists* (1896), by the British writer Lillias Campbell Davidson, was one of the earliest guides for women on the safety bicycle. Davidson, who also wrote the ladies' column in the Cyclists' Touring Club (CTC) *Gazette* and founded the Lady Cyclists' Association in England in 1892, saw cycling as a welcome source of fresh air and exercise, offering "the greatest boon that has come to women for many a long day."[23]

Rather than blaming listlessness and boredom for women's ills, Davidson blamed social limitations, arguing that from the medieval period onwards "the lives of women have been unnaturally cramped and contracted within doors." With few outdoor amusements and little chance for outdoor exercise, with dress styles "calculated to produce palpitation, consumption, and dyspepsia," it was no wonder, she wrote, that women over many centuries had become "physically and mentally degenerated." As she asked, "Who would not have been subject to vapours and hysterics — have suffered from nerves and childish tempers and silly whims?"[24]

Shaping Davidson's praise of the bicycle was her vision of the constricted existence of many nineteenth-century women, not only her readers, whose amusements, she wrote, included the theater, concerts, and accepting invitations for polite visits, but also women who did not belong to the leisure class, who might find their lives "dull and monotonous" and who needed wider interests. For middle- or upper-middle-class women of the 1860s and 1870s — whose sole outdoor ventures, according to Davidson, might be shopping and occasional social visits — daily life was deadened by a monotony made worse by "the strain and worry of nineteenth-century life," for "the tendency of modern life is to become too anxious, too sad, too pessimistic."[25]

Davidson gave high praise to cycling for offering a woman a dramatically widened sphere and teaching her what it felt like to be "perfectly well and devoid of any complaint." Cycling "broadens life for her," giving her "new and fascinating amusement, an interest outside the petty details of her generally narrowed domestic life." Cycling was particularly beneficial to women who led sedentary lives and to women "who have heavy brain work." It was also good for housewives who were "too worried with home cares and anxieties." Cycling would not lead women too far afield, for after cycling, housewives could return home "cheered, refreshed, and braced to take up the burden of daily commonplace life once more."[26]

For women, the journey into the countryside became not just a means of living a more healthy life but a literal and metaphoric ride into a new type of freedom, a venturing away from social restrictions into a much wider universe. Through the bicycle, as *Munsey's* claimed, women could find "a new delight in living, a way to get out of beaten paths."[27]

In its discussion of women and cycling *Munsey's* evoked a nostalgic if not mythic vision of America's colonial past, but without Bottomley's gendered view. To *Munsey's*, women's journey to the country helped them discover the kind of life women had experienced (or so the magazine imagined) in America's early history, a life in which "the healthy, capable, clear headed colonial dame" had the freedom to ride on horseback, visiting and spending time outdoors. But as the magazine suggested, the advance of a "newer and more brilliant civilization" brought "conditions which confined women more and more," and the middle class "chained itself more closely indoors. . . . It was the spinning silver wheels which at last whirled women into the open air," giving them confidence, enjoyment, and strength.[28]

Another important freedom that cycling offered woman was traveling, often unchaperoned, in the company of men. Although nineteenth-century American etiquette books frequently forbade women to ride without chaperones, in actual practice riding with chaperones was rare. Magazine illustrations promoted the idea of outdoor excursions, as in a *Munsey's* illustration titled *A Bicycle Picnic Party*, in which women and men relax on the grass, their bicycles lying on the ground nearby.[29]

But concerns about social decorum were never far away. Though women in nineteenth-century England were encouraged to cycle outdoors, among the wealthier classes chaperones were considered important; indeed, in 1896 the Chaperone Cyclists' Association advertised the availability of female chaperones for women riders wishing to go on excursions and tours. But in her handbook, Lillias Davidson boldly reassured women of the safety and propriety of riding alone in the countryside, though she said that long rides in solitude "pall" and that having a

C. M. Relyea, *A Bicycle Picnic Party*, in *Munsey's*, 15 May 1896. The introduction of the safety bicycle made it easier for unchaperoned women to accompany men in countryside leisure activities.

companion was more enjoyable. Two women could safely travel abroad through France and Holland without courting molestation, rudeness, or annoyance, and through touring a woman could quench her thirst "for longer flights; for the pleasure of going on and on, and never turning back; of feeling like an explorer venturing for the first time into a new country and discovering for herself" a world where she could feel "independent of time or place."[30]

Confronting one of the common objections to women cycling alone in the countryside, the problem of dealing with mechanical breakdowns, Davidson urged women to "take care of the machine yourself." In her 1893 article "Cycling for Ladies," an early version of her handbook, she urged women to "thoroughly understand the use of every nut and cog, and be able to put right any small derangement that might occur in the course of her rides."[31]

(Though she encouraged women to counter some social conventions, however, Davidson was ever mindful of another convention: women's need to keep clean. She urged the woman bicyclist to keep her machine clean in order to also keep "her gown tidy and neat." Reaffirming a popular gender distinction, Davidson added, "Personally, I have never found any man to possess my own views as to the absolute removal of every speck of dirt and grease from a machine.")[32]

Amid the celebrations of women riding with freedom in the open air, cultural debates continued as critics, mostly men, and the press questioned the very idea of women bicycling and raised objections that seemed intended to keep women in their proper, bounded place. In 1896 the cover of *Outing* magazine (formerly *Wheelman*) bore an image of a woman riding at night accompanied only by her running dog as the moon winks at this act of daring, but the fact that women could ride off with men unchaperoned raised fears of licentious behavior. These thinly veiled fears about women crossing social boundaries were sometimes reflected in comic views of women. In 1898 *Harper's Weekly* featured an illustration by Thure de Thulstrup sardonically titled *Repelling Invaders at Camp Black*, in which women on bicycles appear unexpectedly at a military camp in Hempstead Plains, New York, during the Spanish American War, and are about to enter the men's privileged space.[33]

Mindful that its readers, particularly men, might see women cyclists as a social threat, *Munsey's* in 1896 tried to defuse the idea that women were becoming radically changed by representing female bicyclists as embodying more enduring notions of femininity: "If she has ridden her bicycle into new fields, becoming in the process a new creature, it has been gradually and unconsciously. She did not have to be born again in some mysterious fashion, becoming a strange creature, a 'new woman.'

She is more like the 'eternal feminine,' who has taken on wings, and who is using them with an ever increasing delight in her new power."[34]

**Women Celebrate and
Critics Fret**

■ Particularly troubling to many critics of women cycling was the prospect that they might engage in "scorching," or speed riding, a concern that suggested barely submerged anxieties that women on their new machines might be racing out of control. While recognizing the popularity of the new bicycles among women, skeptics belittled the idea of women racing and scoffed at their efforts to develop speed and endurance.

Magazines saw women racers as violating cultural notions of proper feminine decorum. England's *Cycling* magazine wrote in 1892, "This is a go-ahead age, an age when woman is as characteristically go-ahead as man," but "there is a prescribed limit beyond which her modesty and deportment should absolutely forbid her to step." It added, "The record-breaking woman cannot be graceful," for when she centers "all her attention and energy towards the attainment of speed, she cannot fail to be other than an object of ridicule."[35]

Though she championed the benefits of cycling for the independent woman, Lillias Davidson herself clung to cultural clichés about women's frail nature when she rejected the idea of women's racing and insisted that "over-speed is always injurious," adding that "a woman's nervous system suffers a hundred times more than a man's from this excitement and tension from the race itself."[36]

Reflecting these anxieties about women racing or scorching, *Cycling* magazine in its special supplement of November 1893 published a cartoon by George Moore juxtaposing two views of women riding. In one, titled *What We Have Been Used To*, a woman in a skirt smiles confidently as she decorously rides alongside a male companion, who contemplatively smokes his pipe. The companion illustration, titled *What We May Expect*, shows a woman riding in knickerbockers and, like the man riding beside her, bent over in a scorching position as a crowd of people behind gapes. To counter social fears about women racing, bicycle advertisements in nineteenth-century American magazines usually depicted women sitting in an upright position rather than bent over the handlebars.[37]

In a century in which doctors feared that traveling in fast-moving steam-driven trains might cause damage to the brain, fast travel on bicycles was also viewed by some as dangerous, particularly to women's health. Sharing the century's commonly held belief that it was unseemly for women, as delicate creatures, to expend any great energy, *Cycling* in 1893 discouraged women from riding bicycles by citing dangers to their health: "For feats of speed and protracted endurance, she is by nature physically unfit, and bound morally, if she respects her sex, to avoid anything in the nature of deleterious excess of exertion."[38]

As Ellen Gruber Garvey has shown, the actual medical effects of bicycle riding on women's health were hotly debated during the century, and the warnings about dangers were often just thinly disguised efforts to keep women in conventional roles. Often more basic fears underlay the satires of cycling: during the nineteenth century straddling bicycle saddles was often seen as leading to masturbation and excitation, posing a great threat to women's sexual innocence and purity.[39]

Still, by 1895, in the midst of the bicycling mania, many observers shifted their opinion and argued that cycling was good for women's health, a view also promoted in bicycle advertising. An advertisement for Coventry Bicycles in 1895 claimed that bicycling made women more attractive, for it "brightens the eye, puts a flush of health on the cheek, takes you out in Nature, to the pure fresh air."

These mixed reactions to women riding bicycles were reflected in popular art. Legitimating the idea of women cycling, *Scribner's* magazine chose the celebrated illustrator Charles Dana Gibson to create a

Poster for Stearns bicycles by Edward Penfield, 1896. This bicycle advertisement promoted the idea of women as competent riders, contented to ride but not ride too far. Poster Photo Archives, Posters Please, Inc., New York.

version of his famed Gibson girl on a bicycle for the magazine's cover in
1895, and Stearns hired the American illustrator Edward Penfield to
create an advertising poster in 1896 telling women to "Ride a Stearns and
be content," a slogan offering women tranquility but also subliminally
suggesting that they advance forward, but not too far.

Other images clearly presented contrasting views, envisioning
women as feckless and fearless but also needing men's help. In bicycle
posters of the 1890s women are often shown riding with men, and sometimes it is women who lead the way, even riding with only one hand on
the handlebar to signify their confidence and the ease of riding.[40] But
these images of women riding in the forefront were also encoded with
some ambiguity: in an era of bicycle chivalry it was customary for American women cyclists to ride in front of men so that if there were any
mishaps, men could come to their aid.[41] George Moore's comic cover
illustration for *Cycling* magazine in 1905 featured a male cyclist watching
a woman who is learning to ride being aided by another woman, who

S. D. Ehrhart, *The New Woman
Takes Her Husband Out for a Ride*,
in *Puck*, 4 December 1895,
German-language edition.

helps her to steady the seat. The caption ruefully reflects their thoughts: "She: I wish he would offer to help"; "He: I wish they would ask me to assist."[42]

When men and boys were shown being chivalrous, they were sometimes cast as embattled figures struggling with women of huge girth.[43] In *Mrs. Heavytopp's Bicycle Lesson*, a *Punch* cartoon of 1899, a young boy struggles to keep a hefty female cyclist balanced, and in the American illustrator S. D. Ehrhart's cartoon *The New Woman Takes Her Husband Out for a Ride*, which appeared on the cover of the New York comic magazine *Puck* in 1895, an anxious and frowning tiny husband rides perched on the front frame of a man's bicycle as his mammoth wife, dressed mannishly in a polka-dot tie and plaid knickers, steers from behind.[44]

In these portrayals of the woman as Amazon the fat lady cyclist is a source of humor and anxiety, a figure threatening to overpower men yet also tamed by being ridiculous. British and American humorists' jokes about women cyclists with Amazonian physiques often seem to reflect broader cultural fears about women gaining new freedoms and fears about women themselves. Athletic women, including bicyclists, were apt to be depicted as muscular, masculinized, and huge, while men were shown as submissive and feminized, stuck at home to do household tasks (plate 20).[45]

Women cyclists were made to seem less threatening by depicting them as both ridiculously heavy and pathetically delicate, prone to mishaps and falling from their new place. Even *Munsey's* warned that bicycling had the potential to make women vulnerable: "No enemy had ever found out their weaknesses so readily as a bicycle," the magazine wrote in 1896, adding that "the woman who would ride must be patient, watchful, self-reliant," for the smallest "indulgence in indecision" would lead to bruises from a fall.[46]

Alternately celebrating women riders and envisioning them as safely dependent and vulnerable if not low, representations of women and bicycles thus mirrored the mixed feelings in society at large about women participating in the world of modern machines. Photographs also reflected the polarized views about women cyclists. Fin-de-siècle studio photographers created an early version of technology chic: in formal or theatrically staged poses in front of painted backdrops women were photographed straddling a bicycle or perched on the seat, sometimes precariously, sometimes with élan. In this carefully constructed image world the scantily clad London music-hall star Anna Held posed next to her bicycle with her arms dramatically outstretched, reflecting her own confident modernity.

Other photographic images of women and their bicycles clearly signified a social threat, challenging cultural constructions of the proper roles for both women and men. In satirical stereographic images from 1890–1900, comic images of the "New Woman" and her bicycle became popular.[47] In these images female bicyclists undermined traditional gender roles, leaving their husbands at home to do the housework while they pedaled off for the day. In one, a woman tells her husband, "Sew on your own buttons, I'm going for a ride." In these satires the women are usually dressed in bicycle-riding garb, both literally and symbolically wearing the pants in the family. In one image from 1900 a woman dressed in plaid knickers stands in her living room with one foot touching her bike pedal, while her husband, wearing an apron, crouches to adjust her boot as their two young daughters play nearby. Women in these views are imperious: "Mind the children," she commands him, "finish the washing, and have dinner at 12."

Though women's new bicycling outfits became targets of satirical humor, the clothing designs were actually intended as serious efforts to

British music-hall entertainer Anna Held poses with her man's version of a safety bicycle, ca. 1890s. Held moved from London to New York in 1896, became a popular Broadway entertainer, starring in the Ziegfeld Follies, and married Flo Ziegfeld. UPI / CORBIS-BETTMANN.

make bicycle riding, with its potential for putting women in immodest positions, more practical, comfortable, and decorous. Long skirts had proven awkward and troublesome, getting tangled up in machine parts, getting dirty from road debris, and flying up into the wind. Davidson, who wrote in her handbook of women's skirts causing the "deepest anxiety," praised the idea of sewing weights into hems and bicycle dress guards for helping to alleviate the problem.[48]

More radical changes in women's bicycling dress were also recommended. In the *Tricyclist* in 1882 Mrs. E. M. King, secretary of England's Rational Dress Society, following the lead of Lady Harberton, founder and president of the society, suggested that baggy trousers and long jackets be worn.[49] By 1893 and up to about 1898, women, particularly in America, were wearing short, baggy trousers tied or fastened beneath the knee—called "knickerbockers," "zouaves," "rationals" in England, or "bloomers," after the American dress reformer Amelia Jenks Bloomer, editor of *Lily* magazine, who at midcentury had unsuccessfully tried to promote the wearing of skirts somewhat below the knee over baggy trousers, usually trimmed with lace at the bottom. Less intrepid women of the 1890s might also choose to wear divided skirts.[50]

In 1895 the playwright Marguerite Merington reported in a special issue of *Scribner's* magazine on bicycling that while French women were wearing "national culottes" and British women were wearing short cor-

Mind the children, finish the washing, and have dinner at 12, stereograph, 1900. In theatrical settings comic stereographs portrayed the "New Woman" as ready to ride off on her bicycle and leave her husband home to do the chores, but not before he laced up her boots. Whiting Bros., The Whiting View Company, Cincinnati, Ohio. Courtesy of T. K. Treadwell.

duroy skirts with high leggings, Americans had grown accustomed to seeing women bicyclists in a variety of outfits: "The eye of the spectator has long since become accustomed to costumes once conspicuous. Bloomers and tailor-made alike ride on unchallenged." Noting the eclectic styles, she wrote: "No one costume may yet claim to represent the pastime, for experiment is still busy with the problem." And she added that "knickerbockers, bloomers, and the skirt made of twin philabegs, all have their advocates." But in an essay two years later she wrote that the large majority of American women were still wearing skirts of one kind or another.[51]

Fashion-minded American women reading the *New York Herald Tribune*, *Vogue*, and *Godey's* in 1896 and 1897 could see illustrations of European women wearing cycling suits with bloomers and divided skirts. But while the new bicycling outfits brought praise from America's comic *Life* magazine in 1897 for making American girls "look exceedingly well" and demonstrating publicly "that women have legs," a wave of mocking cartoons also presented these women as radical in their departure from conventional modes of dress.[52]

The idea of women wearing outfits resembling trousers not only generated fears on the part of others that women would become manlike or dominate men but also caused women themselves to worry about violating cultural notions of proper female decorum. Mindful of the jibes directed at these outfits, *Munsey's* tried to deflect critics. Though *Cycling* magazine in 1895 had written that an "avalanche of knickerbockers has descended on cycling America," *Munsey's* argued in 1896 that while many women bicyclists had initially worn bloomers, and a few wore knickerbockers, "they were never those whom the great majority would care to imitate." The well-bred bicyclist, like the average American woman, would not discard her whalebone corset and would always be sure "that her dress is dainty and picturesque, without holding out any temptations to the caricaturist."[53]

Still, the magazine also felt compelled to defend the new outfits: "It is nonsense to say the bicycle has revolutionized women's dress, and has glorified Mrs. Bloomer. The most daring costume worn on the wheel cannot approach the ordinary bathing dress as a bold departure from the accepted standards of feminine gowning upon conventional occasions."[54] Wryly suggesting that knickerbockers were hardly radical, the satirical cover of *Life* magazine in 1895 contrasted a woman cyclist in plaid knickers to a woman in a black bathing outfit at the beach, dubbing the woman cyclist "The Voice of Modesty."[55]

In England, Lillias Davidson reported in her column in the CTC *Gazette* in 1894 that women in southern England were wearing knickerbockers, but her replacement columnist, Fanny Erskine, took a skeptical view of the new garb in her own column later that year, writing that women "in their hybrid costume" could well be "classed with the May-flies — ephemerals — things of the day," and she hoped that "their day might be a very short one."[56]

Aware of the attacks against these outfits and sometimes against women bicyclists, women writers offered their own fashion advice to female cyclists. Erskine's handbook *Lady Cycling*, published in England in 1897, suggested ways for women to avoid becoming the butt of jokes and maintain a decorous appearance. Erskine recommended to those who were hesitant about wearing the trouserlike knickerbockers that they wear an "artistically cut skirt, artfully arranged to hang in even portions each side of the saddle." Obviously concerned about women inviting criticism by the excesses of their dress, Erskine advised them that when bicycling in town, they should avoid "ridiculous blouses and flower-garden hats," for "these are unmitigatedly absurd."[57]

As Erskine's comments suggest, guides for women cyclists written by women were often preoccupied with women's appearance in public and recognized that women were often on visual display. These nineteenth-century writers thought that their readers needed to construct a public image that was both decorous and unthreatening. Advocating decorum, Lillias Davidson urged women to avoid making a spectacle of themselves: "The desire of a gentlewoman will be to avoid public notice by her dress in the saddle, as out of it." She advised women to avoid extremes in bicycling costumes, noting that "there have been attempts to foist upon the cycling section of womankind a variety of strange and uncouth costumes."[58]

Mixing pragmatism and a sensitivity to cultural norms, Davidson urged women to wear a skirt "sufficiently plain" to avoid getting caught in the bicycle's machinery but not to wear anything that would draw attention: "It should conform in such measure to the mode of the day as to be inconspicuous when the wearer is out of the saddle. . . . All bright colours are out of place; all flying ribbons, lace, and artificial flowers; all feathers and ornaments beyond a simple broach."[59]

In an era when experiments in moving pictures were already under way, Marguerite Merington's article "Woman and the Bicycle" in *Scribner's* in 1895 used language suggesting the new modes of seeing: "The cyclist is to be thought of only as mounted and in flight, belonging not to a picture but to a moving panorama. If she rides well, the chances are she looks well, for she will have reconciled grace, comfort, and the temporary fitness of things."[60]

Merington, like Davidson, urged women to dress appropriately so that their clothing would not be judged by "the standards of the domestic hearthrug" but instead would reflect "the exigencies of the wheel" and "the mechanical demands of the motion." She warned women that cycling "does not lend itself to personal display" and wrote admiringly of women who rode their bicycles to work while still appearing business-like: "The armies of women clerks in Chicago and Washington who go by wheel to business, show that the exercise within bounds need not impair the spick-and-spandy neatness that marks the bread-winning American girl."[61]

But to women writers, the outer appearance of women cyclists had to do with more than aesthetic issues or decorum. They urged female bicyclists to regard themselves as emblems and as living advertisements for the new sport. To Davidson, female bicyclists had a sensitive role, for their appearance was a crucial factor in maintaining the dignity and legitimacy of the sport for women: "Every woman should look upon herself as, in a certain measure, an advocate, so to speak, of the pastime among the members of her own sex."[62] Davidson emphasized the bicyclist's visual impact: "Upon her it will largely depend whether such women as see her a-wheel will desire to emulate her example, or shun her as a warning against imitation." She warned that "if she rides in a slovenly, awkward style, and sits her saddle ungracefully — if she dashes frantically along hot, dusty, and purple of visage," the bicyclist "will surely not win many recruits to the paths of cycling, but frighten them instead from doing as she has done."[63]

Sensitive to public relations and public images, Davidson said that women should "not dress in a style to excite undue notice, or make their pastime unattractive in the eyes of outsiders," and she continuously reminded women that they were, in essence, promotional public figures: "Every cycling woman who appears in public looking neat, trim, and charming, presents to the public an attractive advertisement; she makes other women want to follow her example, and recommends it in the eyes of all beholders." She too warned women against looking "loud, fast, and simply a fright," for they would do "infinite harm" by "prejudicing all sensible people" against the sport.[64]

In the early years of the safety bicycle female writers urged women cyclists to be practical as well as to maintain their femininity — a theme that would continue to shape representations of women and machines in the coming century. Davidson wrote that although women should seek "comfort, ease, and safety," it was "perfectly possible for a lady to look graceful, elegant, and womanly in the saddle."[65]

Davidson said that one of women's responsibilities was to represent beauty: "It is the duty of every woman to add to the beauty of the world in her own person, not to take from it. The woman who cannot cycle without making a guy of herself had better never cycle at all" or at least ride at night in seclusion so "that the world may be spared the painful and unedifying spectacle." To avoid looking like "a guy," women were advised to dress "suitably, and well, without dressing tightly or injuriously." Davidson also urged women to avoid overdressing, which would only hinder them. Vanity, she added, "is still the enemy of many" and "puts supposed loveliness of appearance before health, comfort, or any other consideration."[66]

Writers of bicycling handbooks thought women had other, equally important responsibilities, including becoming self-sufficient and independent by learning to repair their own machines. Both Erskine and Davidson gave women practical instructions on bicycle maintenance, including tire repair. Erskine insisted that "every lady who cycles should make a point of knowing how to repair her own machine," for this knowledge would give her a degree of autonomy and independence, which "makes cycling so delightful." The delight of bicycle riding, she added, "cannot be appreciated to the full" if women were not self-reliant and "self-contained."[67]

In the preface to her own book *Bicycling for Ladies* (1896) the American writer Maria Ward made clear that by offering women detailed instruction on operation and repair of bicycles, she could help them avoid some of the criticism and ridicule they met when riding. "I have found that in bicycling, as in other sports essayed by them, women and girls bring upon themselves censure from many sources." She added that the censure, "though almost invariably deserved, is called forth not so much by what they do as the way they do it."[68]

To counter ideas about women's mechanical ineptness, Ward sought to teach women the "laws of mechanics and of physiology," telling them that they needed "an intelligent comprehension of the bicycle as a machine" as well as knowledge of "the human machine."[69] Her handbook included specific techniques for bicycle riding and maintenance and the use of tools, as well as suggestions for appropriate dress. But what made her book modern, in a sense, were the instructive illustrations and her insistence that women were capable of mechanical expertise: "I hold that any woman who is able to use a needle or scissors can use other tools equally well."[70]

To demonstrate her point, Ward included illustrations of bicycling techniques based on studio photographs by her New York friend Alice Austen, a photographer living on Staten Island whose lifetime photographic work of eight thousand images charted the lives of New York's middle and upper classes, as well as the city's immigrant groups. Austen's photographs for Ward illustrated techniques for mounting and riding and showed women engaged in a variety of bicycle maintenance activities,

Illustration based on photograph by Alice Austen in Maria Ward's handbook *Bicycling for Ladies* (1896) demonstrating how to dismount a bicycle.

Maria Ward and the gymnast Daisy Elliott, photograph by Alice Austen. Elliott was the stern-looking model whose face in the picture above was replaced by an artist's rendering of a prettier one in Ward's *Bicycling for Ladies*. Alice Austen Collection. Courtesy of the Staten Island Historical Society, Historic Richmond Town, Staten Island, NY 10306.

including holding a wrench, tightening a bolt, using a tire pump, changing a tire, and turning a bicycle over.

Reproduced as illustrations in Ward's book, Austen's images of women cycling gave women a form of visual proof that they were capable of machine mastery. Still, the illustrations also hinted at some of the unresolved cultural ambivalence of the era. Though one of the aims of *Bicycling for Ladies* was to rebut cultural stereotypes about women, the book also made a concession to conventions about female beauty: in Austen's original series of studio photographs the model, the professional gymnast Daisy Elliott, has an unsmiling and rather forbidding face, but for the final published version, bowing to conventions, Daisy's face has been transformed, replaced by a more palatable, commercial image of a pretty woman.

In her memoir about learning to ride a bicycle at age fifty-three, *A Wheel within a Wheel*, the American suffragist and temperance leader Frances E. Willard emerges as a nineteenth-century woman struggling to enter into the modern age. Willard's was an active life: after teaching at the Evanston (Ill.) College for Ladies and working as dean and professor of art and English at Northwestern University, she became one of the most important figures in the Illinois suffrage movement during the 1870s, and served as the national president of the Woman's Christian Temperance Union for twenty years. For Willard, learning to ride a

Frances E. Willard, American suffragist and president of the Woman's Christian Temperance Union, learning to ride a bicycle in England with the aid of three men. Photograph courtesy of Woman's Christian Temperance Union.

bicycle, which she did while visiting England, was clearly a liberating and renewing experience: she tells how through the bicycle her own love of adventure, "long hampered and impeded, like a brook that runs underground," now came "bubbling up again."[71]

But as the photographic illustrations in *A Wheel within a Wheel* show, it was not easy for Willard to cast off cultural stereotypes and, by learning to master the bicycle, create a new sense of self. Although in the days of velocipedes women had been hired by riding halls to instruct other women how to ride, in visual images of bicycle instruction, men are usually shown teaching women how to operate their bicycles. And although Willard's bicycling teachers were women—she disparaged one as "timorous" and another as having "not a scintilla of knowledge concerning the machine"—in her book she included a photograph of herself being supported by men just after first mounting her bicycle: sitting on "Gladys," as she named her bicycle, and somberly dressed in a dark suit with a dark hat, she grips the handlebar while being steadied by the three men who assist. The photograph's caption is revealing, for it recapitulates conventional views about strong men and frail women: "Three young Englishmen, all strong-armed and accomplished bicyclers, held the machine in place while I climbed timidly into the saddle."[72]

After describing her three months of hard work and praising the female instructor who ultimately taught her to ride, Willard ends the book on a euphoric note. In her conclusion she is unabashed in her hyperbole as she happily exclaims, "I had made myself master of the most remarkable, ingenious, and inspiring motor ever yet devised on this planet." And the book's final photograph, signaling her success and self-sufficiency, shows her riding off triumphantly alone.[73]

By the end of the nineteenth century women were also using the bicycle to further their political agendas, including the right to vote. In a *New York World* interview in 1896 the American suffrage leader Susan B. Anthony said of the bicycle, "I think it has done more to emancipate women than anything else in the world." Willard herself relished the boost that the "fascinating and illimitably capable machine would give to that blessed 'woman question.'"[74] Yet in just a few short years, at the beginning of the twentieth century, women in the suffrage movement would be taking emblematic rides in a newer technology as they turned to the automobile as a modern vehicle for change.

WOMEN *and the* AUTOMOBILE

5 IN HER BRASH AND SASSY self-portrait punningly titled *AutoPortrait* (1925) the Polish-Russian French artist Tamara de Lempicka, living in Paris, pictured herself sitting in the driver's seat of a green Bugatti, her hand at the wheel as she coolly and confidently eyes the viewer (plate 21). Like the artist's other paintings of the 1920s and 1930s, *AutoPortrait* joined a machine-age style of abstracted, geometric precision with Lempicka's own provocative and yet dispassionate eroticism. Though the artist herself drove a small, yellow Renault, here she places herself in a green Bugatti, a prime emblem of automotive elegance and modernity.[1] Her vision of herself as a controlled and confident automobile driver clearly spoke to a world seeking images of the independent woman: the painting appeared on the cover of the German feminist magazine *Die Dame* in 1929 and fifty years later in the French journal *Galerie des arts*, where it was captioned "Tamara de Lempicka: Symbol of Women's Liberation, 1925."

AutoPortrait highlighted a potent cultural theme of the era. Engaging us through the automobile window, Lempicka made clear the automobile's central role in providing a mechanism for women's identity, particularly during the early decades of the twentieth century. With the manufacture of American and European steam, electric, and combustion-engine automobiles at the end of the nineteenth century, female drivers found a means not only to free themselves from social and geographic limitations but also to transcend prevailing gender stereotypes about their inherent mechanical naiveté and ineptitude. Through their ability to drive, and sometimes repair, their own automobiles they found a way to forge a new identity for themselves that included the ability to master machines.

Women were among the earliest drivers when steamers and electric automobiles were introduced during the 1890s, though the exact number of women who drove was never easy to establish since women often drove automobiles owned by their husbands or registered in their husbands' names, and the earliest automobile licensing did not take place until 1900 in America and 1903 in England.[2] The actress Minnie Palmer, the first woman in England to drive a car and to own her own car, received delivery of her French-made Rougemont automobile in September 1897.[3] By the end of 1904, the year it was founded, the Ladies' Automobile Club in Britain had more than three hundred members.[4] In the United States, Genevera Delphine Mudge, of New York City, and Daisy Post were driving in 1898, and Mrs. John Howell Phillips, of Chicago, was reportedly the first American woman to obtain a driver's license, in 1899. By 1902 thirty-five Chicago women, most of whom were young and half of whom were single, had obtained licenses.[5]

Breaking New Ground   ■ Stories about these women's early driving experiences often reveal the paradoxes of an era in which women were applauded for groundbreaking boldness yet still seen in traditional gendered terms. In Washington, D.C., Anne Rainsford French was taught to ride a steam-powered Locomobile by her father, a prominent Washington doctor, and in March 1900 she received a steam engineer's license, which was required for driving the automobile in Washington. She became her father's part-time chauffeur and drove alone for visits and errands, traveling at a speed of nine miles per hour, the legal limit for main avenues in Washington.[6]

In a story about French, the *Washington Post* heralded her achievement in becoming the "first duly qualified woman engineer in the Capital City" yet also cast her as archetypically feminine. French, the niece of the famed American sculptor Daniel Chester French and claiming to be the model for the mythic female representation of America in a sculpture by French — one of four created for the U.S. Custom House in New York City in 1907 — was described by the newspaper as "plump and pretty," her shoulders "absolutely flawless from an artistic point of view." French herself kept driving until 1903, when she married and, as she reported, her husband told her, "Driving is a man's business. Women shouldn't get soiled by machinery." But as portrayed by *Life*, French never lost her interest in automobiles. In 1952, at age seventy-three, she was invited to Washington to celebrate the American Automobile Association's fiftieth anniversary, and she was photographed talking to her nephew's young "hot rodder friends" while vacationing in Maine.[7]

The driving achievements of the pioneering British driver Vera Hedges Butler were similarly framed in terms of her femininity. Butler was credited with being the first woman in Britain to pass a driver's examination, having done so in Paris in August 1900 (driving tests were not compulsory in Britain until 1935). In a story about her fifteen-hundred-mile tour from Paris to Nice in 1901, illustrated with Butler's own snapshots, the British magazine *Autocar* attributed her success as much to her automobile as to her own skill. Accompanied by her father, Butler drove her Renault over the Sappey to the Grand Chartreuse, one of the highest roads in France. The road was covered in snow, and the car tobogganed into six-foot-high snowdrifts on the sides of the road several times. The magazine congratulated Butler on her driving feat but also praised the fact that "cars of such reliable and simple character should now be procurable, that a lady can drive one practically from one end of France to the other and back without the least trouble."[8]

In these early years of automobiling the sight of women driving the new steamers, electrics, and gas-engine automobiles was clearly a novelty and a visual sensation. When the wealthy British automobilist Mrs. Bernard Weguelin went for a ride with her husband two days after he bought his De Dion automobile in 1897, wrote *Car Illustrated*, the "plucky *chauffeuse*" drove the automobile home "to the utter mystification of all lookers-on, to whom the spectacle of a lady on a car was altogether new."[9] "Cecelia," a columnist in the same magazine, wrote about a time in the 1890s when she arrived in her automobile in a small British village. The crowds greeted her with excitement. Some "wore a look of curiosity, some of alarm, a good many . . . looked disgusted, and a *very* few seemed envious!"[10]

But in 1907 the *New York Times* suggested that a big shift had taken place. A few years earlier not many women had had "the temerity to make an attempt to operate an automobile," and the sight of "a woman driving a car unattended attracted much attention, and was viewed with undisguised curiosity by all who saw her." Now a growing number of women could "operate an auto as well as many of the best men pilots."[11] But while the newspaper downplayed the novelty of female drivers, the publishers recognized the great interest still generated by the subject and looked for novel examples of women at the wheel.

In an era when electrics were considered women's cars and gas touring cars were associated with masculinity, there was an element of the sensational in stories about women capable of handling heavy gas-engine automobiles. The *New York Times* cited quixotic examples, including that of a ten-year-old New Jersey girl who could drive her father's gas-engine runabout and a ninety-three-year-old woman, a survivor of the San Francisco earthquake, who took a two-hundred-mile automotive tour from New York to Martha's Vineyard, reportedly one of America's oldest female drivers to do so. Adding a note of exotica, the *Times* in 1907 also printed a photograph of Mrs. Charles J. Glidden, a skilled driver, teaching the queen of the Fiji Islands how to steer.[12]

Imaging Women and Automobiles

■ The newspaper and magazine photographs of the era not only capitalized on the novelty and eye-catching appeal of female drivers but also gave the public vivid visual evidence that women were becoming active members of the motoring scene. In 1899 the *New York Herald Tribune* included photographs of "New York's Automobile Girl" driving on a city boulevard, and in the same year the *Automobile Review* featured images of wealthy American society women, including Mrs. John Jacob Astor, in their automobiles.[13] Introducing a new century and introducing its American readers to the new mode of transportation for women, *Munsey's* magazine in 1900 described Paris as "the city of the automobile" and included photographs of French women in their shirtwaists and divided skirts driving automobiles through the busy Bois de Boulogne.[14]

Images of women and their automobiles added decorative and sexual allure to magazine covers and posters for automobile shows, a tacit recognition of female drivers' growing presence. The first color cover of America's new *Motor* magazine in January 1904 featured Henry B. Eddy's illustration *Changing Times*, in which an elderly man in a horse-drawn carriage labeled 1903 appears next to a female automobilist and her son, whose cap is marked 1904. There were also paradoxes in the use of women's images to promote automobiles. Though the Automobile Club of America in New York City barred women from membership, it chose Malcolm Strauss's drawing of a woman in her automobile for its poster advertising the club's first show, held at Madison Square Garden in October 1900.[15]

In an era when automobiles were so costly that they were affordable almost exclusively to the wealthy, Britain's elite *Car Illustrated* magazine included large studio photographs of the "growing list of intelligent and beautiful women who have taken up this new form of sport." These formal photographic portraits of titled and socially prominent women, as well as stage actresses, driving automobiles appeared on the covers of the magazine's earliest issues, starting in 1902, and continued to appear regularly, presenting the women in fashionable dress rather than in the protective, long driving dusters and veiled hats popular at the time.[16]

Seated in their De Dion, Panhard, Gladiator, Renault, and Napier automobiles on the luxurious grounds of their estates, these women exuded elegant glamour. Mrs. Bernard Weguelin, whose appearance in a village had reportedly caused such a sensation, was photographed for the magazine sitting in her Panhard automobile on the manicured grounds of her fourteen-acre estate, Coombe End.[17] Rather than suggesting that female drivers were radical adventurers shaking up the social scene, images like these helped legitimate women's new skill while evoking an aura of stasis and calm.

As early as 1901 an image of fashionable women in their automobile was even considered suitable for exhibition in the prestigious French art salons. In the midst of academic subjects at the salon of the Société National des Beaux-Arts in 1901—paintings of nymphs and satyrs, historical scenes, sedate portraits, and genre themes rife with bourgeois morality—the Philadelphia-born artist Julius L. Stewart exhibited his painting *En Promenade*, later retitled *Les Dames Goldsmith au bois de Boulogne* (plate 22). Moving with his family to Paris when he was ten, Stewart studied with the painter Jean-Léon Gérôme and spent most of his adult life in Paris painting images of fashionable members of French society, yet his close-range portrait of two women riding in the Bois de Boulogne in their 1897 Peugeot accompanied by their dog goes beyond technical gloss to bring a sense of immediacy to the motoring scene.[18]

But there were also signs of unrest at the idea of including painters' images of female automobilists in the French salons. In 1901 the American magazine *Automobile Topics* included a cryptic notice reporting voices of dissension at one of the French exhibitions. An artist named Acheverry had introduced the automobile in art with his painting of a fashionably dressed woman in a gasoline carriage, but "severe critics pronounce against it as an attempt at exploiting social ambitions and the artifice of dressmaking."[19]

Though women represented only a minority of drivers before the 1920s, images of women and the automobile in the early twentieth century often held a certain visual currency and became imbued with much broader and more emblematic cultural meanings. Although many early

An Expert Lady Driver. Photographed on the grounds of their elegant estates, wealthy British automobilists like Mrs. Bernard Weguelin, shown seated in her Panhard automobile, presented an image of dignity and calm. *Car Illustrated*, 18 June 1902.

photographs were ostensibly reportorial, publicizing the social phenomena of women driving, they often went far beyond documentation. Like the newspaper photograph of Mrs. Glidden teaching the queen of Fiji how to steer, images of women and the automobile in both town and country evoked the curio value and exoticism of ethnographic photographs, like those of remote cultures being popularized by anthropologists in the early decades of the century. Women, as the photograph of Mrs. Glidden and the Fiji queen seems to suggest, were not only traveling but traveling far, being transformed by a new technology and, in turn, helping to radically reshape the lives of women afar.

Recognizing women as a lucrative market for automobile sales — and the eye-catching appeal of female drivers themselves — manufacturers like Oldsmobile put images of women in some of their earliest advertisements, including in 1903. A Locomobile advertisement in 1910 presented a female driver transporting a car full of male passengers. Through text and image, these early advertisements visually codified the idea that women were fully capable of successfully operating the machines.

By the 1920s, photographs of women and automobiles had also become a central cultural emblem of women's modernity, independence, and mobility. Studio photographs of Hollywood film stars like Greta Garbo and Marlene Dietrich seated in their automobiles evoked an aura of cool, elegant glamour. In an archetypal image, the photographer George Hoyningen-Huene captured the dark-eyed French dancer Colette Salomon somberly seated in her automobile wearing a white driving helmet; the photograph was captioned "The Modern Woman" when it was published in *Vogue* magazine in 1927.

The automobile also served as a sign of stellar economic and social achievement. The wealthy African American manufacturer of hair-care products Madam C. J. Walker was photographed in the driver's seat of her automobile in 1915 along with three female passengers. Walker was born Sarah Breedlove in Louisiana in 1867 to parents who were sharecroppers and former slaves. An orphan at six, in 1905 she moved to St. Louis, where she developed a hair-growth treatment and cosmetics for straightening hair. Married briefly to the newspaperman Charles J. Walker, she developed her business in Denver and Pittsburgh, and by 1910 she had moved to Indianapolis, where she built a manufacturing company for her hair products that would eventually help make her one of America's wealthier women. Photographed at the wheel of her automobile a few years before her death in 1919, Madam Walker in her hat with sweeping feathers created an imposing image of dignified strength, confidence, and economic success.

Images of women in automobiles also became testaments to women's capacity to control these heavy vehicles. Women themselves rejoiced in the new feelings of power and control. Patricia K. Webster, who was employed as a saleswoman for Franklin automobiles in San Francisco in 1917, praised the allure of the automobile's malleability and the exhilarated "feeling that you are master of the powerful and yet tractable engine and that it will answer your every whim."[20] In its appeal to women, Overland automobiles in a 1924 advertisement in the *Ladies' Home Journal* wrote that women, "with true feminine insight," not only valued safety and the beauty of the car's baked enamel finish but also "thrill to the power of the Overland engine as keenly as any man."[21]

In England, wealthy women who were expert horsewomen wrote of translating their skills to automobiling. In a magazine column on motor bicycling for women the British novelist Mary Kennard—a driver and also the wife of Edward Kennard, the owner of a Napier automobile driven in the Thousand Miles Trial in 1901—wrote that girls who were "accustomed to the hunting field and with sufficient intelligence to take an interest in mechanical matters" would soon overcome their driving difficulties. Writing about a 1902 automobile, she enthused: "She an-

Evoking an aura of confidence and success, the wealthy African American hair-care manufacturer Madam C. J. Walker sits in her automobile with several women, including Lucy Flint, forewoman and secretary at her factory, n.d. Photographs and Prints Division, Schomburg Center for Research in Black Culture, The New York Public Library, Astor, Lenox and Tilden Foundations.

swers to the change speed as does a high mettled hunter to the touch of a spur." She added, however, that women were better suited to handling automobiles than they were to handling horses because horses were liable to become frenzied.[22]

Early photographs of women in their gas-engine automobiles soon became forceful emblems of women's capacity for control. As Michael Schiffer and others have shown, virtually from their inception automobiles were gendered: electric automobiles, because of their cleanliness and ease of handling, were considered particularly suitable for female drivers, whereas early gas-engine automobiles, which required a great deal of strength to turn the crank, were associated with power and were driven primarily by men, becoming a quintessential symbol of masculinity and male prerogatives.[23] (Women may have been less apt to drive early gas-engine automobiles for other reasons as well; for example, they avoided the gas-engine Ford Model T in part because it did not have a front door on the driver's side, requiring women to make the awkward move across the front seat past the gear shift, and also because the Model T adopted the self-starter later than other manufacturers.)[24]

Early in the century, the Ladies' Auto Club in London, hoping to encourage women to drive, promoted an increase in the popularity of gas cars for women.[25] In 1907, a *New York Times* story entitled "Women Autoists Skillful Drivers" announced that American women, not content with electrics, were now using high-powered gasoline cars, and the newspaper publicized the transition with photographs of New York and Philadelphia women in their gas-engine touring cars. The story reported that Mrs. Andrew [Joan Newton] Cuneo, of Long Island, who was "one of America's most energetic drivers" and had driven a steam car in an Atlantic City competition, was now driving a big, gas-engine Rainier touring car in the fifteen-hundred-mile Glidden Cup tour. Another driver, "Miss Griscom," the newspaper added, "drives a high-powered touring car as cleverly as any man."[26]

Like early studio photographs of women on bicycles, images of women and the automobile during the early twentieth century also became photographic emblems signifying daring female behavior and the modern age. At a time when Isadora Duncan, among others, was experimenting with modern-dance idioms, four female members of Albertina Rasch's Celebrated Dancers, from Great Neck, New York, wearing transparent, classically inspired dresses, struck self-consciously artful poses for a photographer as they sat grouped on an automobile in 1926.

Not only was the automobile associated with modernity but it came to embody other important cultural themes as well, including heightened mobility for both urban and rural female drivers and periods of freedom away from the domestic sphere.[27] Virginia Scharff, however, argued in an important disclaimer that the liberation was often more apparent than real: housewives in both suburban and rural areas ultimately spent more and more time driving their automobiles for family chores.[28]

The New York illustrator Ruth Eastman encapsulated the idea of women's wanderlust in the cover illustration for the June 1921 issue of *Motor* magazine. Before setting out on her journey, a female driver in a bright red coat stands next to her automobile near road signs pointing to New York and San Francisco as she ponders her map (plate 23). In texts and images women's capacity to read maps often signified their self-sufficiency and skill, their ability to navigate alone into new territory. In 1909 the British automobilist Dorothy Levitt, in her handbook *The Woman and the Car*, noted that twenty or thirty years earlier "two of the essentials to a motorist—some acquaintance with mechanics and the ability to understand local topography—were supposed to be beyond the capacity of a woman's brain," and she added that there were some who even believed that no woman could understand a road map. But in her own time, she said, the understanding of maps was "a necessity to every active gentlewoman," adding provocatively that "indeed the average woman is probably quicker than the average man in gathering from a map the information it has to offer."[29]

Four members of Albertina Rasch's Celebrated Dancers pose on an automobile, an emblem of modernity, in Great Neck, New York, 1926.

Women on Tour and Touring for Rights

■ During the years when early female aviators were demonstrating their piloting skills through exhibition flying, female automobilists were demonstrating their capacity for endurance and their mechanical skills through cross-country tours. The tours, like so many mechanical feats accomplished by women, served multiple cultural agendas. They were often sponsored by automobile manufacturers not only to promote their products but also to provide proof that long-distance driving was feasible since even a woman could do it.

In 1909 the *New York Times* published photographs of women engaged in long-distance automobile touring, including Alice Huyler Ramsey, who that year became the first woman to drive across the country, traveling from New York to San Francisco on a forty-one-day trip sponsored by the Maxwell-Briscoe Company.[30] In 1910 the automobile company Willys-Overland gave a contract to Blanche Stuart Scott, of Rochester, New York, to complete a similar journey. Scott gained national attention for her cross-country tour in her Overland runabout, which began in May 1910 — four months before her airplane lifted forty feet off the ground, on 10 September, making her a contender for the title of first female aviator.[31]

Leaving City Hall in New York City on 16 May, Scott traveled with a female companion to San Francisco and back, making the fifty-two-hundred-mile trip in ten weeks and returning in July after driving across America's western deserts and weathering rain and storms in the East.[32] In its story before the trip, the *New York Times* emphasized Scott's mechanical abilities with its subheading, "Miss Scott Will Make Trip and Act as Own Mechanician during Long Journey." In the story, Scott told how she had taken up driving five years earlier, when her father had given her a single-cylinder car to improve her health. She had since become a driving enthusiast and owned several high-powered cars. "I became more or less independent by thoroughly familiarizing myself with the general mechanical principles and mechanism of my own car," she said. Quick to defend herself against what might be seen as feminine hubris, she added, "I do not believe I am guilty of egotism when I say that in mastering the different mechanical principles involved I have derived a great deal of pleasure and some little proficiency."[33]

Though proud of her mechanical abilities, Scott revealed her sensitivity to social norms that expected her to rely on men. After stating that "of course I am thoroughly competent to take care of any minor adjustments and small emergencies that may arise," she added that she was more than willing to take offers of help from male mechanical experts along the way. But help from men could also be a nuisance, she said: when making tire repairs on the road, she had "often experienced more embarrassment through the proffered assistance of men who happen along." Almost apologetic, Scott concluded that doing repairs herself "rather magnifies the enjoyment after the little exasperating annoyance is over, don't you know."[34]

Associated with women's emancipation from social confines, the automobile also became linked with women's efforts to gain political rights. Before American women gained universal suffrage in 1920, automobiles were often used to promote the suffrage cause.[35] The machine was already becoming an important cultural emblem of women's quest for autonomy and independence, and for female suffragists it became both a vehicle for transportation and an important sign of their modernity during their crusade to achieve the vote.[36] In their publicity tours and campaigns American suffragists often used gas touring cars instead of electrics because of the gas engine's association with power, though wealthy suffragists were more apt to own electrics.[37]

As early as 1910 American suffragists in Illinois were making local motoring tours, and women also promoted their cause in cross-country travel. Automobiles not only provided convenient, low-cost travel but also helped women gain invaluable publicity. They were used in cities and towns for rallies, parades, formal motorcades, and processions. The vision of women driving automobiles was itself noteworthy: in 1910 the *New York Times* in its story "Suffragists Storm National Capital" gave front-page coverage to women who took petitions to members of Congress engaged in debating suffrage in Washington. Driving in forty-five automobiles decorated with state banners, the women were armed with petitions signed by a half-million people in support of women's suffrage.[38]

Female suffragists capitalized on the automobile as an iconic object, decorating the machine with banners, slogans, and symbolic colors: in England, the automobiles wore the suffrage colors purple, white, and green, and in the United States, they were painted "suffrage yellow." In a huge, sixty-car automobile parade across New York State en route to the Women's Suffrage Convention in Rochester in 1914, suffragists from the city of Binghamton drove an automobile painted blue and yellow with the words "Votes for Women" on every panel.[39] In meetings and parades automobiles served as a useful stage or platform for women delivering speeches. In at least one instance an automobile also became a vehicle on which women literally projected their hopes and aspirations. To publicize their efforts, New York suffragists in 1914 drove what was touted to be the largest motor truck in the state, decorated with a banner "Votes for

Women," to rallies in the city's public squares. Using a stereopticon, the New York women projected humorous and serious cartoons and suffrage publicity on a white sheet covering the end of the truck.[40]

American suffragists took increasingly longer tours to publicize their cause. In August 1915 a radical branch of the National American Woman Suffrage Association sponsored a cross-country trip by three women driving from San Francisco to Washington, D.C., and in April 1916 the *New York Times* reported that two female drivers—Mrs. Alice Burke, of Illinois, and Miss Nell Richardson, of Virginia—would embark on an "automobile suffrage circuit," billed as the first round-the-country suffrage automobile run of the association.

The purpose of the trip, according to the *Times*, was to carry news of the NAWSA's plans to promote the suffrage cause at the Democratic and Republican conventions in Chicago and St. Louis. In St. Louis there was expected to be a huge demonstration outside the convention hall, and in Chicago, 40,000 suffragists were going to parade. Burke and Richardson planned to travel from April to October, going from New York to New Orleans, Texas, New Mexico, Arizona, San Francisco, Oregon, Washington State, through the Midwest, and then back to New York.[41] The women would travel in what the *Times* called Mrs. Burke's "little yellow suffrage automobile," a gas-powered machine donated by the Saxon Motor Company that Burke would drive and service herself.

The women left New York on 6 April, and when they returned in October they had traveled 10,700 miles and visited every state in the country except in New England, making suffrage speeches along the way.

Nell Richardson and Alice Burke in their yellow Saxon automobile during the first cross-country automobile circuit, sponsored by the National American Woman Suffrage Association, in 1916. Photograph provided by and reprinted with permission of the American Automobile Manufacturers Association.

Their automobile became a central part of their enterprise. Serving as a rostrum, it became the locus of their speeches and a platform for the cause. The back of the car bore a sign with the words "Votes for Women," and the car was also covered with promotional advertisements for various states, such as "Eat Raisins in California" printed in large letters. The automobile also became inscribed with personalized testimonials of support as women across the country covered it with their autographs.[42]

According to a report by an official at the National American Woman Suffrage Association's convention in Atlantic City, New Jersey, that year, Burke and Richardson received widespread praise and national publicity not only for their suffrage efforts but also for their technical skill. In each state through which they traveled they received letters praising their endurance and courage as automobilists and their abilities as public speakers. The report accorded the journey almost mythic status: throughout their eventful journey—crossing the desert, driving through the Bad Lands, traveling on the Mexican border during raids, having their car pulled out of rivers during floods—the women's "courage has never faltered" and their trip gave proof of "the well-known fact that you can't discourage a suffragist."[43]

For the Saxon Motor Company, the tour was also a promotional one for their automobiles. In its advertisement reading "Two Noted Suffragists Travel 10,000 Miles in Saxon Roadster" the company included cameo photographs of Burke and Richardson which recontextualized the usual advertising images of women and the automobile, not just providing proof of women's abilities but also shrewdly identifying the company with a political issue central to women. The company's focus, not surprisingly, was not just on the women's political feat and their motoring expertise. It also publicized the auto's ease of operation, telling how throughout the journey Burke and Richardson changed tires when necessary and "personally gave all the slight service that was necessary to keep the car in perfect condition." The ad's ultimate purpose was clear: "This trip furnishes not only convincing evidence of the remarkable endurance of the Saxon Roadster, but also a striking testimonial to the ease with which it is handled."

Stereotypes Surface Anew

■ But while motor magazines and advertisers presented images of women as motorists fully in control, the old stereotypes about women's mechanical incompetence were never far away. The historian Michael L. Berger argued that negative stereotypes of women as poor drivers were minimal during the early period, when almost all female drivers were wealthy women, who posed no threat to the social order. But as autos became less expensive and more readily available to middle-class women, and as the automobile became smaller and simpler to operate due to developments such as the introduction of the electrical self-starter in 1912, the increase in the number of female drivers was accompanied by negative stereotyping. By the end of World War I, during which women had proven their proficiency as drivers, and with the increase in middle-class automobile consumers, the stereotype emerged in earnest.[44]

Though it might not be easy to prove Berger's correlation between the greater number of female drivers, the perceived threat to men, and the increase in negative stereotypes, men's anxieties over female drivers could be clearly seen in comic images such as William G. Stewart's *When Woman Drives*, in which a woman's male passengers look anxious and alarmed.

Even in the earliest days of automobiling, before the machines became more readily available to women, descriptions of female drivers often reflected stubbornly persistent stereotypes about women's inherent nature. The premise of these stereotypes was that women were poor

William G. Stewart, *When Woman Drives*, © 10 August 1915. An early comic view of the alarming woman driver. Library of Congress, Washington, D.C.

WHEN WOMAN DRIVES

drivers because of their inherent nervousness and their delicate physical and emotional constitutions. In its 1902 story on female drivers in America (which was reprinted in England's *Car Illustrated*), *Automobile Topics* reflected many of the prevailing cultural assumptions about women's limitations. As the magazine reported, in Chicago, where licenses and examinations were required, both men and women were tested for color-blindness and to determine the condition of their hearts since "every automobile driver can testify as to the shock that may come to the most unsuspecting and iron-nerved of men who drive these machines through the crush and jam of down-town." The heart test was particularly important for women, the magazine averred, because of "the disposition of women in an emergency to faint and let the machine run where it will," which was "one of the most serious aspects of a weak heart."[45]

In its 1909 story on American women, including Joan Newton Cuneo and Alice Ramsey, who were participating in automobile endurance runs, the *New York Times* sought to dispel notions about women's inherent nervousness: "It has been stated that women are too nervous to become efficient drivers, but this has been greatly exaggerated." It added that "the nervousness would doubtless rapidly wear away as the driver's confidence in herself became established."[46]

The idea of women racing in automobiles was also controversial. Cuneo broke speed records in a race against the prominent racing driver Ralph De Palma in the Mardi Gras races in New Orleans in 1909, but later that year the American Automobile Association banned all women from racing.[47] In England, Winifred M. Pink, in her article "Motor Racing for Women" in *Woman Engineer* in 1928, posited limits on women's ability to drive the early racing automobiles. Pink agreed with the idea that "the average woman is temperamentally quite unsuited to the adequate control of a fast car," adding that "it is only women whose qualifications are far above the average who stand a chance of being a success in any sport." She said that for women, motor racing was more exacting than other sports, and she estimated that women constituted only 2 percent of expert automobile racers. Only a half-dozen women were capable of driving an "abnormally fast car," she said, and few women "have perfect control of a car at anything over 80 m.p.h.," though she added that many women could safely drive standard sports or touring cars.[48]

What Pink's article did not say was that racing cars in the 1920s took a great deal of strength to handle, and for long races on circuits it would have been difficult for either a man or a woman to maintain "perfect control." Women like France's Hellé Nice (the stage name of the acrobat and dancer Hélène Delangle) belied the notion that racing was beyond women's capacity. Nice drove a 1927 Bugatti in a ten-mile Grand Prix near Paris and clocked 118 miles per hour. In the next four years she was the world's only female Grand Prix driver. Driving a Bugatti in 1928, she clocked an average of 198 kilometers per hour, for a record run.[49]

Another persistent assumption was that women were incapable of repairing their own automobiles and that if they became capable, it would diminish their femininity. Maintenance of machinery was assumed to be a man's task because of its complexity and dirtiness.[50] *Car Illustrated* in 1902 was quick to point out that the socialite and titled female drivers it wrote about did not find the idea of automobile repairs "amusing." For example, Mrs. Harold Harmsworth (married women's first names were not given), a "beauty at the helm," always drove accompanied by her male mechanician.[51] Mrs. Selwyn Edge, whose husband was a winner of the Gordon-Bennett Cup, was herself touted as being the first woman in England to drive a De Dion motor tricycle and one of England's first female automobile drivers, "both expert and fearless" in driving her 12-horsepower Gladiator. Still, the magazine, as it generally did in its articles on women, focused on her home estate, her safety record (though she had received a ticket for speeding, she had no recorded accidents), her fashionability (she was "natty and smart" in her white driving veil of chiffon), and her dearth of technical knowledge (she "modestly professes to understand nothing of the mechanism of her motor").[52]

(Showing the persistence of the idea of the mechanically mindless woman, nearly fifty years later, in 1949, *Vogue* magazine published a story that echoed the *Car Illustrated* image of wealthy women who said they knew little or nothing about their automobiles' inner workings. In a photo essay titled "Women and Their Cars" *Vogue* portrayed several wealthy female automobilists, including Mrs. Angier Biddle Duke, Mrs. Henry Ford, and Bette Davis with her orange-red British MG, as happily oblivious to anything mechanical. According to the article, though the automobiles reflected each woman's own sense of style, the fact remained that "these charming and adept drivers know less about the *insides* of their cars than Peter Rabbit." Still, the magazine did not want the women to appear completely frivolous: their lack of concern about machinery was attributed to the achievements of modern automotive engineering, which had provided "foolproof," high-performance cars, so that women no longer needed the services of male mechanics and did not even remember the "mysterious grunts and shudderings from underneath the hood—calling for the proud and secret knowledge of the Male." With modern cars, "the very word 'breakdown' is all but obsolete," allowing the women to be what the magazine affectionately called "mechanically half-witted," women who could now safely buy an automobile simply based on its looks.)[53]

Early writers were often ambivalent about women's abilities. Herbert Ladd Towle, in his article "The Woman at the Wheel," in *Scribner's* in 1915, marveled at the sight of young women in their automobile dusters "manipulating gears and brakes with the assurance of veterans. Not always in little lady-like cars, either." And the accompanying photographs of American and British women in small electrics and large American-made touring cars reinforced his point. But while Towle marveled at the idea of women "sitting cooly at the wheel" commanding cars with six-cylinder engines, he, like so many other writers, attributed their success to improvements in car design that eliminated the need for strength and skill.[54]

Towle embraced conventional cultural notions about women's paltry mechanical abilities: "The average woman has small taste for mechanics; she pushes this lever and pulls that with only a vague knowledge of how the final result is produced; and she is well aware that if anything goes wrong she will have to wait for masculine succor." Towle questioned whether women really felt comfortable with gasoline cars and reported the ongoing search for an "ideal woman's car," one that would demand nothing of a woman driver but pushing a lever and steering. And tire repairs, he argued, still posed a problem for women: with practice, a woman could manage a small tire, but a mid-sized flat tire required masculine help. Reflecting his views, was an illustration showing a woman leaning

A woman photographs her husband making repairs. *Saturday Evening Post,* **July 13, 1907.**

woefully against her car, captioned with words from Towle's text: "If the gallant rescuer isn't at hand, she must wait till he appears."[55]

Even female writers were ambivalent about women having mechanical knowledge. Mary Kennard, who wrote a regular column in *Car Illustrated* under the name "Mrs. Edward Kennard," reassured her female readers that motor bicycles could be ridden by any woman "who will take the trouble, not merely to sail along on her machine like a dummy but learn to understand it." "Instead of feeling hopelessly flabbergasted at the first stoppage," the woman with mechanical knowledge could become "self-reliant and independent." But Kennard clearly had mixed feelings: the motor bicycle, she wrote, was invaluable to women who had husbands and brothers "with a mechanical turn of mind."[56]

American magazines featuring images of grounded women woefully contemplating automobile breakdowns and their wayward machines re-

Spoofing the happily hazardous woman driver, this postcard from ca. 1906 is captioned "Am enjoying the Show. Just ran across an old friend."

inforced cultural notions of women as technological naïfs. In popular magazines repairs were most often shown as being done by men while women stood by as admiring onlookers. In a 1907 *Saturday Evening Post* illustration the woman is a detached observer who photographs her husband at work on a repair. When artists did depict women repairing automobiles, they often showed them as risqué. In the Italian artist Ettore Tito's print *Aide-toi, le ciel t'aidera* (Heaven helps those who help themselves), of about 1925, a woman works underneath her automobile, her stockinged legs jauntily exposed in the rain (plates 24 and 25).

Popular imagery also reinforced stereotypes of women as accident prone and driving recklessly out of control. A mocking American postcard from early in the century shows a smiling woman driving over the body of a fallen man, his hat, cane, and case scattered on the road nearby. "Am enjoying the Show," the card caption reads. "Just ran across an old friend."

Women were also seen as easily distracted drivers whose carelessness, emotionality, and incompetence could lead to catastrophe. Rather than a source of family togetherness and cohesion, they appeared as destructive, shattering social conventions, splitting up families, and fragmenting fragile human bodies and automobile frames. Amidst the auto accidents, carelessness, and moral bankruptcy in F. Scott Fitzgerald's novel *The Great Gatsby* (1925), it is Daisy who kills Myrtle Wilson, leaving Myrtle's mutilated body on the road as she speeds away.

Reflecting social ambivalence — and their own promotional needs — early automobile advertisements were also equivocal: they promoted the idea that women were responsible automobilists, fully capable of handling a car, but their rhetoric of mastery was often undercut by a subtext implying that women were inherently fearful and frivolous drivers, more concerned with fashion than with technology. In this conflicted view women are both confidently capable and incompetent, reckless and fully in control.

Advertising Cars to Women

■ The legitimation of women drivers was reflected — and affected — by automobile advertising illustrations. Seeing wealthy women as a profitable market for the easy-to-drive, clean but expensive electrics, manufacturers like Baker Automobiles recognized that images of women drivers would help widen the market to women, and their advertisements featured illustrations of socialite women at the wheel.[57] Early automobile advertisers also recognized the benefits of luring female consumers with the implied promise of their achieving machine mastery and control. Manufacturers, for example, initially promoted electric models for women drivers since the machines were often enclosed, were considered clean and quiet, and were smaller and slower than other autos.

Advertisers also promoted electrics as being easier to operate than automobiles with combustion engines, which required the strenuous use of a hand crank for starting.[58] Though charging and taking care of batteries could be messy and hazardous because of the chemicals involved, General Electric in its booklet *Charging the "Electric" at Home* in 1911 featured photographs of female models charging their automobiles using a mercury arc rectifier, which the company called the "acme of simplicity" and which, it said, would save women the trouble of going to a garage.[59]

But beginning in 1912, when American and European manufacturers, began replacing the difficult and chancy cranks with electrical self-starters (the Cadillac was the first to make standard use of a self-starter), gas-engine automobiles became more palatable to women. As Virginia Scharff showed, although the new starters were presumably just as appealing to men as to women, they were specially marketed to women drivers, who were considered to have the most anxieties about their ability to control the machine with ease.[60]

Mechanical simplicity and ease of handling continued to be selling points in advertisements aimed at women. Presuming that women were easily flustered by mechanical complexity, the Baker Rauch & Lang Company in 1916 reassured woman drivers that their electric automobile was "free of mechanical obtrusion and confusions." (Years later simplicity was still being emphasized. An Oldsmobile advertisement in the *Saturday Evening Post* in 1941 featured a smiling woman driver, enjoying the automobile's new "Hydra-Matic Drive" with "*no clutch, no shift* driving.")

Making a typical gender distinction, an ad for Lexington automobiles in *Vogue* magazine in 1920 described its auto as a "man's car in power and speed" and "a woman's car because of its luxury, ease of handling, and simplicity of control." While the novelist Mary Kennard had earlier evoked an image of genteel female drivers who translated their ability to control horses to automobiling, a classic advertisement "Somewhere West of Laramie" for the Jordan Playboy automobile, published in the *Saturday Evening Post* and *Vogue* in 1923 evoked the formidable, tough-minded image of the "bronco-busting, steer-roping girl," a woman of the American West who could handle the "sassy pony" with a savor "of laughter and lilt and light—a hint of old loves—and saddle and quirt." This "brawny thing" with its "eleven hundred pounds of steel and action when he's going high, wide and handsome" is the perfect match for the "lass who rides, lean and rangy," and loves the automobile's "cross of the wild and the tame." In the glow of the mythic Western landscape, she is clearly a woman who can handle her machine.

Early advertisers also reinforced a longstanding cultural stereotype of women as preoccupied not with mechanical features but with automotive fashion and style. The Hudson-Essex automobile was advertised in the *Ladies' Home Journal* in 1928 as having "The Qualities Men Admire Made Beautiful for Women." During the 1920s, automobile advertisers urged women to buy fashionably "smart" cars, and a 1920 issue of the Paris *Vogue* featured a woman and automobile on its cover. An Oldsmobile advertisement in 1928 showing a flapper with cloche hat and bobbed hair asked, "What could be smarter?" In other fashion-sensitive advertisements, as Roland Marchand noted, women were often presented in the exaggerated, stylized poses of fashion runway models, reinforcing their role as decorative objects.[61]

While advertisers often geared their ads to women's presumed preoccupation with fashion, some advertisements devalued this fashion-mindedness. An ad by the Lexington-Howard Company, for example, argued that the astute man would not be swayed by superficial appearances. Rather than focusing on automobile exteriors, men picked their automobiles the same way that an employer chose a potential employee: "When you set out to find a man to work for you, you don't hire the most flashily dressed man you meet." Though "many a car has found favor through its paint job," the ad argued, thoughtful men would recognize enduring rather than transient and superficial values.

In some advertisements women were seen as fixated on fashion, but they were also enshrined as the maintainers of family values. Tamara de Lempicka, framed by the window of her automobile, had represented herself as a confident female, the newest of New Women, defining herself in terms of independence, machine mastery, and mobility. But early advertisers reconstructed the frame: in an advertisement for Baker Rauch & Lang electric automobiles in 1916 a woman driver sits with her two small children beside her, the group outlined by the frame of the front windshield (plate 26). Women in these ads sit in a space that reconfirms their place within conventional social frameworks: rather than encasing an image of female assertiveness and independence, the windshield outlines a sphere of protectiveness, security, and family cohesion.

(Images of the cohesive family were also part of a wider marketing strategy. Advertisers recognized that the idea of family unity would resonate with drivers. In an article on automotive sales, *Automobile* magazine wrote in 1914, "Holding the family together is with many families one of the strongest arguments that can be advanced in these days as a divorce mill is grinding overtime. . . . It is an argument that has sold thousands of cars and will continue to sell tens of thousands more.")[62]

The complex and conflicting representations of women automobilists in advertising and the popular press not only mirrored pervasive social attitudes but also helped shape women's own sense of identity and self.[63] It is not surprising that in a cultural climate in which women were perceived as mechanically inept and were socialized to receive less experience in maintaining cars they often defined themselves as incompetent and even mocked their own fastidiousness. Mrs. A. Sherman Hitchcock, writing in *American Homes and Gardens* in 1913, mocked the "motor woman who dislikes machinery—who is afraid of soiling her hands or gloves or gown with a bit of oil or grease," the woman who considers herself "far more elegant when seated like an automaton in the tonneau of the car . . . with a miniature powder-puff always ready to dab her nose."[64]

A Handbook for Women and Ambivalent Views

■ In her rare, early automobile handbook for women motorists, *The Woman and the Car* (1909), Dorothy Levitt tried to counter the clichés about mechanically ignorant females. Wrote Levitt, "I am constantly asked by some astonished people, 'Do you really understand all the horrid machinery of a motor, and could you mend it if it broke down?'" She reassures her readers that "the details of the engine may sound complicated and may look 'horrid,' but an engine is easily mastered."[65]

Levitt's handbook is remarkable not only because it was an early effort written by a woman to help other women become proficient drivers but also because it is a revealing portrait of women's ongoing efforts to become mechanically proficient yet also remain appropriately feminine. A photograph on the frontispiece shows the attractive, dark-haired author smiling knowingly at the reader as she firmly grips the steering wheel, the veil of her motoring hat floating gracefully behind.

Working as a secretary for Selwyn Edge, the noted racing driver and later head of Napier automobiles, Levitt was taught to drive by Edge and

In this frontispiece to her 1909 handbook *The Woman and the Car*, captioned "Her Favourite Photograph," Dorothy Levitt presents herself as a glamorous automobilist.

became one of the first women to participate in speed trials when she drove a Gladiator in 1903 at the trials at Southport, where she won in her class. The trials were often organized by automobile clubs, and manufacturers entered their automobiles to prove the cars' abilities and to provide data to prospective customers. Levitt continued to race for the next six years, not only winning trophies but also helping to boost automobile sales: Edge recognized that if customers witnessed a woman driving an automobile, they might be convinced of the ease of handling the car. The only woman to participate in a 1,000-mile trial at Hereford in 1904, Levitt also gave demonstrations of the automobile to potential customers. Capitalizing on Levitt's success and on the novelty of female drivers, Edge featured photographs of her in automobile advertisements.[66]

At the Brighton trials in 1905 Levitt drove an 80-mph Napier racing car. She came in first in one class, a remarkable feat because the Napier was a large, cumbersome machine that many considered to be too heavy for a woman to steer and disengage the clutch. In 1906 she set a new world's record for women, driving a Napier racing car at 91 mph. Not only was she an expert racer but she was also mechanically self-sufficient. As she wrote in her diary, at the 1904 light-car trials "no mechanic attended to car. Did everything myself."[67]

In his introduction to Levitt's book, the editor C. Byng-Hall carefully established Levitt's credentials as a woman breaking social molds but still framed the details of her life in terms of gender stereotypes. (The book itself, in which Levitt presents no-nonsense instructions on driving and repairing automobiles, is subtitled *A Chatty Little Handbook*.) Worried about the common stereotype that women engaged with machinery are masculine, Byng-Hall thought it was necessary to reassure the reader that even though Levitt could drive a car at a "terrific speed," she was the very essence of a proper female. Writing that "the public, in its mind's eye, no doubt figures this motor champion as a big, strapping Amazon," Byng-Hall described her as "the most girlish of womanly women. Slight in stature, shy and shrinking, almost timid in her everyday life." Byng-Hall described Levitt as feminine in other ways as well, saying that she "passes as a bright butterfly of fashion" and "lives the life of a bachelor girl," complete with giving "many little luncheon parties."[68]

Although portraying Levitt as appropriately ladylike, Byng-Hall was careful to insist that as an automobilist Levitt was no stereotypical female flower: rather than a nervous or timid driver, as women were often assumed to be, she was "quick of eye and sure of hand and nerves troubled her not at all." She was also "intrepid," and "in hill climbs, endurance and speed trials she is alike invincible."[69] (Years later, during the 1950s, the veteran British racing driver and journalist Sydney "Sammy" Charles Davis in his memoir about early racing again portrayed Levitt as quintessentially feminine, describing her handbook as an "amusing little book" and Levitt herself as a woman with a "natural flair for good style in dress." This surprised him, because "popularly the type of woman who muddled with machines was one of the hearty masculine sort" who wore practical rather than ornamental clothes. Levitt, however, wore attractive garb, turning her garden-party dress into a fashionable motoring costume, complete with a huge floppy hat and veil, a long duster, and her black Pomeranian dog, which frequently shared the seat with her when she raced.)[70]

Throughout her handbook Levitt reveals the complexity of transitional times. She both reaffirms and refutes stereotypes about women drivers, telling women that they do have the intelligence to understand mechanical repairs while also advising them on etiquette and on how to choose an automobile with an attractive interior. She tells them that they need not be timid and nervous, yet carefully describes repairs in terms intended not to tax their brains.

The first chapter opens with a revealing image: not Levitt the mechanic but Levitt the feminine automobilist, seated in her auto, her hands at the wheel, fashionably dressed and looking elegant, confident, and composed. "You may be afraid of a mouse, or so nervous that you are startled at the slightest of sudden sounds," she tells her readers, "yet you can be a skillful motorist." And she reassures them that she will "explain everything in the simplest possible manner, without lapsing into confusing technicalities."[71]

Levitt continuously bows to conventional views of female behavior, while also insisting that women can overcome cultural notions of their own limitations. "Twenty or thirty years ago, two of the essentials to a motorist—some acquaintance with mechanics [and] the ability to understand local topography—were supposed to be beyond the capacity of a woman's brain," writes Levitt, but that was simply because a woman had never been exposed to these subjects. "If a woman wants to learn how to drive and to understand a motor-car, she can and will learn as quickly as a man. Hundreds of women have done and are doing so."[72]

Levitt's text is a remarkable mirror of changing times, addressing readers' presumed feminine concerns yet also, untraditionally, giving women no-nonsense technical instruction. Sensitive to automobile fashionability and appearance, she makes suggestions about car color and body (noting, for example, that the Victoria "has the most graceful line"), yet follows with a detailed list of necessary tools for tire repairs. In the chapter "How to Drive," she reassures women that "with the little cars no strength is required" to start the engine, and the accompanying photograph shows a woman reaching for the crank to start the car.[73]

Again confirming and contradicting stereotypes, Levitt assumes that women will start crying at the first sign of car troubles and admits that as a beginning driver she "wept bitter tears" when her car inexplicably stopped and "was so down-hearted that it took me a day to get over it." She advises her female readers to "learn quickly to mend matters and laugh at them rather than weep." Although women can be "really justified in feeling angry" when they have a punctured tire, "if a woman wants to learn how to drive and to understand a motor-car, she can and will learn as quickly as a man." There are many women "whose keen eyes can detect and whose deft fingers can remedy, a loose nut or a faulty electrical connection in half the time that the professional chauffeur would spend upon the work," she adds.[74]

Levitt assumes that women require simple operations (she cites a single-cylinder car as "by far the simplest for a woman to drive and attend to alone"), and her book provides clear, detailed technical information and photographs to illustrate her points. Most of the photographs were

Dorothy Levitt demonstrates how to prime the carburetor to make it easier to start the engine, in this photograph by Horace Nicholls. *The Woman and the Car,* 1909.

taken by the British photographer Horace W. Nicholls, who would be hired as Britain's first official war photographer during the First World War and would help create an important photographic record of women at work in the war.[75] Nicholls's photographs show an appropriately dressed Levitt removing a faulty spark plug, unscrewing the oil cap, and testing the oil with a stick. But the images are more than illustrations: they are testimonials, visual proof of the author's mechanical abilities and evidence that women could indeed drive and service their own automobiles. Through these photographs Levitt offered women a new way to view themselves in the modern machine age.

But though visual and textual representations encouraged women to reconfigure their own self-images and see themselves in new ways, magazines also continued to promote a cultural agenda that insisted that women adapt their images—and abilities—to win the approval, and the approving gaze, of men. In a 1917 article "Little Things about a Car That Every Woman Who Means to Drive One Ought to Know," the *Ladies' Home Journal* noted that male drivers often gave female drivers a wide berth on the road, sometimes driving into a ditch when they saw a car driven by a woman. Their avoidance of women, the magazine added, was due to fundamental gender differences and men's own perceptions: while the average man has an "instinctive sense that an auto is a machine," he avoids female drivers in the "belief that the woman has not yet differentiated a motor from a horse and carriage." Offering women driving and repair rules, the article suggests that if the rules were followed, "every man driver would smile instead of scowl."[76]

Given the cultural emphasis on appropriate gender roles and the pervasive beliefs about women's own inherent lack of mechanical aptitude, it is little wonder that women remained ambivalent about their own capacity to manage an automobile. In a 1915 *Ladies' Home Journal* article "The Girl Who Drives a Car," Ann Murdock, identified as an actress and a skilled motorist, announced optimistically that improvements such as demountable tire rims had "taken most of the labor out of 'tire trouble,'" helping to make automobiles virtually foolproof. Automobile mishaps and forlorn women were becoming obsolete: "those little roadside incidents that stopped the wheels and that used to dot the landscape with woeful feminine countenances, when a carburetor went bad, or a tire blew out . . . are almost as rare as the dodo."[77]

Although Murdock challenged the recurring images of the forlorn woman next to her flattened tire, she herself was equivocal about women's abilities. On the one hand, she threw down the gauntlet to women: "It's your moment," she told her readers, adding that it was up to them "whether you are going to become panicky or prove yourself to be a self-confident, independent, resourceful and intelligent American girl" by mastering the driving techniques described in her article. Illuminating this point is a photograph captioned "Some Girls Have Mechanical Talent," showing a woman dressed in a duster and cap unceremoniously straddling her tire rim while making a repair.

Yet while Murdock described herself as someone with a "certain mechanical talent" who enjoyed "making small repairs and adjustments," she emphasized "the undesirability of doing [repairs] on the part of most girls whose mechanical knowledge is limited to running a sewing machine or pushing little brother around the block in his perambulator." She told women emphatically, "Don't poke around the insides of your machine," advising them instead to go to their auto agency or a "reliable repair man." Setting further limits on women, she also advised them that 12–15 mph was "fast enough for any girl" during her first month of driving.[78]

Women Automobilists in World War I and After

■ Though Murdock expressed reservations about women doing repairs, American, Canadian, and European women were already using their driving and repair skills as they helped out in the war. British women served in Voluntary Aid Detachments (VAD) and the Women's Reserve Ambulance Corps and drove ambulances for the British and French Red Cross.[79] British women who had worked for voting rights increasingly turned their attention to the war effort, like the suffragist Charlotte Marsh, who was a chauffeur for David Lloyd George, then minister of munitions.[80]

With universal conscription in 1916, as British men were called to active service, women engaged in "substitution," working at jobs formerly assigned to men.[81] The Women's Army Auxiliary Corps (WAAC) recruited female driver-mechanics to service army automobiles in 1916, and by 1917 a British employment handbook for women described other job opportunities for women, including working as motorcylists, commercial drivers, and mail van and taxi drivers, and recommended that women have six months' driving experience and a general knowledge of repairs.[82] In 1918 Britain's illustrated magazine *Sphere* publicized women's skills with a full-color painting by Harrington Mann of a female Royal Flying Corps driver.[83] And volunteers in England, like their counterparts in America's Women's Land Army, helped on farms by driving and maintaining tractors.[84]

Before America's entry into the war, wealthy American women, some of whom were recruited by the American Fund for the French Wounded (AFFW), shipped their own automobiles to England and France and volunteered as drivers delivering supplies to hospitals, driving trucks and messenger cars, transporting doctors, patients, and refugees, and working as ambulance drivers with the Red Cross and with British units attached to the French army.[85] Katherine Stinson, who had already made a name for herself as an exhibition flier and who ran a flying school in San Antonio with her sister Marjorie and her brother, was denied permission to fly in combat for the United States in World War I; however, she flew for the American Red Cross, trained American and Canadian male pilots, and later gained permission to go to England and France to work as a Red Cross ambulance driver.[86]

Even Gertrude Stein volunteered as a driver for the AFFW and learned to drive an automobile that had been shipped to her in France. With her companion, Alice B. Toklas, she drove supplies to hospitals, transported ill soldiers in Nîmes, and carried refugees after the Armistice. Photographed with Toklas with the Ford they called "Auntie," painted with the Red Cross sign and the letters A.F.F.W., Stein had the photo printed for postcards, which were sold in the United States to benefit the fund. Back in Paris after the war, the two were decorated by the French government for their war work, and "Auntie" was replaced by another Ford, "Godiva."[87]

In America, the Women's Motor Corps was founded in January 1917, before the country's entry into the war, to recruit women for domestic war service. Interested women had to first demonstrate their skills: to participate in the corps in New York, women needed not only a state chauffeur's license but also a mechanics license, granted by one of the city's mechanics schools. To prepare for these jobs, Chicago women took classes in automobile repair from car dealers, and by 1917 they were

Before America's entry into World War I, a woman at the Westside YMCA in Manhattan is taught automotive repair by a female instructor. *Motor Age,* 19 April 1917.

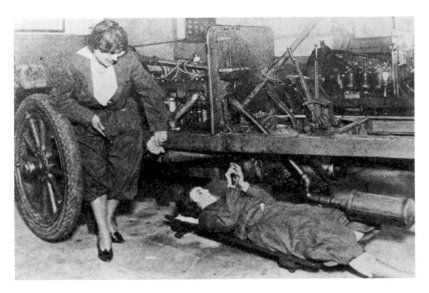

driving ambulances and transporting male recruits from Great Lakes Naval Station to active duty.[88] Female college students like those at Bryn Mawr, in Pennsylvania, took special courses in motor repair.[89]

American magazines and newspapers highlighted the changes that were taking place. A story in *Motor Age* told about women taking driving and repair classes at the Westside YMCA in New York and included a photograph of a woman lying on a "stretcher" underneath an auto chassis getting tips on repairs from her female instructor. Noting the women's special outfits, the magazine noted that they willingly changed their clothes to suit their new roles, finding bloomers more suitable for would-be mechanics, and recognized that "you cannot take an engine apart and explore all its inner mysteries while wearing a delicate georgette crepe waist and a silk skirt."[90]

Women serving in World War I had clearly proven their capabilities as drivers and mechanics, though many lost their jobs when the war ended.[91] But while women had countered gender-role expectations with their wartime expertise, magazine stories written by women during the 1920s revealed their own continuing ambivalence about their abilities and their mixed feelings about flaunting their freedom and being independent of help from men. Beneath their flapper appearance and their sassy, sardonic, jazz-age prose there were signs that they wanted it both ways: they loved the freedom afforded by automobile driving but still wanted male chivalry when it came to broken machines.

Laura Breckinridge McClintock in her 1923 magazine article "Why Take a Man Along?" asked, in a photograph caption, "What woman having tasted the joys of entire freedom would want to go back to man-chaperoned obscurity in the back seat?" The photograph illustrating the story was one bespeaking independence: two women in short skirts stand next to their automobile on a motoring trip, one combing her hair, the other perched on the running board. Yet with the breathy language of a 1920s sophisticate, McClintock was also willing to put men into the picture. She told women not to worry about bringing men along for help on their trip because if necessary they could find helpful men simply by acting very feminine. Rather than taking repair tools, they needed only "a couple of good-looking blouses, a nifty sweater or so, and a pair of curling irons." "This sight," she said, "never fails to bring masculine offers of help from the first car to heave into view." When a man did come along, "we would tell him our tale of woe, and presto! It was great sport." Recommending the "you-are-so-big-and-strong-and-I'm-so-dumb-and-weak" method, she insisted that "mechanics seem to blossom forth into fits of unheard of speed under a fire of ignorant questions and feminine flattery."[92]

Female writers often recommended that women act like children to get men's help. Helen Bullitt Lowry, in another *Motor* magazine article, reasoned that "baby talk may be out in flapper circles, but it is still the best coin when you are hemmed up in a parking space." She added, "One line of baby talk—and some husky young mechanic then will step forward and tug your front wheels by hand to the correct angle."[93] Others, however, took a more chastened view, such as Marie Russell Ullman, who admitted that she was "impressed with the enormity of my ignorance of the car." Regretting her lack of understanding, she recommended that women admit their own incompetence: "We owe it to ourselves and our community to understand our responsibilities" and to have "more than a superficial knowledge of the subject."[94]

Echoing the mixed feelings of its authors, *Motor* magazine presented contrasting visual images of women drivers on its covers and in its stories. Women were still seen kneeling on the ground next to flat tires, but now they were shown repairing them rather than looking picturesquely forlorn. Ruth Eastman's cover illustrations for the magazine in 1921 in-

This photograph of two insouciant flappers was captioned "What woman having tasted the joys of entire freedom would want to go back to man-chaperoned obscurity in the back seat?" Laura Breckinridge McClintock, "Why Take a Man Along?" *Motor*, July 1923.

cluded images of women as bold adventurers, like the map-reading woman en route to San Francisco, and as sportswomen driving with a tennis racket under one arm and the other arm on the steering wheel.

But by November 1921 Eastman was replaced as the magazine's cover artist by Howard Chandler Christy, who presented widely different conceptions of woman and the automobile. His illustrations of women with a wrench in hand alternated with images of statuesque women in neoclassical dress and steamy temptresses sitting on automobile hoods in poses seductive and coy (plates 27 and 28). Combining all three types, a Christy magazine cover in *Motor*'s January 1924 issue featured a pretty but somber muse of the automobile dressed in transparent neoclassical drapery revealing her nude body beneath. Gazing down, she looks reverently at the small model of a touring car she holds in her hands.

Women continued to appear as decorative figures and sexual lures in automotive advertising in the years ahead, and the automobile itself was presented as a love object for women. The horses fondly remembered by

A woman smelling her bouquet of flowers leans suggestively over a "brute of an engine" in this Mustang automobile advertisement, "Six and the single girl," 1966. Used with permission of the Ford Motor Company.

Six
and the single girl.

British women automobilists early in the century and the "sassy pony" of a Jordan Playboy automobile in 1923 were replaced by a Ford Mustang automobile in a 1966 advertisement suggestively captioned "Six and the single girl." A pretty woman smelling a bouquet of yellow flowers leans over a blue Mustang, and the text asks, "What makes a quiet, sensible girl like Joan fall in love with a Mustang? Not simply Mustang's steely good looks or smooth, racy lines." Joan in Ford's formulation is passive, virtually a passenger in a car that seems to drive itself as she places herself in the capable hands of this male machine: "She knew she could trust this husky, suave brute of an engine to squire her around town, drive her to the mountains for a weekend, even drop her off for dinner with the girls."[95]

As in early automobile advertisements, advertisers and automotive designers after 1950 continued to assume that women were mostly interested in comfort, safety, and style and little concerned with mechanical features, even though at least one American survey, conducted in 1979 by *Women's Day* magazine and the National Automobile Dealers Associa-

Margaret Bourke-White, *Breadline during the Louisville Flood,* 1937. During America's Great Depression, the *Life* magazine photographer created this ironic photograph contrasting an advertising billboard image of the American ideal with the harsh reality of victims of an Ohio River flood waiting in line for food supplies from a relief agency. Margaret Bourke-White / Life Magazine.

tion, found that women first considered economy, reliability, durability, and ease of handling.[96]

In an appeal to women's presumed fashion sense, Dodge in 1955 and 1956 produced its limited edition Dodge La Femme automobile, with its 1955 model painted "heather rose" and "sapphire white" and the 1956 model, lavender and white. Both versions came with coordinated women's accessories: the 1955 model featured a matching shoulder bag, rain cape, umbrella, and "dainty boots for unexpected showers." (As a counterpart for men, Dodge also marketed the Coronet Texan, which featured what was considered a more masculine trim.)[97]

Representations by Women

■ Though advertisers, magazine illustrators, and photographers often reinforced cultural stereotypes about women and automobiles, there were women photographers and artists who tried to redefine the terms, revisiting representational conventions and sometimes casting them anew. In contrast to automobile advertisements that envisioned the automobile as a force for family cohesion, picturing family members within the framed boundaries of the automobile windshield, the American photographer Margaret Bourke-White, in the midst of America's Great Depression presented an ironic commentary on the theme of family togetherness in automobile advertising.

In what became an iconic image of America's conflicting values, Bourke-White's 1937 photograph *Breadline during the Louisville Flood* brings together two disparate and discordant images. A giant advertising billboard illustrates an archetypal American family—mother, father, son, and daughter—smiling brightly in the front seat of an automobile,

Fashion models from the house of the French couturier Jacques Heim wearing fashion designs by Sonia Delaunay and Delaunay's coordinated designs for a Citroën B12 in front of the Pavillon du Tourisme at the Exposition Internationale des Arts Décoratifs et Industriels Modernes in Paris in 1925. © L & M Services B.V. Amsterdam 990406. Photograph Bibliothèque nationale de France, Paris.

the entire idealized tableau presented within the frame of the automobile's windshield. This idyllic portrait of harmony, contentment, and middle-class prosperity appears beneath the billboard's bold block lettering, which proclaims, "World's Highest Standard of Living," and to the side appear the words, "There's no way like the American Way." But standing at the base of the billboard, as an ironic counterpoint, is a long line of grim-faced victims of an Ohio River flood carrying baskets and shopping bags to hold supplies from a relief agency. The line of victims cuts across the billboard, in effect negating the optimistic visions of advertisers and the discourse of the American dream.

The French artist Sonia Delaunay engagingly revisited cultural notions about women's fixation on fashion and automotive decor, reinvigorating the conventions with energy and style. Long before the Dodge La Femme, she created her own witty versions of coordinated designs, adapting her avant-garde geometric abstractions to the world of machines.

Born Sonia Terk in Russia in 1885, she emigrated to Paris in 1905, where she married her fellow artist Robert Delaunay, with whom she developed a shared aesthetic and color theories based on "simultaneous contrast": whirling discs and interlocking planes painted in bold patterns juxtaposing complementary colors. Delaunay produced costumes for the Ballet Russes and Diaghilev and applied her patterns to fabric designs for French manufacturers and her own "simultaneous" dresses.

Because of her interests in both art and technology, in 1925 Delaunay was commissioned by a journalist to decorate a Citroën B12, and she applied her dynamic oppositions of color rectangles to coordinated designs for automobiles and women's fashions. In a revealing photograph, models from the design house of the French couturier Jacques Heim wearing Delaunay's patterned dress, hat, and coat ensembles pose with a Citroën painted with Delaunay's design in front of an architectural emblem of the modern age, the entrance to the Pavillon du Tourisme at the Exposition Internationale des Arts Décoratifs et Industriels Modernes. In 1928 Delaunay applied her signature color geometries to dresses designed for Talbot automobiles, and years later, in 1967, she was commissioned to design the exterior of a Matra B530. She covered the sports car's long, sculptural body with bright blue and red geometric squares, transforming it with the lucidity of her designs.[98]

During the 1930s Delaunay also turned her talents to another area of technology, the airplane, as she joined with her husband in painting panels for the Air Pavilion at the Paris World's Fair. Here was a machine that, like the automobile, had become a central emblem of women's quest for new freedoms and new roles, a machine that also became identified with some of women's deepest longings and aspirations.

WOMEN *and* AVIATION

6 TOUCHING THE WING STRUTS of her airplane with one hand and carrying her leather flier's cap in the other, the American pilot Louise Thaden grinned confidently at the viewer in a 1929 photograph, creating a bold emblem of the modern machine-age woman. Thaden won the first Woman's Air Derby (mockingly dubbed the "Powder Puff Derby" by Will Rogers) that year and, with Blanche Noyes as copilot, won the prestigious Bendix Prize in 1936, the first women to win that prize. She became one of a growing number of women who found in aviation a welcome feeling of power and control, a way to transcend the boundaries of traditional social roles and assumptions about women's limited capabilities. As she wrote in her book *High, Wide, and Frightened*, "Flying is the only real freedom we are privileged to possess."[1]

Thaden's photograph belies the notion of a dainty, "powder-puff" pilot. Framed by the sharp diagonals of the wing struts, which splay out like the ray lines in modernist paintings of the period, she stands with her goggles around her neck and one foot planted firmly on the airplane's wheel. At a time when women were still dogged by stereotypes about their frail gentility, she and other notable female fliers would set their own records for endurance, speed, and altitude.

Though women initially found it hard to gain entrance into flying schools, by 1930 there were two hundred licensed female pilots in the United States, and by 1935 there were between 700 and 800, and many women flew without licenses.[2] During the 1920s and 1930s celebrated women aviators, including Amelia Earhart, Thaden, and Ruth Nichols, worked as test pilots, transport pilots, flying-school operators and instructors, company pilots, aerial chauffeurs, and in aircraft sales and advertising. Manufacturers were particularly interested in hiring female salespeople to demonstrate that flying was safe and that airplane designs had improved.

Photographs of the period capture the excitement felt by many of these pioneering female pilots, happy at the chance to learn how to fly. A

1915 photograph shows Marjorie Stinson grinning broadly as she holds the wing struts of her instruction plane, a Wright Model B biplane. In her diary a year earlier she had worried that "girls might be discouraged in this so-called hazardous undertaking," and she herself was at first cautious about even writing for information about flight lessons at Orville Wright's school in Dayton, Ohio. Yet after Wright showed her the landing field and introduced her to her instructor, Howard Rinehart, she learned how to fly after only four and one-half hours of flight time over a period of a few months. Later she became an exhibition flier, a member of

Louise Thaden, winner of America's first Woman's Air Derby (the "Powder Puff Derby"), stands in front of the left wing of a Beechcraft Travel Air ca. 1929. National Air and Space Museum, Smithsonian Institution (SI Neg. No. 83-2145).

the famous "Flying Stinsons," and operated a flying school with her sister Katherine and her brother.[3]

Fliers like Stinson were sensitive to the role of photography in helping them establish their careers. In her 1914 diary she reported that when a cub reporter for the *San Antonio Light* called on the Stinson family for a story, she was disappointed that in the published photograph she only appeared as a small figure in the background, adding, "So far I have not been purposely photographed alone for the newspapers."[4] But her exploits as a flier would be well covered by photographers during the course of her career as newspapers recognized that female fliers, like female automobilists, were a novelty worthy of special attention.

In both text and images newspapers charted the records being set by these pioneering female pilots, including Harriet Quimby, who in 1912 became the first woman to fly across the English channel, and Ruth Nichols, who in 1930–31 set the women's records for speed (210 mph), distance, and altitude (28,743 feet). In 1930 the British flier Amy Johnson became the first woman to fly solo from England to Australia. In 1932, in a heroic feat, Amelia Earhart, flying in her Lockheed Vega, became the first woman to fly solo across the Atlantic, and in the same year she became the first woman to complete a solo, nonstop transcontinental flight, flying from Los Angeles to Newark.[5] (Five years later, Earhart's

Learning how to fly, Marjorie Stinson holds the wing struts of her instruction plane, a Wright Model B biplane, in 1915. Stinson Papers, Library of Congress, Washington, D.C.

career would be tragically cut short when she attempted a round-the-world flight in her twin-engine Lockheed Electra, and was lost, with her navigator Fred Noonan, en route to Howland Island.)

Photographs also became an artful testament to the ways aviation was helping women reconfigure their own identities. In the thousands of images of Earhart taken over the course of her celebrated career, photographers often caught her close personal identity with aviation and the airplanes she flew. Framed by the outlines of the cockpit window, she, like Tamara de Lempicka in her signature *AutoPortrait*, found in the machine the parameters of her own sense of self. In one striking photograph Earhart is shown on a diagonal next to her Lockheed 10-E Electra's right "Wasp" engine. With her left hand on the propeller tip, her right hand on her hip, and her arm stretched the length of the propeller, Earhart is at one with her aircraft, her identity barely separated from that of the machine. In other photographs of Earhart the airplane joins the automobile as a central emblem of the modern, as in one layered image in which Earhart stands proudly in front of her new Cord Phaeton automobile, which is parked in front of her new Lockheed 10-E Electra.

Female artists in the early part of the century also found aviation an alluring emblem of modernity. The French artist Sonia Delaunay and the Russian painter Natalia Goncharova shared the modernists' infatuation with mechanical imagery. Goncharova's 1913 painting *Aeroplane over Train* fuses a cubist fracturing of images with the futurists' love of dynamism and speed as it celebrates the airplane's ability to surpass the older transportation technology.[6] In 1937 Sonia Delaunay joined with her husband, Robert, in designing mural decorations for the Air Pavilion at the International Exhibition of Arts and Technics in Paris. As her own contribution to the project, she painted three murals in gouache — of an abstracted airplane propeller, control panel, and engine — all with the Delaunays' signature "simultaneous" color contrasts.[7]

In the work of other women artists aviation not only figured as a sign of modernity but was also closely linked to issues of female identity. Paradoxically, the French artist Marguerite Montaut apparently masked her own identity.[8] Montaut, the wife of the famed French automobile illustrator Ernest Montaut, is widely believed to be the artist whose signature "Gamy" appears on a number of early hand-colored lithographs celebrating feats of aviation and automobiling.[9] Her prints added a painterly elegance to the daring exploits of early aviators. In her image of Henri Farman winning the Grand Prix d'Aviation in Europe's first 1-kilometer circuit over Paris in 1908, stark black silhouettes of the biplanes contrast dramatically with the sky of pink and blue washes.

Gamy not only celebrated the feats of male fliers but also created telling images contrasting the allure of aviation with women's conventional roles. In her lithographed poster "Circuit Européen," celebrating Beaumont's triumph in the European Circuit in a Blériot monoplane in 1911, the event is witnessed by a woman in traditional peasant clothing and wooden shoes carrying heavy water buckets on her shoulder. The woman gazes up at this new world of aviation, a world that soon would offer women a way to transcend the burden of their traditional gender roles.[10]

The lure of aviation also appeared in Berlin artist Hannah Höch's modernist mapping of female identity. During the years of the Weimar

The famed American flier Amelia Earhart creates an image of unity with her aircraft as she stands in front of the right engine of her Lockheed 10-E Electra. National Air and Space Museum, Smithsonian Institution (SI Neg. No. 71-1050).

Republic, 1918–33, Berlin dada artists, including Höch, George Grosz, and John Heartfield, created photocollages combining fragments of advertising, clips from newspaper and magazine photos, and bits of text. Höch's photocollage *Öhne Titel* (Untitled) (1920) creates an emblematic version of a female in the modern machine age (plate 29). The collage joins a photograph of a woman's face, the eyes directly engaging the viewer, with a halftone photographic fragment of the body of a dancer with bare legs and thighs, dressed in shorts and high heels. The composite figure stands amidst other images of technological modernity, including a coffee grinder and a ball bearing, while the woman's entire figure is superimposed on an aeronautical map.

In Höch's photocollage the intersecting lines of the aeronautical map become a new matrix for the modern, situating the woman among contemporary machines. For Höch, the map also illuminated the art-making process itself. In an interview in 1972 she suggested that the imagery of aerial photography, along with that of microscopy and radiology, may have inspired the making of the collages during World War I.[11]

But while women artists seized on aviation as a central emblem of modernity and female fliers eagerly embraced the airplane's potential to lift them into a freer realm, women in aviation were still dogged by a cultural climate of skepticism that seemed intent on keeping them rooted in more conventional, earthbound roles. The American flier and manufacturer Glenn Curtiss and the noted flier Claude Grahame-White in England were reluctant to help women learn how to fly, and Bessie Coleman, the famed African American flier, had trouble getting flight instruction.[12] The American pilot Helen Richey was briefly hired by Central Airlines to fly as a commercial pilot, but she resigned when she was not allowed to fly in bad weather and she was rejected by the all-male pilots' union.[13]

In her article "Engineering and Aviation" (1927), Lady Sophie Mary Heath, the noted British flier and promoter of flying for women, writing under her former name, Mrs. Elliott-Lynn, referred to the "chilly aloofness" of the Royal Aero Club toward admitting women members, and she noted that among a minority of commercial pilots there was a "feeling against women in aviation," which she attributed to the men's fear that women would interfere with their "good times" after a day's work. But she added, in a conciliatory tone, that "we do not want to oust any men from any position they hold," insisting that "we women are entirely against any sex warfare."[14]

There were other important signs of skepticism. In early stereotypes female fliers, like early female bicyclists and automobilists, were often portrayed by men as incompetent operators of their machines. These skeptical views continued even as women were setting world records and successfully displaying their piloting skills. As Amelia Earhart remarked in *The Fun of It* (1932), women "are considered guilty of incompetence until proven otherwise."[15]

In his essay "Milady Takes the Air," which appeared in the *North American Review* in 1929, the American pilot Bruce Gould argued that most women lacked the strength, focus, and endurance needed to succeed in aviation. He cited the prejudices of male pilots, who believed that "women are by nature impulsive and scatter-brained" and "don't like to mess around machinery," which kept them from giving "motors and controls the meticulous attention" they deserved. Summing up these views, Gould concluded that female fliers were "all right in the rear cockpit, or the upholstered cabin; anywhere, in fact, except in the pilot's seat."[16]

Even in stories praising women's flying aptitude male writers often revealed their low expectations about women's abilities. In his article "Teaching Women to Fly," in *Aero Digest*, Fred L. Hattoom evoked lingering and pervasive cultural notions about the psyches of early female aviators. Hattoom, a flying instructor at Standard Flying Schools in Los Angeles, saw differences between the women of 1928 and those of 1930. In 1928, he said, female students were a rarity and were primarily interested in being stunt fliers. Typically, he claimed, this kind of woman was an "aggressive, dare-devil type whose sole purpose in learning to fly was for the novelty and thrill she got out of it. Stunts delighted her, and she invariably went in for them after her first solo." By 1930, he added, women were flocking to the training school and "showed flying ability far exceeding the expectations of their instructors."[17]

Although he praised women for their skills, including their ability to gain a good understanding of engine mechanics, Hattoom nevertheless assumed that his female students were dainty figures who needed delicate instruction. He wrote that being a flight instructor required patience and the ability to impart knowledge "in a way that appeals to women students." These students, he added, "are quite sensitive and cannot be corrected harshly," and he urged male instructors to forgo expletives such as "hells" and "damns," which "must be eliminated entirely in the training of the ladies." The censorship of swearing was necessary, he explained, not because of good etiquette but because it would dampen the women's enthusiasm and retard their progress, and he urged instructors to make their corrections and explanations "in a softer manner."[18]

Interwoven in Hattoom's article were assumptions about gender differences in instructional needs: "Generally, girls require a little more dual time than do men and are apparently not in such a hurry to solo. They require constant explanation and repetition of actual flight maneuver and are a little slow to think quickly in a tight pinch."[19] (Presenting an alternate view of gender differences, Louise Thaden in 1931 reported that women were slower to learn landing but that they were better in air work and quicker in learning takeoffs.)[20]

The photographs illustrating Hattoom's article on training women pilots reinforced his view that the women were, after all, decidedly feminine and shy with their male instructors. In one image the student Elizabeth Kelly stands with helmet in hand and smiles as she pats her hair in place after making a solo flight; in another photograph Marion Bowen, dressed in flying pants, goggles, and helmet, shakes her instructor's hand even as she minimizes the fact that she is taller than her instructor by posing in a model's S-curve.

Women were welcome at Hattoom's Los Angeles flight school, if only for economic reasons. As Hattoom noted emphatically at the end of his article, "It is time to adopt the proved policy of the automobile dealer — Cater to the women."[21] Yet although some instruction schools may have welcomed female students and even though women were proving that they could master flying and compete with men in endurance

Working for a transport license, Marion Bowen strikes a model's pose with her flight instructor at the Standard Flying Schools in Los Angeles in 1930. *Aero Digest,* March 1930.

and speed, they were often faced not only with stereotypes about their abilities but also with very real restrictions and limits.

Though Hattoom described his early students as "dare-devil" types looking for thrills, he neglected to note that in the early years of flying women were excluded from many air competitions, denied jobs as pilots on regularly scheduled commercial airlines, and prohibited from serving as pilots in World War I. Looking for flying opportunities, they were often confined to stunt flying and exhibition flights. American women could not compete with men in major races until the mid-1930s.[22] For many female pilots, including America's Matilde Moisant, Marjorie Stinson, Helen Lunde, and Florence Lowe "Pancho" Barnes, among others, stunt flying became an important source of employment and a way to display their flying skills. Blanche Stuart Scott, who, along with Bessica Raiche, was credited with being the first American woman to fly, left flying in 1916. She complained that "there seems to be no place for the woman engineer, mechanic or flier. Too often, people paid money to see me risk my neck, more as a freak—a woman freak pilot—than as a skilled flyer. No more!"[23]

In England, as the columnist C. Griff reported in 1927 in her regularly featured column in the British journal *Woman Engineer*, there were not many opportunities for women pilots to earn a living apart from stunt flying and film work. She even suggested that if women fliers wanted work, they might go to Australia and Canada, where she foresaw opportunities in aerial transport, air surveying, and commercial photography.[24] Female stunt fliers, however, were a big public attraction and a novelty. Dressed in a linen frock, soft white pull-on hat, silver leather jacket, and goggles, Lady Sophie Mary Heath performed in her de Havilland Moth before the Prince of Wales, and the flier Sicéle O'Brien wrote about her own stunts including, "looping-the-loop" and spinning nosedives and flying over the Crystal Palace in London.[25]

Restrictions and limits on female fliers were particularly evident during wartime. In England women could not serve as military pilots in World War I. Ruth Law and Katherine Stinson volunteered to be pilots in 1917, but the United States Army turned them down. Looking for a way to participate, Marjorie Stinson trained Canadian pilots for the British Royal Flying Corps, and Hilda Hewlett, who was the first British woman to earn a pilot's license, instructed men as fighter pilots.[26]

During World War II, American women serving as Women Airforce Service Pilots, or WASPs, were allowed to transport planes but not to fly them in combat, and aside from Helen Richey's ten months flying for Central Airlines in 1934, it was not until 1973 that American women were given jobs as pilots in commercial aviation (Emily Howell Warner was hired by Frontier, a regional airline, and Bonnie Tiburzi, was hired by American Airlines). And it was not until 1993 that the first female pilot was assigned to combat duty by the United States Navy.[27] Though American women were precluded from working as combat military pilots until 1993, Russian women were flying in military aircraft in 1938, and Soviet women flew fighters and bombers in combat missions in World War II.

Transcendence and Grounding

■ Reflecting social ambivalence about female fliers, cultural images of women and airplanes were often equivocal about women's place in the new order of aviation. Photographs celebrating the high-flying exploits of feckless female aviators coexisted with popular poster images showing women as simply decorative figures or onlookers in a world of male fliers. In these contrasting images women were championed for their capacity to transcend cultural limitations, yet they were also visually grounded, brought low for their audacity and envisioned as firmly planted in their conventional roles.

This tension between women's longing for freedom and transcendence and the social pressures for them to remain grounded became one of the central themes in early-twentieth-century representations of women and aviation. The cover illustration for the American composer Irving Berlin's witty song *When Katy the Waitress (Became an Aviatress)*, of 1919, featured a young woman in her open cockpit looking down happily at a row of admiring men waving to her from the ground. In contrast, the cover for the American fox trot *Daughter Come Down*, ten years later, shows a young woman and a young man soaring up in an airplane and heading home. In the song's lyrics, the mother, fretting about scandal, pleads, "Daughter, I'm saying come down, Your place is here on the ground. They call you a flapper, but I call you a clown." But in a comic reversal, the mother finally has second thoughts, and tells her daughter, "Push back the clouds, clear a path to the sky, Your mother's woke up and she's gonna fly."

Women were also portrayed as anxious onlookers, worrying about the safety of men in their airplanes overhead. In the 1920s America's airmail service hired mostly men as pilots. A poster celebrating the service shows a woman with bobbed red hair looking up anxiously as she reaches out to a male pilot flying through lightning in a dramatic, storm-swept sky (plate 30).

In other posters and programs celebrating men's flying achievements women served simply as adornments. In Ernest Montaut's poster for an air meet at Reims in 1909 an elegantly drawn belle époque woman with a fashionably curved hat and graceful, serpentine body waves up to the fliers in their biplanes overhead.

The famed American illustrator Charles Dana Gibson named one of his famed Gibson girls "The Ace" in the program for the First International Aero Congress in Omaha, Nebraska, in 1921. Dressed in her middy outfit, this female "Ace" reclines fashionably rather than demonstrating any flying skill.

But though women in aviation illustrations were often portrayed on the ground, in the early era of aviation women fliers provided real-life displays of transcendence and grounding as they engaged in highly dangerous feats to demonstrate their daring and skill. Blanche Scott and Marjorie Stinson were both stunt fliers who took dangerous dives. Barnstorming during the teens and twenties, Stinson, whom one magazine in 1916 cited as the youngest flier in America, dazzled crowds by first going high into the air, then turning the plane sideways, plunging it precipitously groundward for fifteen hundred feet, and then suddenly righting herself and flying calmly over their heads.[28]

(Offering an intriguing psychoanalytic insight into the possible lure

In Ernest Montaut's belle époque poster advertising Aviation Week in Reims, France, 22–29 August 1909, the woman's serpentine figure serves as an elegant adornment. Science Museum / Science & Society Picture Library.

of stunt dives for both men and women, Michael Balint suggested that people engaged in daring feats, whom he called "philobats," were often displaying a form of psychological regression and mother-dependency. Their search for death-defying thrills became an attempt to demonstrate their omnipotence. Their fantasy, he argued, was that the whole world "is a kind of loving mother holding her child safely in her arms," and therefore external dangers and hazards could be tolerated and mastered safely.)[29]

While death-defying women were being celebrated for their aerial exploits and their ability to transcend cultural conventions, women continued to be hounded by cultural images that represented them as cultural lightweights. In 1929, amid America's infatuation with airplanes and flight, the satirical American magazine *Life* (not to be confused with the later, more famous American news magazine) mocked women's fixation on glamorous images of flight. In a cover illustration titled *Air Minded* an elegantly dressed woman with bare shoulders contemplates a wall covered with pictures of pilots in their flying gear (plate 31). (Being "air minded" in the 1920s meant being enthralled with this latest version of modern transport, but for this soignée and image-conscious woman it simply means contemplating a wall of male fliers.)

As another sign of cultural ambivalence, newspaper photographs honored female fliers as serious, competent, and courageous, but popular images at times represented them as frivolous creatures preoccupied with the latest in flying fashions. In fashion magazines of the 1920s and 1930s the women were sometimes portrayed as vacuous and airheaded. In the *Vogue* magazine story "Madame, the Aeroplane Waits" flying was simply a pleasurable pastime for New York female socialites, who use their airplanes for shopping outings from Long Island to New York City. These are women for whom flying vies with motoring as "a smart and popular diversion." Written with the insouciance meant to please the magazine's wealthy readers, the article addressed those mildly naughty and daring women who wished to distance themselves from social obligations, leaving behind "poor stay-at-homes," their lovers and husbands. "With a ship of your own, you'll be able to forget Theodore in a little spiral or two after dinner."[30]

Illustrating the article were drawings by the French designer Paul Iribe, who created glamorous outfits for fashion-conscious fliers, including mink-trimmed suede suits and cashmere scarves. Offering advice to would-be female airplane owners, *Vogue* recommended Orville Wright's "most alluring little aero coupé" and also the small airplane, the "Oriole," which the magazine praised not only for its mechanical merits but also for its good looks, recommending it "to brunettes whose favourite colours will contrast agreeably with the orange fuselage."[31]

If the airplane in the pages of *Vogue* served as an attractive complement to women's needs for haute couture, American fashion photographers during the 1940s appropriated images of aircraft as high-fashion props. During these wartime years plastics were being promoted as an exciting and fast-developing technology, and fashion imaging often conflated plastics, airplanes, and femininity in advertising the latest in women's clothing designs.

In 1940 *Fortune* magazine presented its own conflation of plastics, aviation, and an archetypal emblem of femininity. In its story "Plastics in 1940" it heralded the material as a $5 million-a-year industry, noting that in 1940 alone 3 million tons of plastics were being used for products like the nonshattering Lucite and Plexiglas bubbles used for cockpits and bomber gun turrets. Encapsulating plastic's modernity, the magazine illustrated its story with a surrealistic photomontage it called "An American Dream of Venus," which coupled images of femininity and technology, like a transparent plastic woman's shoe overlaid on the yellow plastic rudder of an experimental airplane.

More revealing, the montage also included an armless female torso made of clear acrylic set amidst a jumble of plastic products: doorknobs, telephones, playing cards, a toothbrush, and celluloid strips of movie film. *Fortune*'s plastic Venus, photographed in a surreal subterranean world of blue ocean and swimming orange goldfish, embodied the world of utopian promise and hopeful dream. As the magazine self-consciously commented, "Only surrealism's derangements can capture the limitless horizons, strange juxtapositions, endless products of this new world in process of becoming."[32]

In his book *American Plastic* Jeffrey Meikle saw *Fortune*'s plastic Venus as embodying a "carnival of material desire" and as a forerunner of the postwar "expansive culture of impermanence," brought about by the introduction of unstable thermoplastics "capable of being melted and reshaped."[33] Yet more telling about *Fortune*'s conception of the "American Dream of Venus" is its hallucinatory merger of plastic femininity and the sophisticated military machines of war. The magazine's dreamlike conception of women and technology is decidedly one of transformation: nylon hosiery "can turn into parachutes," and Venus, the "transparent lady," also "serves as the nonshatterable window on bombing planes."[34] In this vision of metamorphosis Venus as timeless beauty merges into the time-shattering world of bombs and death. Ultimately, though, the plastic Venus as nonshatterable protective window, like the nylon parachute, is also an emblem of virtue and benevolence, a lifesaver in the theater of war.

Glamour in the Skies   ■ *Fortune*'s Venus suggested the intricate weaving of patriotic virtue, mechanical imagery, and femininity that helped shape cultural representations of women in wartime. In a world of competing agendas — a world in which commerce and advertising turned aviation into an attractive prop for fashion photographers and women were encouraged to develop their mechanical abilities while maintaining norms of femininity — women pilots and defense workers often struggled to maintain a complex fusion of multiple identities, presenting themselves as serious professionals while being sensitive to both fashion and femininity.

The tension women experienced as they tried to meet the practical needs of flying while remaining feminine had a long history. In the early days of flying, women's long dresses proved to be a hindrance. On a demonstration flight in France Wilbur Wright had trouble with three British female passengers, whose skirts had to be tied around their knees. Mrs. Hart O. Berg, whose husband was Wright's representative in Europe and who became the first American woman to fly as a passenger when she flew with Wright in 1908, had to tie her skirt around her ankles to avoid problems with the wind.[35]

The rough conditions of early flying often required female pilots to wear what was traditionally considered men's clothing — breeches, leather jackets, leather helmets, high boots; and during World War II women wore the khakis and helmets required of flight personnel. In a Columbia University oral history interview, Ruth Law (Oliver), who had starred in early flight exhibitions and broken several American altitude and long-distance records, laughingly remembered when her flight clothing nearly caused a crisis of identity. Preparing for a cold-weather flight in 1916, she put on heavy gloves and several layers of clothes, including a leather mask, coat, and breeches. As she remembered, "I didn't look any more like a woman than anything at all. A man, a workman with his lunch pail, came hurrying over, stretched out his hand, and said, 'Well, good luck, young feller. I hope you make it.'" She did make it: she completed the 590-mile trip from Chicago to Hornell, New York, in five hours and forty minutes, which she cited as a record in the United States at the time.[36]

Photographs of early fliers revealed women pilots' sensitivity to maintaining a feminine appearance. In 1926 Amelia Earhart was photographed sitting in an open airplane cockpit in Boston wearing helmet and goggles, elegantly fingering her trademark pearls. In the *Harper's Bazaar* article

"Plane Clothes" she addressed the question, "What does the woman aviator wear?" She said that early in her career, when she had become the first woman to fly as a passenger on a cross-Atlantic flight, she wore "clothes suitable for roughing it" rather than a "ladylike ensemble." But as planes became more comfortable and clean, female pilots could wear more fashionable sports costumes, like the one she was wearing in the *Harper's* photograph that showed her standing in front of an airplane wearing an Abercrombie and Fitch tweed coat, a close-fitting felt hat, and a large, boldly patterned scarf.[37]

Will Rogers may have comically dubbed the women's 1929 flying meet the "Powder Puff Derby," yet women found nothing incongruous about being concerned with cosmetics in the midst of rugged flying. Photographs revealed female pilots' preoccupation with cosmetics as well as clothes as they tried to fulfill dual (and dueling) agendas. Pioneering female aviators were photographed putting on makeup shortly before take-off. In a 1912 photograph Harriet Quimby, who was known for her purple satin flying outfit, was photographed powdering her nose before flying from Dover to cross the English Channel in a Blériot monoplane (three months later she would be tragically killed in a crash at the Harvard-Boston aviation meet).[38] Jacqueline Cochran, who won the prestigious Bendix Prize in 1938, worked as a beautician while she was taking flying lessons, and by the 1930s she not only had her own beauty salon in Chicago but was a successful businesswoman with a cosmetics lab in New Jersey.[39]

Even earlier, Britain's Lady Heath, who had made the first solo flight from the Cape of Good Hope to Cairo, was known for wearing fashionable frocks and keeping her nose powdered while she was flying. Heath, however, seems to have taken a wry view of her use of cosmetics. Piloting a transcontinental flight from London to Capetown, Lady Heath told Amelia Earhart, who accompanied her on the journey, that "one of the absurdities of her journey was pulling out a powder puff and powdering her nose over the South African wastes."[40] Women were also represented as being concerned with cleanliness. In its appeal to a female market, United Airlines in a 1940s advertisement "Why Women Fly" said that women flew because they liked speed, as well as "the sensation of freedom and clean, effortless flight."[41] Even earlier, Earhart had in "Plane Clothes" reassured her readers that with improved aircraft, "one has a clean face at the journey's end." She added that women could contribute to aviation by insisting on better conveniences for women and "demanding comfort," just as they had successfully promoted the creation of gas station conveniences.[42]

Women in Aviation in World War II

■ During World War II the women who worked in airplane manufacturing, like other women engaged in war work, continued to try to balance conflicting roles and images, balancing work responsibilities with their roles as mother and wife and maintaining signs of their femininity while working in jobs sex-typed as masculine.

During the 1930s and early 1940s American women were employed making wing and fuselage covers for airplanes, but in 1941 they constituted only a small fraction of aircraft-factory workers. However, with increased airplane manufacture after America's entry into the war, starting in 1942 airplane manufacturers like Grumman Aircraft, on Long Island, began increasing the number of women hired. During the peak of wartime employment more than 30 percent of Grumman's workers and more than 40 percent of Los Angeles aircraft workers were women. By November 1943 nearly half a million women, or 36.7 percent of the American airplane industry work force, were employed nationwide, helping to mass-produce parts like clear Plexiglas nose cones for light bombers being manufactured at a Douglas Aircraft plant in Long Beach, California.[43]

At plants like Grumman women were considered especially adept at assembling delicate parts, and they largely predominated in the electrical department. Grumman also became the first major manufacturer to have women serve as test pilots for the newly manufactured military planes. Many women working in the aircraft industries later told of enjoying the respect they had received and said that their work experience had changed the way they thought about themselves, giving them a new sense of pride in their competency.[44]

During the war, women also found jobs as civilian pilots flying for the military. In Britain they served in the Air Transport Auxiliary (ATA) ferrying airplanes, worked as flying instructors, and also served as ground crews and mechanics. Starting in 1940 a group of eight headed by Commander Pauline Gower began flying in the women's section of the ATA ferrying airplanes. Eventually more than a hundred women from Britain and other countries, including twenty-two American women recruited by Jackie Cochran and the celebrated British pilot Amy Johnson, were flying aircraft for the ATA, including heavy bombers. (Johnson, who had broken records flying from England to Capetown in the thirties, died in 1941 while ferrying a plane.)[45]

By 1942 Nancy Love had organized America's own Women's Auxiliary Ferrying Squadron (WAFS), in which women also served as civilian pilots flying for the military. To prepare women as pilots in the WAFS, Jackie Cochran, back from England, organized a Women's Flying Training Detachment, which moved from Houston to Avenger Airfield, in

Sweetwater, Texas, in 1943. That same year the two groups were merged into one organization, the Women Airforce Service Pilots.[46] After being taught navigation, mechanics, weather, and other subjects, the women engaged in a variety of aviation activities, including flight testing, but their chief function was ferrying planes from factories to air bases. By the time the program ended on 20 December 1944 more than a thousand women had graduated.

Among the many photographs of WASPs taken by the U.S. Army Air Forces, an iconic image dramatically captures the women's feelings of camaraderie and wartime excitement: four women wearing leather jackets stride forward as they leave their Boeing B-17 aircraft named "Pistol Packin' Mama" at the engine school at Lockbourne in 1944. (Though women were only allowed to ferry the large bombers, such as B-17s, the airplanes themselves, like "Pistol Packin' Mama," were sometimes given admiring feminine names. One of the first B-17s to be used in Britain during the war, attached to a bomber group at Molesworth in Huntingdonshire, was named "Fast Woman.")

Women Airforce Service Pilots (WASPs) leaving their Boeing B-17 aircraft "Pistol Packin' Mama" at the engine school at Lockbourne in 1944. U.S. Air Force Photo Collection (USAF Neg. No. 160449 AC), National Air and Space Museum, Smithsonian Institution.

Many of the jobs held by women in the war had formerly been held by men, and writers at times showed amusement at the ambiguities of gender identity that took place as the women assumed these new roles. In an article publicizing the activities of women training at the Army Air Forces's Women's Flying Training Detachment at Sweetwater, where Jackie Cochran served as director, *Pegasus* magazine began by coyly feigning puzzlement at the sight of "an animate bundle" wearing khaki overalls and a leather helmet, carrying a cumbersome parachute, and walking "straightlegged like a man." But as the person's helmet is lifted, a surprising change of identity takes place: "Long, red hair frames the face," and "a girl, of all things, says 'Hi!'" Yet the transformation was more than superficial; in the magazine's bemused view, the real change for these women was the setting aside of their conventionally feminine preoccupations: "They forgot about men and dates and hairdo's," and "they worked and flew like Jackie knew they could."[47]

Like other wartime articles and photographs publicizing women at work, the *Pegasus* article portrayed the women at Avenger Field as bold in tackling untraditional roles yet never losing their essential femininity. The magazine praised them for their competence yet also marveled at "the appearance of so much femininity in an activity about which there is such grimness and determination." Even the airfield's atmosphere reflected this fusion of femininity and grit, for, as the article noted, the "pungent odor of grease blends incongruously with the essence of milady's perfume and powder." The students are portrayed as going about their work "with a totally ungirlish drive."[48]

Reflecting wartime photographic and cinematic conventions, the article's photograph of a female trainee climbing into the cockpit is taken at a low-angle view, her body looming upward to create heroic iconography, and she appears imposing in her leather flying helmet and khaki coveralls. As a reminder of the women's essential femininity, *Pegasus* also photographed them sitting on the ground playing with their cat Flip, the group's feline mascot.

Reflecting a familiar theme, the magazine praised the women for their determination while also grounding them in their traditional, gendered sphere. *Pegasus* reassured its readers that the women serving as pilots had not forgotten their domestic roles, nor were they getting notions about continuing flying after the war. At the article's end Jackie Cochran is quoted as saying, "The principal boost they can give aviation is as mothers and wives," and she added that she expected all the women at Sweetwater "to raise families and make stable, happy homes for our men when this war is over. And I know that's what they want, too."[49]

Through text and image stories like these heightened women's stature yet also limited them with cultural grounding. The tension wove its way through stories about women who took on other roles in the field of aviation during wartime. In a chapter titled "Schoolmarms of the Sky," from her book *Girls at Work in Aviation* (1943), the American war photographer and correspondent Dickey Chapelle, writing under the pseudonym Dickey Meyer, described the work of American women employed as instructors in civilian flying schools under contract to the Army and Navy. As part of the government's War Training Service Program, the job of these schools was to ready men for active military duty.

In her book Chapelle — who had studied engineering at MIT, learned to be a pilot at the Curtiss-Wright airport near Milwaukee, and worked as a writer and photographer for the publicity offices of Transcontinental and Western Air Lines — voiced the women's purported inner thoughts, expressing their pride in their own abilities as teachers as well as their awareness of cultural attitudes that might limit their achievements.[50]

As described by Chapelle, the women themselves embraced some gender stereotypes: they "have always known that as women they pos-

A female trainee in the Army Air Force's Women's Flying Training Detachment at Avenger Field, Sweetwater, Texas. *Pegasus*, **November 1943.**

sessed an extra share of the elusive ability to teach." But they were also clear-eyed about the limits they faced. Although they felt that they possessed a special teaching ability, "they have also known that its free expression could be earthbound by another feeling equally inherent in the hearts of men," the "belief that women were somehow not quite adequately prepared for military aviation." Chapelle cites the example of women who trained to become instructors at Tennessee's Aviation Research Instructor School for Women and had to prove their mettle through studying even though they came with high credentials (several were former ferry fliers, and one even had a degree in aeronautical engineering).[51]

During the 1910s and 1920s Coles Phillips's magazine illustrations had pictured women kneeling forlornly next to their automobile flat tires, their faces charmingly smudged. Years later Chapelle wrote that for the women aviators and factory workers the smudges were at once both fanciful and real. She sardonically recounts an earlier period when the print media was intent on constructing images of female workers as quaintly dirty: "The first women in the factories had to smear grease on their faces and pull their hair down to satisfy the art editors of the newspapers."[52]

Yet, as Chapelle points out, having dirt on their hands and faces was part of the reality of being airplane mechanics. The caption to a photograph in Chapelle's book of a woman tackling repairs on a huge airplane engine describes her as "dressed in overalls, her face smudged and her fingernails in most unlady-like condition." Rather than being artifice, rather than being quaintly cute, as in Coles Phillips's images, this woman's unladylike dirt is a sign of her seriousness of purpose.[53]

Issues of femininity and competence, transcendence and grounding, also were present in stories about British women working on balloon crews during World War II. The women's tasks included inflating, hoisting, repairing, and bedding down the huge balloons that were used for barrages and surveillance as well as for guarding the British coastline and industrial sites. Dame Laura Knight's painting *A Balloon Site, Coventry* shows a crew of three women and a man, overseen by a female supervisor, tugging at lines launching huge white balloons in air clouded with factory smoke (plate 32).

Women in eighteenth- and nineteenth-century caricatures, such as in Daumier's lithographs, were sometimes pictured as flighty creatures, made airborne in their huge inflated crinoline petticoats and balloonlike dresses; the caption to one of Daumier's lithographs reads, "These are not women but balloons." But the women who worked on balloon crews during World War II were represented as being thoroughly down to earth as they tackled their job. In his pamphlet *Eve in Overalls*, describing British women's war work, Arthur Wauters included a chapter titled

"Beauty and the Blimps," in which he argued that England's Balloon Barrage Defense was one of the toughest jobs taken on by the Women's Auxiliary Air Force because the pulling of mooring ropes required great "dexterity, method, co-ordination and physical strength."[54]

Still, Wauters took pains to preserve an image of the women as safely feminine and young rather than formidable. After all, "not very long ago their pretty hands must still have been playing with toy balloons." While he praises the women's current expertise, he also portrays them as both erotically captivating and safely maternal: "They wear exactly the same uniform of their male comrades, but they have not lost any of the allure of their sex." Suggesting their maternal nature, Wauters writes that the women had to contend with the type of balloon that, like a wayward child, tore "impatiently at its gigantic umbilical cord."[55] Wauters himself seems to have been intent on keeping these women, with their potential for reaching great heights of expertise, safely contained. With their hair tucked beneath caps and steel helmets, they "very much resemble miniature captive balloons." And beneath the guise of the stalwart worker, they are reassuringly feminine. When one young woman anxiously asks whether she is becoming too mannish, the author firmly establishes her femininity, evoking her "magnificent eyes, lovely hair," "delightful mouth," and her "bewitching smile."[56]

Throughout the war, photographs and texts continued to mirror a wider cultural ambivalence that championed women's roles as pilots and industrial workers while also presupposing their fundamental place in the home. Even earlier, female writers like the American poet Muriel Rukeyser had reflected some of this same ambivalence, seeing in the airplane an apt metaphor for her own longings for an expanded self yet also her clinging to the more familiar, and more grounded, social roles. Rukeyser, in her long poem *Theory of Flight* (1935), found in aviation imagery the embodiment of her own conflicted longings.

In her poem Rukeyser forges a dialectic of opposing impulses: the pull toward the seductively warm, earthbound, grounded domestic roles of wife, lover, and mother and the wish to find freedom and power through her imaginative identification with airplane pilots and flight. The speaker's quest is for a lusty, elemental knowledge of earth as well as for the experience of a world without boundaries. Her poem opens with the invocation, "Earth, bind us close and time," yet the speaker also resists her own longings for the ground, for the fastening lips of earthly lovers. She becomes free by imaginatively allying herself with the plane and pilot who lift off "in one access of power." In the poem the airplane itself becomes a gleaming "metaphor most chromium clear" that embodies her quest for expansion, for distancing, for "Power, electric-clean, gravitating outward at all points." Yet ultimately her centrifugal pull outwards becomes too problematic, and in the poem's final section she again feels the earth's pull:

> Flying, a long vole of descent
> renders us land again.
> Flight is intolerable contradiction.
> We bear the bursting seed of our return[57]

A Distanced View   ■ The allure of airplanes has long been linked to their ability to provide detachment, distance, and disconnection from earthly care. Lady Heath wrote euphorically that "the women who have flown will never give it up, for they have realised the joy and happiness that comes from being so disassociated with the world."[58]

But women writing in the early twentieth century, like male writers, also celebrated another aspect of distancing: the new perspectives offered by aerial views. In *Everybody's Autobiography* Gertrude Stein described her first experiences in an airplane. In 1934 she flew from Paris to England to the United States. En route to Chicago she talked with the pilots and stewardesses and was even allowed to sit in the pilot's seat, where she "made the wheel move a little."

But for Stein one of the real delights of flying was the way it brought clarity to her vision of America. She enjoyed "the looking down and finding it a real America" and seeing the states' clear boundaries, the "straight lines and quarter sections," which validated her vision of the American map. The clean shapes of the American terrain also validated her view of cubism: "the mixed line of Picasso coming and coming again and following itself into a beginning was there, the simple solution of Braque was there," as well as the "wandering line of Masson," and "it made it right that I had always been with cubism and everything that followed after."[59]

While travelers like Henry James in *The American Scene* (1907) had written about the distancing effect of landscape images seen from the window of a rushing train and Matisse in his painting *Through the Windshield* (1917) had envisioned the landscape through the defining framework of automotive glass, Gertrude Stein in *Picasso* (1938) privileged the world seen from an airplane as providing the most wondrous views. "One must not forget that the earth seen from an airplane is more splendid than the earth seen from the automobile," she wrote. Here again she marveled at "all the lines of cubism made at a time when not any painter had gone up in an airplane."[60]

While Stein was fascinated by the abstracted view of the landscape below, for the American painter Elsie Driggs in *Aeroplane* (1928) the

Plate 21. Tamara de Lempicka,
AutoPortrait, oil on board, 1925.
The artist is shown at the wheel
of a green Bugatti. © Kizette de
Lempicka Foxhall.

Plate 22. Julius L. Stewart, *Les Dames Goldsmith au bois de Boulogne,* oil on canvas, 1901. Stewart's painting was one of the earliest paintings of women and an automobile to be exhibited at a Paris salon, in 1901. Compiègne, Musée de la voiture et du tourisme. © Photo RMN-Arnaudet.

Plate 23. A woman driver maps out her cross-country journey in Ruth Eastman's cover illustration for the June 1921 issue of *Motor* magazine.

Plate 24. A teasing view of a woman servicing her automobile in a jigsaw puzzle titled *Start Something,* manufactured by Tryawhile, n.d. Photograph © Anne D. Williams.

Plate 25. A woman repairs her automobile, undaunted by the rain. Ettore Tito, *Aide-toi, le ciel t'aidera,* hand-colored line block print, ca. 1925–30. Victoria & Albert Museum, © V&A Picture Library.

Plate 26. Framed by the front windshield of her electric automobile, a woman and her children form an image of family cohesiveness in this Baker Rauch & Lang Company advertisement in *Life* magazine, 13 January 1916.

Plates 27 and 28. These contrasting magazine covers by the American illustrator Howard Chandler Christy show women as self-assured automobile repairer and decorative sex object. *Motor* magazine, November 1923 and January 1922.

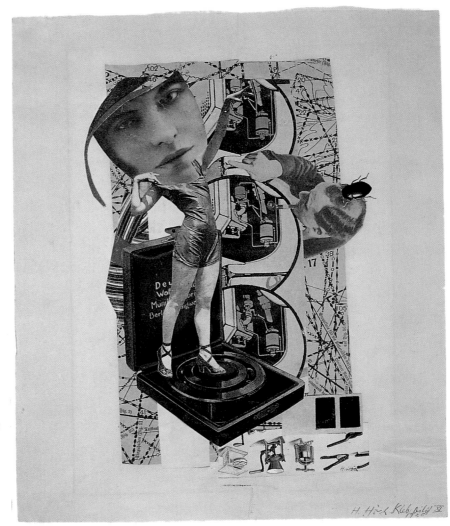

Plate 29. In her 1920 photocollage *Öhne Titel* (Untitled) the Berlin dada artist Hannah Höch presents a female figure standing on a ball bearing against a matrix of an aeronautical map, creating an iconic image of the modern woman and the machine. Courtesy of Morton G. Neumann Family Collection. © 2000 Artists Rights Society (ARS), New York / VG Bild-Kunst, Bonn.

Plate 30. In a familiar motif, an anxious woman holds out her hand to a male flier in a storm-swept sky in this poster for America's airmail service, ca. 1924. National Air and Space Museum, Smithsonian Institution (SI Neg. No. 98-20433).

Plate 31. The cover illustration titled *Air Minded* for the American humor magazine *Life* spoofed an elegant woman's fixation on images of fliers. *Life,* 8 November 1929.

Plate 32. Dame Laura Knight,
A Balloon Site, Coventry, oil on
canvas, 1942. A balloon crew and
their female supervisor in Britain
during World War II. Women's
tasks included inflating, launching,
repairing, and bedding down the
huge balloons. Used with permission of the Trustees of the Imperial
War Museum, London.

Plate 33. Audrey L. Flack, *Spitfire,* acrylic with oil glazes, 1973. In her air-brushed painting of model airplane kits Flack presents a glittering look at the allure of military war games. Courtesy of National Air and Space Museum, Smithsonian Institution, and Louis K. Meisel Gallery.

Plate 34. Shannon W. Lucid, 1996. Completing her mission, Lucid floats in the tunnel that connects the space shuttle Atlantis's cabin to the Spacehab double module in the cargo bay of the Mir docking station before the September 1996 return to Earth. Courtesy National Aeronautics and Space Administration.

Plate 35. Edward F. Skinner, *Women Workers Cutting Files*, 1917, oil on canvas. Here women are shown cutting rat-tail files at the Cammell Laird Company's Cyclops Steel and Iron Works. Science Museum / Science & Society Picture Library.

Plate 36. Edward F. Skinner, *For King and Country*, oil on canvas. Women manufacture shells at the Grimesthorpe Steel and Ordnance Works. Workers wearing yellow are carrying highly explosive shells. Used with permission of the Trustees of the Imperial War Museum, London.

Plate 37. A female defense worker drills holes in an airplane fuselage frame in this cover illustration for *Fortune* magazine, April 1942.

Plate 38. Dame Laura Knight, *Ruby Loftus Screwing a Breech-ring,* oil on canvas, n.d. Used with permission of the Trustees of the Imperial War Museum, London.

Plate 39. A doughty defense worker in J. Howard Miller's poster "We Can Do It!" (1943). National Archives, Washington, D.C.

Plate 40. Benton Clark, *Keeping Up the Good Work,* oil on canvas, 1944. Featured in a Coca-Cola advertisement in 1944, this female riveter—whose yellow sweater is alluring to male admirers rather than a sign of explosive danger as in World War I—works at a wartime shipbuilding site. Courtesy of the Coca-Cola Company.

Plate 41. "The more WOMEN at work the sooner we WIN!" In this 1943 recruitment poster, with a photograph by Alfred Palmer, a woman is shown putting the finishing touches on the nose section of a B-17F Navy bomber at the Douglas Aircraft Company, in Long Beach, California. Library of Congress, FSA-OWI Collection.

Plate 42. Jean-Léon Gérôme, *Pygmalion and Galatea*, 1890, oil on canvas. In Ovid's tale, Venus brings Pygmalion's sculpture of a beautiful woman to life. The Metropolitan Museum of Art, Gift of Louis C. Raegner, 1927 [27.200].

Plate 43. The pink plastic, digital Venus next to the gallery janitor Milo in Donna Cox's computer animation *Venus & Milo*, 1990. Courtesy of the artist.

experience of being in an airplane seemed to evoke a sense of isolation and disengagement. Driggs, whose work became associated with that of the American precisionist painters, was among the very small number of early-twentieth-century female artists painting images of industry and machines, in particular images of airplanes. The daughter of an engineer who worked in a steel factory in Sharon, Pennsylvania, Driggs moved with her family to New York, where she studied at the Art Students League. From 1924 to 1932 she exhibited her early paintings of flowers at the Daniel Gallery in New York, a gallery that also presented paintings of flowers by artists who would later share a fascination with industrial imagery and be dubbed first the immaculates and then the precisionists.[61]

Intrigued by industry, in the 1920s Driggs returned to Pennsylvania to visit the Jones and Laughlin steel mills in Pittsburgh. She returned to New York to paint her own version of towering smokestacks in *Pittsburgh* (1927), a painting whose clarity and classical distillation of imagery was like that of the work of other precisionists, including Charles Sheeler. On an airplane trip to Detroit from Cleveland in 1928 to make preparatory drawings of the Ford Motor Company's River Rouge Plant (a subject also painted by Sheeler, who had gone there a year earlier), Driggs sat next to the pilot and made sketches that she used for her painting *Aeroplane*.

In choosing to paint an airplane Driggs created her own version of a machine that was already becoming a modernist icon. In an essay entitled "The Machine Aesthetic" the French painter Fernand Léger described seeing airplanes in the French Salon d'Aviation and admiring the "beautiful, hard, metallic objects, firm and useful" and the "geometric power of the forms."[62] And at the Machine-Age Exposition in New York in 1927 a Curtiss airplane engine and an airplane propeller were displayed as examples of modern design. During the 1930s the airplane's streamlined, tapered body and sweptback wings inspired American engineers and designers. Industrial designers, including Norman Bel Geddes and Raymond Loewy, who admired the aerodynamic engineering of airplanes such as the Boeing 247 and DC-3, produced their own streamlined versions of railroad locomotives and automobiles, as well as more quotidian objects, from toasters to pencil sharpeners. The streamlined style, wrote Sheldon and Martha Cheney in 1936, was "borrowed from the airplane and made to compel the eye anew."[63]

In Driggs's painting the airplane is isolated for visual inspection and crossed with hints of ray lines, those radiating lines that so often appeared in modernist renderings of skyscrapers and the Brooklyn Bridge as emblems of a dynamic age. Like other precisionist paintings, *Aeroplane* presents a clear, classicized vision, uncluttered with human presence, social commentary, or the shifting impact of dirt, decay, and time.[64] Still, with its flat, uninflected grayness and clear light, her painting also becomes a psychological analogue for disconnection and evokes an eerie feeling of isolation. Driggs's precisely rendered airplane becomes a gray phantom flying in a void, cut off from any contextual landscape.

Driggs's *Aeroplane* and her interest in machine and industrial iconography were unusual for an early-twentieth-century female artist, a fact noted by Thomas C. Folk in a 1991 exhibition catalog in which he considered her machine images in unabashedly gendered terms, oddly echoing a critic's comments about Angèle Delasalle's French Salon forge painting a century earlier. "Her Precisionist paintings do not suggest her sex," wrote Folk, adding that her American contemporaries "tended to do well in the decorative and overly feminine subjects." Marveling at what he considered Driggs's masculine interest in industry and technology, he concluded, "Driggs's early art rarely indicates that it was created by a beautiful, young, and very feminine woman."[65]

Driggs's *Aeroplane*, detached from the social landscape, a machine free from any associations with its uses in war. But in the work of other artists and writers, the image of the airplane became associated with the psychic distancing afforded by modern technologies—the troubling type of distancing that might keep bombardiers, for example, separated from a full awareness of the damage they were inflicting on the ground.[66]

Concerns about the destructive role of aerial bombing itself had actually emerged early, as in poems about aviation written by female writers included in Stella Wolfe Murray's anthology *The Poetry of Flight*. Justifying her decision to place the poems by women in a separate section, Murray explained that it reflected a "natural desire to give them prominence and to show that despite their lack of opportunity, women feel all the poetry of flight and are fully alive to all that progress in aviation means to the world."[67]

One poem in Murray's anthology, "The Gift of Flight," by a poet who identified herself only as "Astra," contrasts the initial pride felt about aviation with the use of aerial bombing in World War I. Aerial bombing was then only in its inception, but to the poet it already seemed that the airplane had become an indiscriminate source of destruction, bringing

> "the terror that flies by night,"
> Bombing the baby at the breast, the old, the sick,
> Widows and orphans, like a poison'd hail, on the just
> and the unjust.

For this poet, airplanes brought benefits—"commerce and prosperity," "tonic joy flights"—but bombing represented the "base use of the gift of flight," and she denounced its "death-dealing / deviltry." Instead of "This machining through the air," she ultimately longs for a redirection of flight and for "peace, eternal peace."[68]

During World War II, however, the British painter Laura Knight, working in the service of the government, understandably took a more sanguine stance, in her oil painting *Take Off* meticulously detailing the mechanism of the bombardiers' mission. Working for the War Artists' Committee, Knight was commissioned by the government to create a painting of the crew of a Stirling bomber at Mildenhall in East Anglia. She visited the site where the Stirling bombers were being built and created a watercolor painting of the planes being constructed. She also studied the cockpit interior of a Stirling aircraft and the faces of crew members about to leave on one of their nightly missions to Germany.[69] In Knight's painting, produced amidst antiaircraft fire, the bombardiers' preparation for a mission is shown with authority and force, capturing the intensity of the navigator scrutinizing his charts and the huge interior space of the aircraft, displaying the intricate mechanics of war.

In her own participation in a bombing mission, however, the American photographer Margaret Bourke-White offered a more probing view of aerial warfare, going beyond reportage to a contemplation of her own distanced feelings. Bourke-White had long been fascinated by aviation. "Airplanes to me were always a religion," she remarked.[70] In her early career she worked as an industrial photographer, and in 1934 and 1935 she was hired by several aviation companies, including Pan Am, Eastern Airlines, and TWA (then Transcontinental and Western Airlines), for whom she photographed phases of their operations.[71]

Bourke-White's early photographs of airplanes reflect her own aesthetic distancing, as well as some of the visual conventions of machine-age photography, in which machines were admired for their clean lines, classically elegant geometries, and the dynamic interplay and patterns of machine parts. Her airplane photographs of the 1930s endowed the machines with elegance and monumentality, such as in her closely cropped detail of a TWA aircraft, its soaring vertical tail juxtaposed with nose and propeller facing a huge block of empty sky.

Photographing the towering concrete spillway of the Fort Peck Dam in Montana for the cover of *Life* magazine's first issue, on 23 November 1936, Bourke-White became one of four staff photographers first hired by the magazine, and during the 1940s she became a war correspondent for the magazine. On a photographic assignment for *Life* she received permission from Major General James Doolittle early in 1943 to become the first woman to fly on a military combat bombing mission over enemy soil, and she accompanied thirty-two planes on a mission to bomb an Axis airport — the El Aouina airport, near Tunis — where German planes were being kept. (She flew in the lead plane, "Little Bill," which was piloted by General Atkinson. The copilot was Paul W. Tibbets Jr., who in August 1945 would pilot the *Enola Gay*, the plane that dropped the world's first atomic bomb, on Hiroshima.) In a story titled "Life's Bourke-White Goes Bombing," in March 1943, the magazine touted Bourke-White's milestone: "It is no common thing for a war correspondent to accompany Army fliers on a combat mission in the western war theater."[72]

As a photographer reporting about the American armed forces, Bourke-White was aware not only of her role as a photographic image maker but also of herself as a glamorous icon. *Life's* Tunis story included a photograph of her in a high-altitude flying suit, complete with fleece-lined leather jacket, heavy pants, and boots. In her autobiography, *Portrait of Myself*, she wrote: "I was flattered when this picture in my high-altitude flying suit became popular as a pin-up." But she noted wryly that her popularity was fleeting, her photograph being quickly displaced by "my rivals in their bathing suits (or less)," whose images were hung on the walls of dugouts, barracks, and Nissen huts.[73] In the Tunis story, for which she photographed preparations for the bombing and the group of thirty American B-17 Flying Fortresses heading along the Mediterranean coast, Bourke-White revealed her reactions to the Tunisian bombing mission as a mixture of emotional detachment and engagement. Flying four miles up, she took photographs from the waist-gunner's port, the radio operator's window, and even the waist-gunner's hatch of the lead plane, wearing an oxygen mask as she moved about the plane in sub-zero temperatures.

Focused on getting photographs rather than on the danger of the mission, she surprised the bombardier who was pulling safety pins out of the bombs and momentarily storing the tabs in his mouth. When she took a shot and the flashbulb of her camera fired, he cried out in alarm, "Jesus Christ! They're exploding in my hands!" In a revealing commentary in her autobiography, written years later, Bourke-White, ever conscious of gender issues, seemed less concerned with the explosiveness of the moment than with reflecting on her own feminine presence in the plane. She saw the bombardier's startled surprise as evidence that her presence had been forgotten, belying "the old wives' tale I had been pestered with ever since my early days in the steel mills and coal mines: the legend that if you bring a woman into a spot where men are doing dangerous work, her very presence will so distract them that they will be careless and have accidents."[74]

In her personal notes about the bombing mission written at the time, some of which were later used in *Portrait of Myself*, Bourke-White revealed that hers was an aesthetic view, the view of a photographer watching from afar. As they approached the target, she wrote in wonderment: "The majesty of it, the rendezvous. Ancient Carthage lay below. Catching the glint of the backlight on the salt lake, now a marsh. We were wheeling in now the bomb run." Even when she could see smoke columns from the blasted airfield, with its bombed planes and fuel dumps, she reacted with the detachment of a professional craftsman, "I felt a little cheated. No fighter attacks within camera range."[75]

Only later did she ponder her own detachment and her disturbing feelings of being distanced from acts of war and destruction. As she wrote in *Portrait of Myself*, "I don't believe I ever thought of this expedition for which we were all preparing as a mission of death. The impersonality of modern war has become stupendous, grotesque." She added, reflecting on her own sense of distance, that "even in the heat of the battle, one

The *Life* magazine photographer and war correspondent Margaret Bourke-White dressed in the high-altitude flight suit she wore when she accompanied American military fliers on a bombing mission to North Africa in 1943. She became a pinup for American servicemen during World War II. © Margaret Bourke-White Estate / Courtesy Life Magazine.

human being's ray of vision lights only a narrow slice of the whole, and all the rest is remote — so incredibly remote." In her notes she also mused, "The remoteness between killer and killed in this kind of warfare doubtless has to be there if those of us who deal with war are going to carry on." But, she wondered, "what are one's responsibilities? What should we feel? . . . What can we do?"[76]

In her notes Bourke-White also further probed the issue of the distancing permitted by new technologies and modern war machines, a subject that seemed to haunt her. She remembered speaking to Paul Tibbets in Kansas after the war and asking him about his response to the Hiroshima bombing, including how he had felt about the people down below. "You don't think about anything like that," he replied, adding, "They're so poor and miserable it probably helps them as they'd only die anyway." Calling this a cliché, "for which we must forgive him," Bourke-White wrote that although Tibbets was intelligent and kind, his response was "a striking example of the remoteness that was troubling me." About her own feelings on flight missions she said, "I was distressed that I could feel no emotion."[77]

After the war Bourke-White's fascination with flying — and her aesthetic distancing — continued, though she was by no means distanced from danger. In 1951 she risked her life by dangling from a helicopter to create a series of photographs of the earth that appeared in a 1952 *Life* article, "A New Way to Look at the U.S.," images that captured the mosaic of fields and abstract aerial patterning.[78] Also in 1951, Bourke-White revealed the continuing lure of aviation adventure when she took on a *Life* assignment for a photographic essay on America's Strategic Air Command. The assignment permitted her to fly in the bombardier's seat of an experimental B-47 bomber being tested at SAC headquarters in Kansas, the first woman to fly in one of these experimental planes (the testing project was again headed by Paul Tibbets). Her photograph of the bomb bay of a B-36 bomber at an airfield in Texas uses the sharp symmetries framing the central figure of the crewman to draw us into the body of the plane.[79]

During the 1960s and 1970s, airplane imagery continued to appear in the work of women artists who explored issues of distance and immediacy, detachment and raw immersion, in the harshness of war. A chilling view of military aircraft is presented in the precisely rendered drawings and paintings of the New York artist Vija Celmins, whose childhood memories of aerial bombing in Latvia during World War II helped shape her grim, gray representations of airplanes hovering ominously in a high and distant space, removed from any view of the targets below. Meticulously created, her images have an uncanny realism and presence yet also have an air of the surreal.

Celmins and her family fled to Germany during the war and then settled in California in 1948. Revisiting her memories of the war, she began building little model airplanes and buying war books. In 1966 she produced four paintings of warplanes — three of American bombers and one of German military aircraft. In *Flying Fortress, Suspended Plane,* and *German Airplane* the familiar is made strange; in each painting a lone, near-photographic rendering of an airplane fills the pictorial space and spreads over a gray, undefined landscape of a troubled dream.[80] Celmins's version of the airplane is detailed yet without reflective surfaces, ominously absorbing light rather than reflecting it. The artist's mordant eye endows the machine with a dull, heavy cast, making its malevolence unmistakable.

Taking a less mordant view, the New York artist Audrey Flack in her 1973 painting *Spitfire* brought a cool precision to her images of plastic toy airplane model kits, evoking the seductive allure of aircraft and America's fascination with war games (plate 33). Using projected 35 mm slides and an airbrush, Flack created a precise, photorealist's rendering of hot images — the packaging for models of military airplanes, including a Curtiss fighter bomber and the Spitfire, the airplane that played such a prominent role in Britain's air battles during World War II.

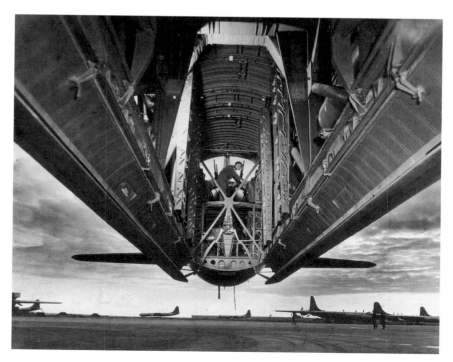

Margaret Bourke-White, *Bomb Bay of a B-36 Bomber, Carswell Field, Fort Worth, Texas,* 1951. Bourke-White produced a photographic essay on America's Strategic Air Command for the *Life* magazine issue of 27 August 1951. © Margaret Bourke-White Estate / Courtesy Life Magazine.

In interviews Flack emphasized her formalist agenda, her concern with light and color, her love of "reflections and glitter." *Spitfire* includes not only shiny images of airplane models and strings of metallic beads but also multicolored tubes of paint, the references to color and paint and the process of making art often found in her paintings from this period. Commissioned by the art collector Stuart Speiser to produce a painting of an airplane, Flack chose to paint models of military aircraft and passenger planes because, she said, as a child she had enjoyed making airplane models out of balsa wood and she liked the "volumetric aspect of them."[81]

Still, *Spitfire* rises above Flack's professed aesthetic agenda: the painting's glossy images and metallic blues chillingly encapsulate a consumer culture that both glamorizes and trivializes deadly military machines. In her painting, the airplanes' plastic surfaces reflect the hard-edged destructiveness of war. Their surface reflections, and the artist's detached clarity of vision, produce a provocative visual and political tension: behind the apparently benign and impersonal view of toy aircraft lies a reminder of the insidious boxing and merchandising of war, a merchandising that transforms the reality of aerial bombing into a game of play far distant from the harsh, brutal realities of combat.

Four years later, Flack's fascination with airplanes and glamorous merchandising turned to paintings probing women's quest after a glamorous identity—and by extension, a society that encouraged them to do so. In her photorealistic painting *Marilyn (Vanitas)*, of 1977, a dripping paintbrush hovers over a table laden with images of Hollywood glamour, cosmetics, and time: a framed photograph and its mirrored image of Marilyn Monroe, a tube of lipstick, rouge, eyeshadow, a gold compact filled with powder, an hourglass, and a clock, images suggesting not only the transient nature of beauty but also the tragic implications of a female identity that is transformed, covered over, made up.

In her own work, the New York artist Nancy Spero created a confluence of airplane imagery, an exploration of technological distancing, and feminist issues. Early in her career, during her years in Paris, where she lived with her husband, the painter Leon Golub, from 1959 to 1964, Spero produced her murky "Paris Black Paintings." Her expressionist oil paintings of lovers, prostitutes, mothers, and children focused on women in their traditional maternal and sexual roles. Yet as Lucy Lippard has noted, women artists who participated in the antiwar movement during the Vietnam War years merged their political and feminist agendas, feeling that they too were among the oppressed, the powerless.[82] Spero said that in her early years as an artist she felt confined, frustrated, and silenced: "My anger was in that people weren't listening to what I had to say, meaning that people weren't looking at my work."[83] Her paintings such as *Female Bomb* (1966), showing tiny faces with tongues sticking out, merged her concerns about nuclear explosions and her fury against war with her own struggles to discover her own voice and be heard as both a woman and an artist. "I wanted to depict women finding their voices, which partly reflected my own developing dialogue with the art world, that somehow I had a tongue and at least a part of the language of that world."[84]

During the 1960s Spero joined protests against the Vietnam War and began producing tortured and tormented images of airplanes as war machines. Airplanes in the work of Flack and Celmins were separated from any human context, stripped of their deadly associations or removed to a nightmare space far removed from the plains of war. But in her own work Spero creates a highly charged, expressionist outcry; in her paintings of military aircraft produced during the Vietnam War she engages in a piercing, painful jabbing aimed at eliminating any sense of detachment. Rather than evoking remoteness, her paintings reassert a direct connection to human torment and suffering.

In her "War Series" paintings (1966–70), which she first exhibited in 1983, Spero revealed her concern with what she has called "technological obscenity," an obscenity made manifest not only in the gross gore of blood and excrement but also in the mechanical neutrality, the uncomprehending eye of monster machines processing their human victims with a ghastly, joyless promiscuity. Her cries to articulate her own voice coalesced with the impassioned cries of open-mouthed victims of war, and it was this merger that helped make her images of violence and suffering so memorable. In the War Series, she pierced through to the visceral and more explicitly sexual components of the rage and violence unleashed during the war. Airplanes emerge as emblems of mechanical oppression, and military helicopters become ravenous destroyers, uncompromising death machines.

In Spero's work the helicopter is central to the artist's vision of the ravenous destruction and cruelty unleashed by the tactical and rhetorical weapons of war. Helicopters were used in the Vietnam War for fighting, dropping napalm, and dropping and picking up soldiers. Painted with delicate gouache and ink on paper, Spero's helicopters have a delicacy that belies the fierceness of her vision. Drawn with slashes of black, pale orange, yellow, and steel-gray gouache, monstrous silver military helicopters engorge human victims and spew forth naked bodies smeared with dark red blood. Mushroom-shaped bombs with male, female, and androgynous torsos explode with cataclysmic spasms of military and sexual fury.

Typical of Spero's War Series is *Gunship and Victims*, a small work with an explosiveness that erupts off the paper, producing sharp shrapnel shards of anger that pierce complacency. The gunship is a large silver airplane-monster with huge black propeller blades and a solitary eye that stares like Leviathan with unblinking gaze. Hovering in empty space, the monster machine vomits out small human figures outlined in black and covered with dark red blood while the plane's tail end releases its excrement of tiny human skulls and bones.

In her painting *Search and Destroy* Spero eliminated the possibility of distancing by infusing the work with the visceral feel of intrusive aircraft — the horrendous, unwanted feeling of intimacy created between victim and machine. Her work dives in low, immersing us in technological menace.[85] The helicopter, with its stylized, exaggerated nose cone, jabs at its human victim, who is attempting to crawl away on the ground. On its impersonal search-and-destroy mission the helicopter becomes a mechanical tormenter, directed at the man who makes a futile effort to escape. With a few deft strokes, the artist evokes all the pain and suffering of war.

Helicopters often appear as grotesque, serpentlike or reptilian creatures in Spero's paintings. In *S.U.P.E.R.P.A.C.I.F.I.C.A.T.I.O.N.*, for example, the helicopter is an androgynous monster that is both serpent with a phallic fuselage and a cruel maternal figure. As the helicopter hovers, a row of nude human figures — prisoners dangling before being dropped below — cling to ropes that also look like rows of breasts hanging from the plane's underbelly. The prisoners hang like docile, helpless children,

In Nancy Spero's *S.U.P.E.R.P.A.C.I.F.I.C.A.T.I.O.N.* (gouache and ink on paper, 1968), prisoners hang from the mother machine. Photo © David Reynolds, courtesy of the artist.

dependent on the mother machine for nurturing, caught in the paradoxical condition of symbiosis and alienation.[86]

While artists like Spero probed the menacing nature of aircraft in time of war, there were also signs that for women like the American scientist Shannon Lucid the world of aviation and space exploration offered a welcome sense of expansion and release. During the 1990s, photographs and video stills made by the astronauts themselves helped reconceptualize not only the idea of space itself but also the cultural space occupied by women. Rather than portraying women as constricted in their aspirations, grounded, and circumscribed by cultural conventions that limited their roles, these images of women like Lucid, a biochemist flying in space-shuttle missions, suggested the widening circumference of women's roles.

In 1996 Lucid set records for endurance in space during her six months aboard the Russian space station Mir, where she studied the impact of a space environment on living tissue. The photographs of her in the Mir docking tunnel before heading home on the Atlantis shuttle are revealing. Though the Soviet crew reportedly envisioned her in conventional roles—an officer in the Russian space program is said to have assumed that Lucid would help the space station be neater because "we know women love to clean"—the image of Lucid floating in the surrounds of the Mir docking tunnel shows her encircled but not confined, a figure who has clearly transcended women's once circumscribed roles (plate 34).[87]

WOMEN *in* WARTIME

From Rosie the Riveter to Rosie the Housewife

7 IN A CARTOON published in the *New Yorker* in 1942 a female factory worker dressed in overalls and carrying an oil can cautiously approaches another human figure whose identity is hidden by welder's gear, including a protective face-covering hood. "Madge?" the overalled woman anxiously inquires of the welder. As this image coyly suggests, women working in defense jobs were appearing in a new guise and experiencing new ambiguities of identity. Leaving their more conventional blue- and white-collar jobs or their roles as students or housewives, these women were

working in jobs that were socially defined as men's work, such as machine-tool operators, riveters, welders, mechanics, airplane pilots, truck drivers, gas-station attendants, and railroad workers. In the process, they were also dramatically reconfiguring, if only for a time, their own identities as women and their sense of self.

Female Workers in World War I

■ The issue of changed identities actually was not new: in World War I women had worked as automobile and ambulance drivers, welders, mechanics, machine-tool operators, and at myriad other skilled and unskilled jobs usually considered men's occupations. Working in munitions factories, women donned masks, caps, coveralls, and sometimes trousers and tunics, their new uniforms signaling their changed role. Recruiting British women for munitions making, a poster by Septimus Scott of a woman pulling on a factory duster announced, "These Women Are Doing Their Bit." Instead of being shown forlornly contemplating a flat tire, women were pictured wearing army coveralls and using wrenches to repair their own wheels. In text and image these women were shown as surprisingly transformed.

To a writer in the *Sphere*, one of England's earliest photographic magazines, the sight of these alterations was astonishing: a "modern day Rip van Winkle" would rub his eyes in amazement "at the war-time change which has come over our English Eve." In nearly four years of war, "she has discarded her petticoats, she has a vote," and as a "munition girl" she is "turning out shells and handling complicated machinery with as much calm and aplomb as if it were a sewing-machine." Highlighting the change, the magazine's cover in 1918 featured an official Ministry of Munitions photograph of a woman making cartridge cases at a Taylor and Challen drawing press, its gigantic gears looming over her and rising twenty feet into the air.[1]

American magazines also marveled at women's new roles. In 1917 a story in the *Ladies' Home Journal* by Harriet Sisson Gillespie reported on women driving electric factory trucks down to the docks at Bush Terminal in Brooklyn. The story contained themes that would reappear in reportage about women industrial workers during both world wars. Gillespie described the women as both attractive and competent, fashionable in their "natty seersucker suits" yet also adept as they drove their loads of cotton bales "in a most businesslike manner." Guided by male instructors, they had the machines firmly under control: "although a big burly engineer stood beside the feminine operator to lend timely assistance," the trucks moved "slowly forward toward the docks to the pressure of a girl's hand on the lever."[2]

In 1906 magazines like the *American Machinist* had featured photographs of female munitions makers rolling tubes for blank cartridges and inspecting paper bullets at a U.S. arsenal, but during wartime in 1916 the same magazine was publishing images of white-capped women in more technical jobs, working as machine-tool operators and bench operators in French munitions plants.[3] In 1918 the *Scientific American* described women learning welding and assembling huge airplane wings in an American naval aircraft factory where 1,000 of the 3,600 employees were female, and stories in popular magazines, such as Mary Brush Williams's "Industrial Amazons," in the *Saturday Evening Post* in 1917, chronicled British women working at new skills in airplane, shipbuilding, and automotive plants.[4]

Williams's aim in "Industrial Amazons" was to ease the fears of employers like a Pittsburgh manufacturer who had a "note of anxiety in his voice" when he thought of hiring women to replace men in factory jobs. To reassure American employers that female workers were responsible and capable, she visited fourteen different types of industries

This cover illustration from the *Sphere*, 4 May 1918, shows a British woman making cartridge cases at a huge drawing press rising twenty feet high. Used with permission of the Trustees of the Imperial War Museum, London.

in England, including an airplane-manufacturing plant in Warwickshire, where women worked at welding and sewing fabric on airplane wings, and her description of the airplane plant included many of the themes found in descriptions of women's war work.

At first she described the scene as "nothing novel," a comfortable domestic tableau in which the women worked in familiar feminine roles: "The spectacle was like a great sublimated quilting bee, to which the neighbors had been invited, to help and have cake and coffee." But on closer inspection her vision changed, and as would so often happen in narratives about women working in wartime, she focused on a woman's hands: "A woman was tacking with a hammer! She held a handful of small nails in her left palm . . . and without the assistance of her right fingers she managed to get them one by one between her left thumb and forefinger in position for driving into the framework of the machine." Williams added, "My grandmother and her neighbors could not have done that."[5]

Reflecting her own fears about electricity as a menacing power, Williams also visited an electric generating plant in London, where, to her surprise, a confident young woman was fearless and authoritative in handling "sinister-looking levers and handles." Tackling the recurring anxiety that women engaged in war work would lose their femininity, Williams described the women as dressed in trousers and in tunics but also "red-cheeked, feminine and twenty-two," wearing earrings, bracelets, and ropes of imitation pearls around their necks. "I think it safe to say," she concluded, "that with femininity unimpaired the women of England are marching on." To Williams, these capable yet still feminine women represented a new breed, "transfigured women, altered in point of view and purpose from their mothers" and from women of the past.[6]

Still, for all of her rhetoric of transformation, and with the ambivalence that often characterized stories about women and machine work, Williams reported the skeptical views of British employers who insisted that "women are by no means adepts at managing machinery." As a manager of a velvet factory argued, "Women plod along and gauge and measure and work, but they have not acquired sympathy for machinery. A man will hear the squeak of a machine and stop it, but a woman will let it run along with something wrong with it until a man, rasped at the feeling of permanent damage being done, runs to its rescue." One department head in the British Ministry of Munitions told Williams that women excelled at repetitious work but tempered his praise with rueful remarks: "'Women will keep on doing the same thing forever,' he observed. 'A man will invent a machine to do it. Man is lazy; woman is industrious.'"[7]

In her own report, *Women and War Work*, Helen Fraser introduced American readers to British female drivers helping in the war effort, and her book featured a striking visual distillation of women's new roles. One photograph shows a woman in a dress and hat driving a large steam-roller, her imposing figure silhouetted against the sky. She stands with her back to the viewer, her figure paralleling the machine's steampipe, a timeless image of woman at work.[8] Americans were also presented with more light-hearted views: spoofing the transformation in women's work, a drawing in America's comic weekly *Life* in 1919 featured a woman in overalls and wielding a rivet gun, dubbing her "The Girl Who Used to Drive a Nail With Her Hair Brush."[9]

Images such as these highlighted women's new roles, which were considered crucial to the war effort. In England, as growing numbers of men left for military service with the advent of universal conscription in May 1916, and with the country's acute shell shortage and greatly increased need for other war material, women increasingly took over men's jobs in what was called "substitution." In addition to recruiting munitions workers, by 1916 Britain's Women's Army Auxiliary Corps was advertising the need for female automobile drivers and mechanics, and by the beginning of April 1918 the Women's Royal Air Force (WRAF) began recruiting thousands of women to work as drivers, welders, fitters, rig-

"The Girl Who Used to Drive a Nail With Her Hair Brush." *Life,* 1919.

gers, riveters, typists, telephone operators, telegraphers, railroad conductors, and engine builders.[10] Members of the WRAF overhauled and serviced automobiles and also worked as airplane mechanics during the war, though none were allowed to fly military planes.[11]

The recruitment was a success. In 1916, the British journal *Automobile Engineer* reported that "almost every kind of machine is worked by women in some part or other of the country," including nearly every operation in the engineering trades, whether mechanical, electrical, or chemical. The journal was particularly impressed by the women's skills, noting that whereas before the war women had only rarely worked at manufacturing internal-combustion engines or operating machine tools to produce parts to precise measurements, now, through "dilution of labor" or "distribution of unit skill over multiple hands," they were manufacturing automobile and marine combustion engines and showing their prowess at precision grinding of airplane engine parts as well.[12]

Photographs of women working during the war were used not only to publicize this new phenomenon but also to help dispel doubts on the part of employers and the public about women's capacity to operate machines. The photographs became a form of visual proof that women could do the job. In his 1916 article for the *American Machinist* I. William Chubb included photographs of British women backing off milling cutters in the lathe, grinding valve-seating cutters, and running shapers. The images

Mechanics of the Women's Royal Air Force working on the fuselage of the trainer airplane AVRO 540 ca. 1919. Used with permission of the Trustees of the Imperial War Museum, London.

were clearly intended to certify women's abilities, for as Chubb wrote, "Any engineering employer in Great Britain who has doubts as to the possibility of female labor can often get them set at rest by consideration of official photographs." He noted that several hundred images in the collection of the Ministry of Munitions were available to firms for examination in the ministry's offices.[13]

But while the photographs in the *American Machinist* may have visually validated women's capabilities, the article itself presented a contradictory subtext, sometimes suggesting that the female machine-tool operators had only limited capabilities. The machine-tool manufacturing processes were described as simplified for the women: the shell-turning process was broken down into units, "each of a simple character," said Chubb, though he added that "given a simple machine tool, an ordinary woman does not require much in the way of training to become efficient."[14]

Probing the specific skills of women, Chubb repeated the commonly held assumptions that women were good at repetitive, detailed, precise work and were "content" doing one job all the time, making them "unlike the skilled workman" and young men, who objected to being kept to one process. For the repetitive work of turning drills he found "the dressmaking and milliner type of girl" to be "specially suitable," and the women were also good at precision work like grinding, checking the dimensions with a micrometer.[15]

But while Chubb found the women's work "entirely satisfactory," he added that they often "cannot get beyond a somewhat uncertain limit, though whether this is inherent in the sex or is simply due to an insufficient period of training is not determined." Still, as his article implied, women's most significant limitation was one that was imposed on them: the presumption that they would return their jobs to men at the war's end. Chubb summed up succinctly the central paradox of women working in war industries: "In an engineering sense they have no past to help or hamper them, and in many instances they can have no future."[16]

To illustrate women's new jobs, British and American magazines published documentary or relatively unmanipulated photographs of serious, unsmiling women working at machinery in crowded factory settings. Yet writers and illustrators also went beyond providing proof of women's abilities and transformed the often gritty, tiring, and even dangerous work into prettified, romanticized views, elevating the female machine workers into the realm of theater, spectacle, and art.

In his highly propagandistic book *Our Girls*, Hall Caine, with the sanction of the British government and using photographs supplied by the Ministry of War, presented a heroic view of British women in the munitions industries, which he numbered at half a million, spending long hours making fuses, filling shells, turning the shell surfaces, and riveting brass plugs into high-explosive shell bodies. In Caine's book vision itself is consistently contested, and he often presents a gauzy, filtered view of the women's work. At first Caine himself questions what he sees, for the very sight of women working at Woolwich Arsenal is astonishing: "There is at first something so incongruous in the spectacle of women operating masses of powerful machinery (or indeed, any machinery more formidable than a sewing-machine), that for a moment, as you stand at the entrance, the sight is scarcely believable."[17]

Caine praises the women for their appearance as much as for their endurance and skill, and clearly these are women whose value is heightened by men's gaze. The women are dressed in no-nonsense khaki-colored overalls with caps to protect their hair from machinery, but, says Caine, their outfits have the "not altogether negligible advantage, in the eyes of the male creature, of being extremely becoming." Caine wonders "if there is any man in London who can pass through the workshops of Woolwich without thinking he has been looking at some thousands of the best-looking young women in the world." With unabashed idealization, he describes the women as unaffected by the difficult working conditions: "Their hard work does not seem to be doing much harm to their health, for their eyes are bright, their cheeks are fresh, and there is hardly any

British women manufacturing munitions at Woolwich Arsenal in World War I. Photograph in Caine, *Our Girls.*

evidence of fatigue among them." The "clamourous and deafening noise of the machinery," he adds, is "like music in their ears."[18]

Observing the women, Caine takes an aesthetic view. At midnight, when the dinner whistle blows, the girls in their trim overalls and caps still look "fresh and bright," and when he thinks about the three thousand women working at the plant, with their "human hearts capable of being filled to over-flowing," he wonders, "What more does any artist want?" With unflagging rhetorical flourishes he casts the women into a variety of exotic roles, likening them to Arab women with their protective face veils, to nuns moving noiselessly, and to mythic "Brynhild, the warrior maid." When he first looks at an older woman boiling lead and pouring it into shot molds, he thinks of the witches in *Macbeth*, but on closer inspection he softens the view, seeing a face "such as Rembrandt loved to paint."[19]

But if Caine's transforming eye saw the women as bright-eyed and beautiful, working cheerfully without fatigue, the reality was a different story. The women in World War I were indeed transformed, but not in a painterly way: packing cordite, phosphorus, and TNT into shells, their faces became yellow from the TNT, earning them the nickname "canaries." In an oral interview conducted years later one British female munitions worker said, "We had bright yellow faces, because we had no masks." She said that sometimes when the women boarded packed trains they would be ushered into the first-class carriages, where the officers "used to look at us as if we were insects. Others used to mutter, 'Well, they're doing their bit.'"[20]

As women worked in dangerous, potentially explosive conditions, their appearance changed in other ways as well: because they were entering sterile explosives areas in the factories, before donning their uniforms they were asked to remove their corsets, shoes, hairpins, wedding rings, and any other metallic objects that might get embedded in their bodies in the event of an explosion.[21] Rather than working without fatigue, they told of the exhausting work.[22]

In her study of British female workers in World War I, Gail Braybon argued that propagandistic social commentaries and stories in the British press did women a large disservice by presenting them in a patronizing way and even exaggerating their abilities. These portrayals led many, including feminists and suffragists, to believe that men's conceptions of women had greatly changed and that radical social changes for women would take place after the war. This sanguine view was dramatically dispelled, however, by the postwar expulsion of women from men's jobs, demonstrating, Braybon argued, that "in reality the old prejudices and ideals still existed" among employers, trade unionists, and government members.[23]

But rather than representing women as dramatically changed, writers like Caine continued to see women in conventional and stereotyped terms, which served as a reminder that no profoundly dramatic change in views had taken place. To Caine, these were women who still had limitations, including limited physical strength and the lack of long, formal mechanical training. Although the women "have done almost miraculous work in the munitions factories," he insists, they still "have their limitations, and it would be madness to forget it."[24]

For the women in munitions factories, working with explosives and being subjected to unexpected air raids were terrifying experiences. Years later, in oral interviews, former munitions workers remembered being "frightened to death" at their first sight of factory gas masks and petrified that they would get killed by an explosion. One woman found even the sight of an overhead lathe frightening. But in Caine's account the emotional reactions of the female munitions workers are melodramatically heightened. Caine said that when Zeppelin air raids threatened, the women cried hysterically and fainted, though they soon started singing and dancing in the dark to dash their fears, a sight he found "thrilling."[25]

Writers like Caine and L. K. Yates exaggerated the female war workers' stereotyped feminine trait of emotionality, yet they saw the women as not only charmingly emotional but also admirably doughty. Yates, in his book *The Woman's Part*, praised women for their "grit and endurance," even citing one woman who went back to work after losing her thumb and finger in an explosion. Still, for Yates "the femininity of the worker, however, has its drawbacks," and he described the women as highly emotional creatures who at the initial stage in the munitions shops had "to overcome the instinctive fear of the machine" and required delicate handling.

Yates also added an erotic note to his vision of the women at work, seeing them as girlish and charming as they developed a love relationship with their machines. With proper coaching and comforting by male supervisors, they could be coaxed to lose their fears of a "monster machine" and "soon get attached to the machines they are working, in a manner probably unknown to men." Soon, too, an understanding arose "between the machine and the operator which amounts almost to affection," and said Yates, injecting a hint of sexuality into the workplace scene, "I have often noticed the expression of this emotion in the workshops; the caressing touch of a woman's fingers, for instance."[26]

Artists' images of women in the war also reflect some of these same conflicting signals, dignifying women's war work and praising their abilities while also casting them as charming if somewhat limited creatures, best portrayed in prettified terms. In 1917, after the British government decided to create a national war museum, soon to be named the Imperial

War Museum, a Women's Work Subcommittee was formed to gather documents and images of women's efforts in the war. In 1918 the subcommittee commissioned Britain's first official female war artist, Victoria Monkhouse, to create artworks of female war workers. There is a light charm rather than seriousness in her pencil and watercolor sketches done in 1919 of sweetly girlish women in untraditional jobs, including a bus conductor and mail van driver.

Magazines like the *Sphere* lent dignity to women's new occupations while softening the impact of change. In its stories and images chronicling women at work the women's special abilities are cast in comfortably familiar, gendered terms. In its May 1918 issue, a special issue focusing on women in the war, the magazine included a photograph of a woman installing electrical wires on board a government warship, noting, "It is a branch of mechanics at which women should do well, as it necessarily demands a certain deftness of touch which they can supply."[27]

Updating the centuries-old pictorial theme of men laboring in fiery forges and ironworks, the magazine in the same issue also included *Women Charging a Gas Retort*, by D. MacPherson, a drawing in subdued oranges, blues, and browns. The scene of women working in the darkened, cavernous space of a South London gasworks amid "cascades of fire" and tubes filled with gas-emitting burning coal is described in the language of Satanic gloom once used to describe early ironworks like those at Coalbrookdale, shrouded in smoke and fire: the women labor in a setting that "at first sight seems to represent a passage from Dante's 'Inferno.'" But while earlier paintings of forges and ironworks, such as Philip James de Loutherbourg's *Coalbrookdale by Night* (1801), had been cast in the highly charged aesthetic of the picturesque and the sublime, the awe-inspiring structures of industry evoking a sense of wonder, in the *Sphere*'s description of *Women Charging a Gas Retort* the wonder was directed not at industry but at the "daring and successful experiment" of women doing men's work, as the "heavy iron lids of the retort tubes, glowing red, are lifted by the women with dexterity."[28]

In other paintings during World War I artists transformed industrial scenes by adding an element of softness, beauty, and order, casting a sentimental glow over images of women at work. While Braybon saw the prettifying of women workers in written accounts as a sign of the authors' patronizing views, the prettifying of women workers in art was more likely a reflection of contemporary pictorial conventions and propaganda needs. Here again was a subtle tension between the heightened respect for women's new expanded identity and the gauzy, soft-focus romanticism that often surrounded representations of them at work.

From 1917 to 1918 the British artist Edward F. Skinner was commissioned by the London office of the Cammell Laird Company, a Sheffield armaments manufacturer, to create paintings of men and women processing steel and producing munitions in the firm's steelworks in Sheffield. The paintings were later reproduced as postcards, which were sold to raise money for the British Red Cross. In his thirteen commissioned paintings Skinner lent order, balance, and a graphic elegance to the factory scenes. His painting of women cutting rat-tail files at the firm's Cyclops Steel and Iron Works presented a sweeping row of women in blue and lilac skirts all bent over their work in the same way, their muscular arms extending from rolled-up sleeves (plate 35). With their identical hair styles — their hair drawn back in a single braid — and their elegantly classicized faces, the women echo the comfortable traditions of nineteenth-century British sentimental realism in painting.

In *For King and Country*, Skinner's painting of women doing shell work at the Grimesthorpe Steel and Ordnance Works, he again brought an ordered symmetry and graphic clarity to the scene, featuring receding rows of white-capped women with a central, smiling figure walking toward the viewer carrying a shell (plate 36). For all of its prettiness, however, the painting's color scheme becomes a reminder of the very real danger of the work, for some of the women are wearing yellow overalls, a sign that they are carrying highly explosive shells.

The good looks of Skinner's women, however, were more than an aesthetic conceit, for female workers themselves and the British press were sensitive to the issue of women staying pretty and attractive as they worked in the world of machines. A story in the London *Daily Mail* in 1916 described women in workshops of the Women's Service Bureau as wearing overalls, leather aprons, and goggles and yet "displaying nevertheless woman's genius for making herself attractive in whatsoever working guise."[29] But there were also serious-minded admonitions to women, including housewives. In 1917 a column in the *Sphere* undercut the idea of glamour, noting, with a rhetorical flourish, "No one is proud now of having lily-white hands and almond-shaped, highly-polished nails. Rather are we proud of the honourable scars and blemishes acquired in washing up at our pet canteens or doing our own housework. No one is too dainty for a rough job."[30]

The dual themes — and often conflicting perspectives — of glamorized personal appearance and no-nonsense work are often evident in the photographs by two men hired to document women's war work in Britain during World War I. In 1918 the Women's Work Subcommittee of the new museum reported that it was creating a series of photographs — "for record purposes" — of women working in "substitution" employment to free the men for military service.[31]

Though many of these specially commissioned photographs were released to the British press, most were not used to recruit women into the war effort but rather for other purposes, including documentation, pamphlet illustrations, or for sale in the shop run by the new propaganda department, the Ministry of Information in London. The photographs were also shown at an exhibit of women's war work at the Whitechapel Art Gallery in London in 1918, an immensely popular exhibit seen by 82,000 visitors, including the queen, and sponsored by the Ministry of Information and the Women's Work Subcommittee. The exhibit also included plaster models of women engaged in work and technical models of factory production produced by the women, including examples of munitions sent by twenty-six firms.[32]

Horace Nicholls, who was known for his photographs of the Boer War and Edwardian society, was hired in 1917 by the Department of Information (soon renamed the Ministry of Information) to photograph women working in munitions factories. He was given the title Official Photographer of Great Britain and was the only photographer employed full time on the home front. In 1918 he and the photographer G. P. Lewis, also working for the Ministry of Information documenting home-front employment, were engaged for the Women's Work Subcommittee documentary-photographs project. Nicholls photographed London factory women, while Lewis photographed factories outside the city; within eighteen months the two men had produced 4,600 photographic images.[33]

While both photographers documented the skilled and arduous work being done by women, Lewis often showed the gritty aspects of factory work. Accompanied by local inspectors, he often focused on health and safety issues and photographed women doing heavy work. His tightly cropped images of unsmiling women lent intensity to the work. He, like Nicholls, also had a flair for the theatrical. He used dramatic foreshortening in his 1918 photograph of a female worker operating a filter press in a glucose factory in Trafford Park, Manchester. With one foot raised high on the machine's wheel, she tugs mightily at a large lever to tighten the press. Lewis was not above sentimentality, transforming the toughness of women's work into softly lit scenes of tranquility in which women appear more notable for their loveliness than for their muscularity. In his picturesquely posed photograph of a woman hoisting barrels of oil onto rail trucks at an oil and cake works, the pretty young woman wears the winsome smile of a dairymaid as she lightly touches the barrel with her head slightly tilted and foot pulled gracefully back.

At their best, Nicholls's photographs blended promotional punch and intimacy, such as in his image of a lone woman lying on her side repairing an automobile. At the other extreme, his work also captured the

George P. Lewis, *Screwing Up the Filter Press*, 1918. A British woman operates a filter press at a glucose factory at Trafford Park, Manchester. Used with permission of the Trustees of the Imperial War Museum, London.

Woman making tire repairs in World War I. Horace Nicholls, *Chauffeur: Trouble on the Road*, ca. 1917–1918. Used with permission of the Trustees of the Imperial War Museum, London.

huge scope of war operations. For his image of women operating over-head gantries in a shell-filling factory in Chilwell, Nottinghamshire, he photographed the workers high in the air, far above the long, receding rows of shells below, the heightened perspective conveying the enormity of the industrial scene.[34] Though Nicholls, like Lewis, was charged with creating a documentary record, his photographs also became artfully staged tableaus, echoing the compositional and lighting conventions of academic painting. In his photograph *The Woman Blacksmith* a female blacksmith is the central figure, working in dramatic light and shadow and flanked by men on each side. Grasping a hammer, she works as a child looks in through the open window, perhaps a reminder of the woman's dual roles.[35]

Olive Edis Galsworthy, known as Olive Edis, was commissioned by the Women's Work Subcommittee to photograph women in 1919 engaged in the war effort in France and Flanders after the Armistice. Edis produced a series of photographs of members of Queen Mary's Army Auxiliary Corps (formerly the WAACs) who were attached to the American Army Expeditionary Force stationed in France, showing them working in a variety of roles, including as office workers and teachers.[36]

Images of Women in World War II

■ During World War II, with its acute labor shortages and huge increase in war materials production, there was again a large influx of women into American and European war-related industries, and artists and photographers captured signs of the transformation, however transient, of women in their new roles. Photographs taken for America's Office of War Information told stories of identity shifts in a world of war, such as one of a woman in overalls putting air in automobile tires, captioned "Former Beautician's New Occupation."[37] While nineteenth-century women shaped their bodies with the latticed frameworks of steel-cage crinolines, *Fortune* magazine's cover photograph in 1943 was a carefully crafted image of a woman whose identity is seen through a new framework: she is seen drilling holes in the gridded metal frame of an airplane fuselage (plate 37).

In its 9 August 1943 issue *Life* magazine reported that of the 65 million American women who were employed overall, one-quarter were working in war industries. To highlight changes that had taken place, the magazine published its photo essay "Women in Steel," focusing on women working in a Gary, Indiana, steel mill. Offering what it called "the first comprehensive picture of women toiling in heavy industry," the magazine told a tale of transformation: before the war, women had only worked as sorters in the mills, but now they were being employed as welders, crane operators, tool machinists, and engine and furnace operators. And although the "concept of the weaker sex sweating near blast furnaces" had long been accepted in England and Russia, *Life* argued, it had "always been foreign to American tradition"; only the need to replace the diminishing supply of manpower had "forced this revolutionary adjustment." Taking a role in legitimating the women, the magazine insisted that they were "not freaks or novelties" and that they had become accepted by management, unions, and "the rough, iron-muscled men they work with day after day."[38]

To illustrate the story, *Life* enlisted its stellar staff photographer, Margaret Bourke-White, whose photographs showed a sampling of the women in their new roles, including images of women flame-cutting steel slabs with four-torch machines and using pyrometers to measure the temperature of steel in an open-hearth furnace. The magazine's cover featured Bourke-White's dramatic closeup photograph of Ann Zarik, a "flame burner" using an acetylene torch to cut out pieces of armor plate (for ballistics tests) as molten metal sprays out.[39] Reflecting Bourke-White's experience as an industrial photographer, the steel-mill images, with their high-contrast lighting and darkened backgrounds, also captured the clean geometries of soaring factory silos, massive ladles, and furnaces without allowing them to dwarf the female workers, who staunchly hold their own.

To personalize the workers, *Life* also published cameo portraits of individual women, grounding them in family life by mentioning their husbands and children. As a sign of the national effort in wartime, the magazine also indicated the racial mix of the workers, citing one woman as Mexican American and including a central photograph of an African American blast-furnace worker, as well as a photograph of Mrs. Rosalie Ivy, described as a "husky Negro laborer" doing the work of "pan men," mixing mud in the mill.[40]

The 9 August 1943 issue of *Life* not only celebrated women's new roles in industry but took a revealing look at the multiple versions of feminine identity — and social agendas — that women were being asked to embrace. In the magazine's pages, the women in wartime were presented in a wide variety of guises — as workers in traditionally male roles, as glamour girls, and ultimately as contented housewives in a postwar world. In a story titled "Hollywood Girls Are All Things to Servicemen" the magazine presented pinup images of women as glamour girls and sex objects, women chosen by men in the military services to fulfill their fantasies. In captions rife with sexual innuendoes, Betty Grable was chosen as the "Girl We'd Like to Fly With in a Plane With an Automatic Pilot" by Ferry Command, while a submarine crew chose Paulette Goddard as "The Girl We'd Like to Submerge With." Maxine Barrat was chosen by bombardiers as "The Girl Whose Hair We'd Like Most to Have in Our Bombsight."[41]

The emblematic glamour girl baring her body and the woman welder camouflaged by goggles and overalls in the *Life* stories represented two central female archetypes promoted by the media, advertisers, and government recruiters during the war. The two images also suggested the complex social agendas women were being asked to fulfill. In a world at war, government agencies and advertisers intent on recruiting female workers had to convince women that they had the stamina and skill to operate equipment and to do jobs formerly held by men. But the women also had to be convinced that while working as riveters, welders, mechanics, and pilots they could still retain an aura of glamour and femininity. These dual agendas became dominant themes in representations of women at war.

But women were also being asked to keep in mind a third national agenda — that they were expected to return to their domestic duties as housewives and mothers after the war. To emphasize this point, the same issue of *Life* that offered images of women as steel workers and Betty Grable also included images of women cooking in its article entitled "Kitchens of Tomorrow" under the heading "Postwar Living." In contrast to women working amid towering steel furnaces and cranes, the article presented a visionary image of a well-dressed woman in the "ultra-modern kitchen" of the future, a sparkling clean kitchen of gleaming smooth surfaces, a kitchen that would turn gritty industrial scenes into a distant memory.[42]

Designed by the industrial designer H. Creston Doner, *Life*'s experimental kitchen, built by Libbey-Owens-Ford Glass Company in Toledo, Ohio, had a streamlined look, with a built-in toaster, hidden pots and pans, and lidded utensils, all designed to disappear so that when work was done the kitchen could double as a playroom. With its air of easeful living, the kitchen offered the recurring cultural fantasy that by using ultramodern, electrical appliances, women who returned to their domestic duties in the postwar world would have less work to do. Illustrating the effortless nature of the kitchen, the designer's wife was photographed paring potatoes and roasting meat while posing gracefully in her elegant attire, including high heels.

During the war, the myriad visual images of women working as riveters and welders, operating heavy machinery, driving automobiles, and building and repairing airplane engines not only publicized their increased presence but also helped validate their new roles. The images ranged from the quasi-documentary photographs taken for U.S. and British government agencies to more overtly propagandistic and promotional photographs presented by private industry, magazines, and advertisers. Then, too, there were Norman Rockwell's iconic illustrations of Rosie the Riveter that appeared on the covers of the *Saturday Evening Post* in 1943, as well as other serious and satirical Rosie images during the course of the war. Few, if any, of these images could be considered unmediated or documentary: in a wartime environment, with its own propagandistic imperatives, visual illustrations and written texts frequently served, whether overtly or subtly, to promote the war effort. These images of women engaged in war work often tellingly illuminated central and frequently conflicting cultural assumptions about women's more fundamental roles and identities.

A revealing series of photographs by the American photographer Russell Aikins of women working for the United States Steel Corporation during the war gave an undoubtedly unintended hint of women's problematic and contested roles. After working as a newspaper photographer for the *New York Times* and as a freelance magazine photographer, Aikins in 1937 shifted to photographing business and industrial workers at Chrysler, General Electric, and U.S. Steel, among other corporations. In 1941, before America's entry into the war, corporations like U.S. Steel mounted public-relations campaigns to promote their products and to show that they were ready for the war effort. In 1941 Aikins was hired by U.S. Steel to produce a large-scale photographic project documenting the company's myriad plants and production jobs. He eventually produced 1,150 photographs for the company, which were at first used for traveling exhibits that appeared throughout the country in libraries and department stores. Reflecting changes in company recruitment as men left for war, a number of Aikins's images showed women at work.[43]

In a particularly suggestive photograph a worker smiles at her own reflection in the mirrored surface of a tin-plate sheet that she is holding up for inspection in a tin-plate mill at the Tennessee Coal, Iron & Railroad Company in Birmingham, Alabama. In this carefully constructed image the woman smiles self-consciously at her mirrored self, perhaps aware of playing an iconic role as female war worker in this industrial setting.

But Aikins's other images were also reminders that female workers' jobs were largely temporary. In one photograph a Marine in full dress uniform, complete with medals, shows a female worker, Mrs. Gifford, how to make a trigger part using the milling machine he used to operate. Behind her protective glasses the seated Mrs. Gifford looks up respectfully at the standing, smiling soldier—a promotional reminder that she owes her job, and her training, to the exigencies of war.

During World War II the American and British governments also created their own records of women engaged in war work. In England,

Russell Aikins, Woman sorting tin plate at Fairfield Works, Tennessee Coal, Iron & Railroad Company, Birmingham, Alabama. Russell Aikins U.S. Steel collection. National Museum of American History, Smithsonian Institution.

Russell Aikins, photograph of a Marine aviation machinist instructing Mrs. Frances Gifford how to make a trigger part (a seer) for a carbine on the milling machine he used to operate at the Gerrard Company, Chicago, Illinois. Russell Aikins U.S. Steel collection. National Museum of American History, Smithsonian Institution.

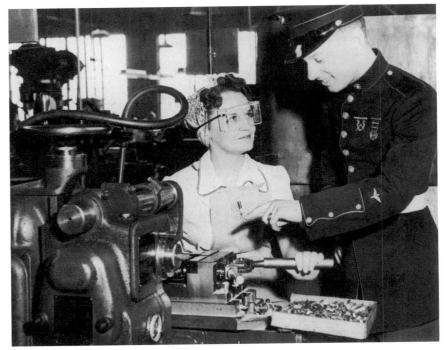

the Ministry of Information, as well as the British press, photographed women engaged in work for the home front and taking on jobs previously held by men, including the reportedly 50,000 women employed by the British railroads. Photographs in London's *Daily Herald* showed women working as conductors on London suburban lines, repairing railroad signals, and doing the heavy work of coupling the passenger cars.

British photographers, including Cecil Beaton, best known for his *Vogue* magazine fashion photography and celebrity portraits, contributed their own images. In the preface to his overtly propagandistic book *Air of Glory*, issued by the Ministry of Information in 1941, Beaton wrote that he wanted to create a "truthful picture," but his artifice was evident, as in his photographs of female workers taken with the exaggerated perspective of a bird's-eye view. "I wanted to show the various groups as vignettes," he added, and indeed the female defense workers in some photographs emerge as figures in an aesthetic tableau. In his photograph *The Web (WAAFs Working on a Balloon)*, the dramatic, weblike pattern of the mammoth balloon is central, dwarfing the three small figures of women grouped in statuelike stillness below.[44]

In America during the 1930s and through 1941, documentary images

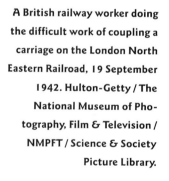

A British railway worker doing the difficult work of coupling a carriage on the London North Eastern Railroad, 19 September 1942. Hulton-Getty / The National Museum of Photography, Film & Television / NMPFT / Science & Society Picture Library.

created by photographers hired by the Farm Security Administration (FSA), headed by Roy E. Stryker, captured a sense of urgency and moral concern about the plight of America's urban and farm workers displaced by the Great Depression. But in 1942 the FSA Historical Section—Photographic was superseded by the historical section of the Office of War Information (OWI), which during the period 1942–43 produced photographs that were often more overtly propagandistic and included images of women recruited to work in wide-ranging occupations—in steel mills and aircraft factories, as bus drivers and gas station operators—during the course of the war.[45]

Used in magazines and newspapers, in museum exhibits, and for sale to people who wrote asking for them, the OWI photographs told a story of women and industries in transition. The photographs' captions—neatly typed by OWI staff members and often transcribed from photographers' handwritten notes on the back of the images—were also encoded with value-laden notions about gender and the women themselves. Working at a midwestern vacuum cleaner plant that had been converted to the production of gas masks, the woman in a 1942 OWI photograph operates a crimping machine fastening eye pieces to the

Ethel Peterson, a forklift operator at the Paraffine Company in Emeryville, California. Office of War Information photograph by Ann Rosener, 1943. Library of Congress, FSA-OWI Collection.

masks. The caption describes her as a former beauty-shop operator who "finds that the dear dead days of bobby pins and curler clamps have left her with skillful, speedy fingers and a natural tendency to eliminate wasteful hand movement."[46]

Among the OWI photographers were eight women, including Dorothea Lange whose images documented women working in California shipyards and whose 1936 iconic FSA photograph of a California migrant mother and her two children, with pained eyes and lined face, provided a timeless image of hardship in the midst of the Depression. Among the others were Marjorie Collins, Marion Post Wolcott, and Ann Rosener. In a story in *Parade* magazine in 1942 Rosener's photographs of women in training become emblematic of the important transformation that was taking place, a transformation evident in the magazine's reference to the synecdochic image that often appeared in representations of women and machines, a pair of women's hands: "Hands that used to rock the cradle today run the lathe and the punch press."[47]

Reflecting the values of war, women who had switched roles were praised for their new sense of purpose. A story about Rosener's work in *Minicam Photography* in 1943 included her photograph of a former actress

Marjorie Karwaske works at a crimping machine fastening eye pieces to the bodies of gas masks at a vacuum cleaner plant converted to the production of gas masks in America's Midwest. The original caption says that Karwaske, a former beauty shop operator, left behind "the dear dead days of bobby pins and curler clamps" to work at her new job. Office of War Information photograph by Ann Rosener, July 1942. Library of Congress, FSA-OWI Collection.

doing aircraft part casting, noting approvingly that the former child star of silent films, once known as Baby Dorothy Phelps, found her new role "more satisfying and useful."[48]

Going beyond documentary record, many of the visual images of female war workers reflected the widespread need to recruit women for war industries and to publicize new job opportunities. The American and British governments made great efforts to produce promotional material, and the American government often coordinated its efforts with magazine editors and commercial advertisers. By 1942, as American men left their jobs to fight in the armed forces and as industrial production was going into high gear, the new War Manpower Commission and the OWI actively recruited women for work. With labor shortages in industries such as aircraft and steel, shipping, and small arms ammunition assembly, the number of women working in war industries soared. According to one government report the number of women in aircraft-engine plants went from 600 in December 1941 to 4,000 in June 1942, and by the end of 1943 women made up 40 percent of the nation's aircraft workers.[49]

During this period the War Manpower Commission, joining with the OWI, pondered ways to convince women to join the war effort. Working with advertising companies, these government agencies launched concerted campaigns to urge women to leave their conventional roles as domestic caretakers for new roles, telling them that they were sorely needed in wartime work. To promote recruitment, the two offices supplied magazine editors and advertisers with photographs and stories of women working in factories. Several agencies were established specifically to create publicity and recruit women: in 1942 the Magazine Bureau was formed to supply publishers with reports and to promote articles that it wanted to publicize, and the OWI's Bureau of Campaigns produced a guide for advertisers detailing promotional methods. In 1944 a national "Women in the War" campaign was planned, a joint venture of the OWI and the War Advertising Council.[50]

But as Leila Rupp, Maureen Honey, and others have noted, these joint ventures of American government agencies working with advertisers produced inaccurate visual and written images that promoted the idea that the typical female war workers were housewives and single women working for the first time. The typical worker was presented as a young, white, middle-class woman who was temporarily working to help with the war effort and would happily leave at the end of the war to resume her role as wife and housemaker.

These recruitment campaigns were often aimed at housewives with little or no work experience, and they often tried to convince these women that their participation as workers was a patriotic necessity, freeing men to fight for their democratic freedoms. Newsreels, propaganda films, and journalistic stories promoted the idea of women making heroic sacrifices, leaving behind their glamorous jobs as actresses and beauticians for the more sober work in defense factories. By joining the war effort, women were told, they would help assure the men's military success and also help bring them home sooner and more safely.[51]

The reality was different. Although a significant number of women working in war industries were indeed middle-class married women who had previously been housewives, the workers were more diverse and included widows and students. Also, a large number of women had already been working and had changed jobs not only for patriotic reasons but to earn more pay. According to one set of statistics, in 1944 one-third of female defense workers had formerly been full-time housewives and one-fifth were recent students, but nearly half of the women in war industries had previous work experience (50% had five years' experience and 30% had ten years' experience).[52] And although some women worked for patriotic reasons, most worked because they needed the wages to supplement their family income or to support themselves.[53]

Contrary to the much-publicized image of the white female worker, women doing war work were ethnically diverse, with African American women making up a significant number of the work force. During World War I, African American women were recruited to make shells, airplane parts, and gas masks, but many of their gains were lost after the war. In 1939 a report written by the Women's Bureau, U.S. Department of Labor, stated that "the opportunity to tend a machine usually is denied the Negro woman," who in factories was usually employed "at such dirty, hot, or heavy [jobs] as cleaning and sweeping, sorting old rags, or pressing finished garments," though these women were finding work in the African American beauty shop industry.[54] During World War II a significant number of these women left domestic service and farmwork to work in war industries. Although the numbers of minority women in the work force increased to 18 percent, they were rarely represented in posters and advertisements, and with some exceptions, such as Palmer's image of an African American woman operating a hand drill (see plate 4), African American women were not often photographed working in the aircraft industries.[55]

Still, efforts were made to suggest that black and white women could work together.[56] In their promotional material American corporations attempted to show that American employees, with all of their cultural diversity, could band together in a common national enterprise. In *Steel in the War* (1946), a yearbook published by U.S. Steel, three grouped cameo photographs by Russell Aikins—of an African American, a Caucasian, and an Oriental woman—implicitly testified to the diverse ethnic range of women workers and that, in the words of the accompanying caption, an "All-American Team of American Women Help Speed Victory."[57]

African American, Chinese American, and Native American women were also represented in photographs created for the OWI, with typed photographic captions making note of their ethnic background. Still, photographic images of African American women in the work force belied a different reality: African American industrial workers, for example, were often placed in segregated housing, and protests erupted when there were efforts at integration.[58]

Publicizing Jobs and Women's Capabilities

■ Paging through magazines like *Life*, *McCall's*, and *Harper's Bazaar*, American women during the war could see stories and advertising that publicized the new wartime occupations and tried to convince women that they could master technical jobs. Advertisements featured photographs of models and women representing purportedly real-life case studies to help promote the idea that women could handle complex mechanical work. In 1943 the Barbara Gould cosmetic company, advertising in *Harper's Bazaar*, told women that exciting jobs in designing and engineering were open to them and that "if you have aptitude or training for this type of work, the Government urgently needs your service." Illustrating a woman at work riveting, another Gould advertisement offered more encouragement: "Don't hesitate to take a job in a war plant just because you've never been 'mechanically-minded.' It's surprising how quickly you learn."[59] In the OWI recruitment film *Wanted: Women War Workers*, the narrator tells the country's women that war jobs were no harder than household tasks. "Most war jobs are as simple as operating the appliances you use in your home," adding, "any woman who can operate a vacuum cleaner can join these women at this job."[60]

The issue of whether women indeed had the capacity to develop mechanical expertise was not a trivial one. In her *Shipyard Diary of a Woman Welder*, Augusta Clawson, a former government worker who went to work as a welder in Oregon, reported a sense of accomplishment, as well as some conflicted feelings. On the one hand, she wrote with pride that "now a door in the poop deck of an oil tanker is hanging, four feet by six of solid steel, by my welds. Pretty exciting!" But tempering her excitement, she added, "What exhausts the woman welder is not the work, nor the heat, nor the demands upon physical strength. It is the apprehension that arises from inadequate skill and consequent lack of confidence." Ultimately, she convinced herself that her doubts "can be overcome by the right kind of training," and she recorded her feeling of accomplishment: "I enjoyed the work today because I *could do it*."[61]

In America and England artists also created images of women that reinforced the notion that women had the capacity for difficult work with machines. In England the War Artists Advisory Committee, which had been formed in 1917 and was reestablished in 1939, commissioned Laura Knight to produce paintings of members of the Women's Auxiliary Air Force (WAAF) and other records of women in the war. Her painting *Ruby Loftus Screwing a Breech-ring* shows the twenty-year-old Loftus doing one of the most difficult jobs in an armament factory: cutting the threads of a breech ring for a Bofors gun, a job she learned in seven months rather than the years required of male apprentices (plate 38).[62]

Advertisers and government agencies also tried to convince women that they had the heft and muscle power to do strenuous work. Women were presented as formidable figures in posters and ads, such as J. Howard Miller's widely circulated poster "We Can Do It!" produced by the Westinghouse Corporation for America's War Production Coordinating Committee, in which a determined-looking female defense worker displays her powerful arm (plate 39). In an advertising layout popular during the war, Camel Cigarettes featured a page of small square individual photographs, ostensibly small documentary portraits of women working in factories during the war. In one small portrait Helen O'Brien, described as a steel tester, works while being observed by men. The text of the advertisement proclaims, "Roaring Furnaces . . . giant ladles of molten steel . . . it's a *tough* business. A man's business, but women are in it too."[63] An advertisement for the Pennsylvania Railroad presented an artist's conception of "Mrs. Casey Jones," a hefty railroad worker with a formidable figure like those found in OWI photographs taken by Marjorie Collins of real-life female workers employed to load railroad cars during the war.

In advertising, images of female pilots appeared often as emblems of women's mechanical know-how and expertise. During the war, female civilians working in Britain's Air Transport Auxiliary served as test pilots, flying instructors, and ferrying-service pilots and also worked as aviation ground staff, repairing and tuning up engines. American women were similarly trained for air-transport services. Though the actual percentages of women engaged in these flying services were small, the novelty of their jobs may have contributed to their frequent photographic appearance in ads for products ranging from Elgin watches to cigarettes. An advertisement for Chesterfield cigarettes in 1944 presented an artist's illustration of a generic woman pilot for the Women's Auxiliary Ferrying Squadron. Taking a broad-shouldered, he-man stance, the woman lights up her cigarette, and the text proclaims, "Theirs is the man-sized job of ferrying war planes from factories to air-bases for Uncle Sam. Expert flyers, each and every one . . . THEY ARE THE BEST."

Rosie the Image. In the war years of the 1940s, government agencies and advertisers alike worked at convincing women that even as competent industrial workers they could still fashion themselves into glamour girls. Probably the most popular American wartime image was that of Rosie the Riveter, as seen in Norman Rockwell's painting *Rosie the Riveter,* which appeared as the cover illustration for the 29 May 1943 issue of the *Saturday Evening Post.*[64] Rockwell's is a muscular Rosie, a more genial version of the mammoth, mythic female sculptural figures presiding over nineteenth-century industrial fairs.

But for all of her muscularity, Rockwell's Rosie also shows signs of femininity: in her pocket is a powder puff, and she wears, however incongruously, nail polish and rouge. A more feminine image of Rosie appeared in a promotional painting by Benton Clark commissioned by Coca-Cola in 1943. Here the attractive shipyard worker is an object of sexual interest as she stands, soft drink in hand, surrounded by an admiring group of muscular men during lunch (plate 40). In contrast to the munitions workers in E. F. Skinner's World War I painting, who wore yellow as

Norman Rockwell, *Rosie the Riveter,* oil on canvas, 1943. Rockwell's painting of this saucy, muscular Rosie with a compact in her pocket and her foot on a copy of Hitler's *Mein Kampf* appeared as the cover illustration for the *Saturday Evening Post* issue of 29 May 1943. Photograph courtesy of The Norman Rockwell Museum at Stockbridge. Used with permission of The Norman Rockwell Family Trust. © 1943 The Norman Rockwell Family Trust.

a sign of explosive danger, the Coca-Cola Rosie, dressed in a bright yellow sweater, is safely cute and coy.

Rosie the Real. The complexities of the Rosie images—female factory workers who were both glamorous and tough, women who could successfully fulfill cultural definitions of both men's and women's roles—reflected both the rhetoric and reality of women in defense work. Women were employed in jobs using heavy machinery as well as in jobs calling for qualities more traditionally linked to women's special abilities: dexterity and precision.

American government pamphlets reflected the conviction that women could excel at conventionally masculine as well as feminine skills. A pamphlet published by the Women's Bureau of the U.S. Department of Labor in 1942 noted that some American women in war factories were operating heavy machine tools, including lathes, grinders, and milling machines, while thousands of others were "skillfully doing work requiring a delicate touch, manipulative dexterity of a high degree, as well as extreme accuracy in measurement."[65]

Beautiful Hands. The emphasis on women's delicate touch and manual dexterity again revealed the emphasis placed on women's hands. In 1942 the National Association of Manufacturers arranged a cross-country trip for fifteen female journalists working for American newspaper and press services so that they could report on women working in midwestern and West Coast war plants, including those involved in aircraft manufacture, shipbuilding, machine crafts, munitions manufacture, and tank and jeep assembly. The reports of seven of these journalists appeared in the association's *American Women at War,* published in 1942.

In one of the seven published reports, "Victory in the Hands of Women," Janet Owen, of the *New York Herald Tribune,* focused on a central image of women working with machines: "The new symbol for victory in the war effort is a pair of woman's hands." Owen was one of several of the journalists who portrayed women as nimble-fingered, patient, and glamorous.[66]

Owen's report, billed as "unedited" by the National Association of Manufacturers, still managed to idealize the women and see them through the eyes of management. Wrote Owen of a female worker, "As the mind's eye of industry sees them, her hands are small but flexible, sensitive but strong, delicate, perhaps, but very agile. They are neat, and not gaudy with lacquer talons. They are eager. They are efficient." Taking a polemical stance, she added, "Some leaders of war industry will tell you they are more efficient than the bigger, brawny hands that preceded them," and she quoted an aircraft-plant executive as saying that "woman's dexterity in sub-assembly work is superior to man's." Still focused on fingers, Owen summed up, "The byword is 'women's nimble fingers.' . . . Their 'nimble fingers' make them far more skillful at all forms of precision work."[67]

In her own report, "Factory Housekeeping," Mary Hornaday, of the Washington bureau of the *Christian Science Monitor*, wrote of the concessions made by manufacturers to women's requests, ranging from creating day-care nurseries to allowing female workers into the lunch cafeteria earlier than men so that the women would not get hurt in the lunchtime rush. Still, as Hornaday reported, the manufacturers drew the line at what they considered frivolity. Again women's hands emerge as the prime emblem of femininity. Reacting to rumors that beauty parlors were being installed in factories, one manufacturer stated, "We're glamorizing cartridges, not women." Hornaday reported further, " 'We use our red nail polish here,' he said, picking up a cartridge with a red-painted tip identifying it as a tracer bullet."[68]

Cosmetics, Femininity, and Patriotism

■ While *American Women at War* saw women's dexterous, unpolished hands as a sign of their virtuous commitment to war work, the idea of no-frills fingernails posed an obvious challenge to cosmetics manufacturers, who were confronted with the task of reconciling patriotism and the national interest with their own need to promote their products. At a time when wearing too much makeup might be considered frivolous, cosmetics companies developed marketing strategies that emphasized moderation, glamour, and patriotism. The American cosmetics manufacturer Elizabeth Arden in 1943 developed a "war face" to help women look "radiantly healthy" and natural, a face with light rouge, no exaggeration to the shape of the lips, and a deeper than usual powder base to give their faces a "healthy glow."[69]

Conflating goals, governmental agencies, which needed to recruit women into war industries, worked with cosmetics manufacturers and women's magazines, developing a symbiotic relationship as they tried to convince women that they could maintain glamour and femininity even in a factory setting. Continuing a fifty-year-old tradition in advertising, manufacturers equated makeup with female identity, an identity that came under siege as women took on men's roles.

Promoting femininity while giving government jobs a human face, an advertisement for Cutex nail polish featured Ruth Gray, a female trainer of Pan American World Airways transatlantic pilots, who said of her nail-polish color, "Wearing Cutex YOUNG RED is like going into a glamour spin. It keeps me looking feminine even though I have a man-size job." In its advertisement "Soprano over the Atlantic," DuBarry cosmetics included an illustration of Eunice Damant, microphone in hand, working as a "sky cop," or traffic controller, at the Navy's Floyd Bennett Field. The soprano, the ad proclaimed, "is doing a man's job for the duration," but with DuBarry beauty preparations, "she's keeping pretty and womanly as ever."[70]

Reassuring women that they could still achieve feminine goals, such as getting married and keeping alluringly clean even in the midst of factory work, Pond's Cold Cream ran a series of advertisements in a campaign called "She's Engaged!" in 1943. In one version, Susan Tucker Huntington, who is pictured drilling holes in metal castings as she trains for aircraft production, uses the company's cold cream to "slick off every tiny little speck of machine dirt and grease," leaving her face "soft as a glamour girl's." While citing Susan's mechanical expertise — her fiancé "would be surprised if he could see how mechanically exact" she was getting to be — Pond's also reaffirmed her more conventional feminine goal of marriage, including an inset photograph of her emerald-and-diamond engagement ring.[71]

The conflation of mechanical skill, aviation, and cosmetics had an analogue in the career of Jackie Cochran, one of America's star aviators, who helped found the Women's Auxiliary Ferrying Squadron and headed the WASPs in World War II. Born in Georgia, Cochran later became a partner in a Pensacola, Florida, beauty salon and worked at an exclusive salon in New York's Saks Fifth Avenue. By the early 1930s, thinking of becoming a cosmetics-company representative, she learned to fly and earned a commercial pilot's license in 1933. By 1935 she had established her own cosmetics company, Jacqueline Cochran Cosmetics, but was focusing on flying, taking part in air competitions.[72]

Wartime cosmetics advertisements also promoted the idea that makeup would improve women's morale. Early in the war, wearing makeup was strictly prohibited in women's service units in England, but British women rebelled by insisting on wearing lipstick, and the ban on makeup was lifted several months later. Supporting British women, a *New York Times* writer reporting from London in 1942 argued that "if all the cosmetics suddenly disappeared from Britain, the morale of both sexes probably would slump severely." But writers were also quick to defend the patriotism of British and American women. Faced with the possibility of new restrictions, these women would readily comply. "No one needs to teach British and American women patriotism," wrote one American commentator, who added that there would also be no question of "lipstick versus bombers and bullets."[73]

After America's War Production Board rescinded its restrictions on cosmetics, American women were encouraged to curtail their cosmetics use. The country's cosmetics industry, however, mounted campaigns to keep working women wearing nail polish and face powder by linking cosmetics and patriotism. In its advertisement "A Woman's Fighting Face!" the cosmetic-color manufacturer H. Kohnstamm promoted the wearing of makeup during wartime. The advertisement showed a woman in blue work overalls and a blue hairnet standing against a backdrop of painted white stars, and the text insisted that wearing makeup was essential to boost American women's morale and this boosting was essential for successful factory production: "It is essential for every woman to look her best," the text argued, citing the case of Britain, which had found that morale and production dropped when it attempted to ban cosmetics at the beginning of the war. "When the ban was lifted," Kohnstamm claimed, "up went the production curve." Using Germany as another example, the text also evoked the name of Hitler, who was said to have made a futile effort to order the abolition of makeup, not recognizing that, as the ad's inflated rhetoric insisted, cosmetics could help promote social order: "In the height of the bombings, powder and lipstick banished panic and confusion."[74]

In her autobiography the photographer Margaret Bourke-White told of an actual incident in which cosmetics seemed to help maintain troop morale during World War II. On assignment as a war correspondent for *Life* magazine, she was given permission to travel on the flagship of a convoy en route to the North African coast, a ship carrying 6,000 American troops as well as 400 nurses, WACs, and Kay Summersby, working for SHAEF (Supreme Headquarters, Allied Expeditionary Force) as General Eisenhower's personal driver. When the ship was hit by a torpedo, Bourke-White managed to get into a lifeboat, but two WACs stayed behind on the ship. She was later given a "shining account" of the two women, who not only "laughed, sang, told jokes," and made sandwiches for the men but also boosted the men's morale in an unexpected way: "They brought out their vanity cases on deck, flapped powder puffs, applied lipstick, until the troops were saying, 'If a girl can use a lipstick in a spot like this, you feel there isn't much for a boatload of fellows to be afraid of.'"[75]

The image of stalwart women also appeared in advertisements like "War, Women and Lipstick," in which the American lipstick manufacturer Tangee, in the *Ladies' Home Journal* and *Photoplay* in 1943, promoted wearing cosmetics as a way for women to maintain their own morale while also doing their patriotic duty. Raising the banner of national chauvinism, Tangee mixed politics, gender, and morality and made bold claims for cosmetics as a means for women to balance conflicting roles. In its photograph of a smiling, waving female flier climbing into the pilot's seat, the ad touted American women as "the loveliest and most spirited in the world," capable of doing "double duty" by piloting and still carrying on their "traditional 'woman's' work of cooking, and cleaning, and home-making."

Tangee even linked democratic values and women's delicate balancing act: "It's a reflection of the free democratic way of life that you have succeeded in keeping your femininity — even though you are doing man's work!" In the Tangee advertisement lipstick becomes the very embodiment of American women's patriotism, a symbol of "this fine, independent spirit" and women's "courage and strength." With stirring marketing logic, Tangee even portrayed lipstick as the embodiment not only of women's political rights but of their gender rights as well, saying that it "symbolizes one of the reasons why we are fighting . . . the precious right of women to be feminine and lovely — under any circumstances."

One of the more intriguing aspects of the Tangee advertisement was its equation of masking with morale. Cosmetics are seen serving a crucial role, helping women become stronger and more stoic by camouflaging their feelings: "A woman's lipstick is an instrument of personal morale that helps her conceal heartbreak or sorrow; gives her self-confidence when it's badly needed." By referring to concealment, Tangee entered the tricky world of artifice and authenticity found in wartime cosmetics ads. Through promoting products long associated with artifice, surface, and masking, advertisers also worked at investing their products with an aura of authenticity and verisimilitude. The technique they used most often was to include documentary material: mini-biographies and photo insets of actual women working in factories and flying, whose presence and names, coupled with their wartime credentials, not only provided product testimonials but also encoded the advertisements with moral authority.

At times, however, the distinction between the genuine and the generic, the (real) individual and the invented self, was blurred. Advertisers freely mixed mini-biographies of real-life female workers with generic versions of women at war.[76] The distinctions between artifice and realism were also occasionally blurred in photojournalistic stories purporting to be about women actually employed in the war. *Harper's Bazaar* cast its photo essay "Hands on the Job" as documentary realism and focused on the elegantly polished hands of women operating machines. The women are wearing a new shade of nail polish, Cutex's "On Duty." Rather than associating the new nail polish with artifice and glamour, the magazine took a more subdued tone: "On Duty" was defined as appropriately modest, "an unobtrusive shade for work." In its own advertisement for "On

Duty" in *Harper's Bazaar* Cutex featured photographs of women fliers wearing the new shade, including an active member of the female flying group the '99ers, who said, "With extravagance out for the duration, no wonder it's so popular."[77]

Emphasizing naturalism rather than artifice, in "Hands on the Job" *Harper's* made a point of insisting that it had used actual working women rather than models and that the images were spontaneous and unposed: the photographs of women working at machines such as a drill press had been chosen "at random from batches of snapshots," and the women were "typical, not posed for Bazaar."[78] But in spite of these disclaimers, the photographs in the *Harper's* story retain the elegance and sheen of high-fashion photography. The story's most dramatic image is carefully constructed: in an enlarged color closeup a pair of women's hands, slightly smudged with dirt, are shown operating a drill press, the fingers elegantly draped to display long, shiny, red nails.

The women in the *Harper's* story are identified by name as workers for Wright Aeronautical Corporation and Consolidated Aircraft, but rather than being randomly picked, they were apparently chosen for their fashion-plate good looks. They are shown delicately and elegantly hold-

In a photo essay titled "Hands on the Job," in the February 1943 issue of *Harper's Bazaar*, Mrs. Geneva Hansen, looking like a glamorous movie star, with dark glasses and manicured hands, works at Consolidated Aircraft welding parts for a B-24 Liberator. Hansen, the magazine said, was the first woman to pass the Army and Navy tests for steel welding.

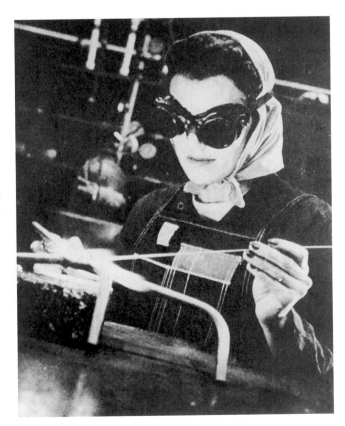

ing their tools in tightly cropped, posed images with darkened backgrounds that add to the theatrical air. Mrs. Geneva Hansen, wearing dark welding glasses, is shown working on parts for B-24 Liberators at Consolidated Aircraft, her long, red fingernails gently touching the welding rod. Her identity masked by the dark glasses, Hansen looks glamorous, barely distinguishable from a Hollywood star.

Investing its story with high moral purpose, *Harper's* praised the women for their war efforts while deftly transforming them into images of fashionable femininity. Geneva Hansen is described as the first woman to pass the Army and Navy tests for steel welders, and both she and the other women in the story are praised for their "feminine charm" and the "grace with which they go about their highly skilled tasks." Clearly, these are women in a temporary role, women whose elegant lives are not far away. Their fingernails are "as beautifully manicured as if their owners had spent the morning at home and had nothing more on their minds than an afternoon of bridge or a cocktail party."[79]

The theme of beautifully manicured women capable of great mechanical skill also appeared in articles in aviation magazines that emphasized the femininity of women in military air-transport services. In its 1943 story "Pilots in Nailpolish," New York's *Pegasus* magazine, published by the Fairchild Engine and Airplane Corporation, highlighted the activities of women in Britain's Air Transport Auxiliary, who were engaged in flying planes, including bombers, from factories to airfields.[80] Evoking the theme of masked identity, the magazine's opening lines are disingenuous and coy, feigning surprise when the person flying the plane turns out to be a woman: the Spitfire flier has fingertips "tinted a pleasing red," a mouth "discreetly rouged," and her hair under the flight helmet is "curled in the waves of a 'permanent.'" But after establishing the woman's feminine credentials, the story notes that the women who have received training as test pilots and instructors are not shy flowers but both competent and heroic. They have surmounted the "considerable opposition to their doing such arduous, dangerous, and responsible work," and their "magnificent endurance and skill proved their ability."[81]

As portrayed in *Pegasus*, these women represent the best of both masculine and feminine worlds: they are "fit without looking tough," "light-hearted but never light-headed, emancipated yet entirely feminine." Still, as was often the case in journalistic accounts of women warriors, *Pegasus* felt compelled to present reassuring images of the women's fundamental femininity and reminders that the women had not forgotten their maternal role: one woman saves her pay for her baby's education, another goes home every night to be with her baby.

But while the women are described as preparing for flights in a professional manner—studying meteorological charts and flight routes, packing and testing their parachutes—they are also portrayed as engaging in a stereotypically feminine pursuit, rushing into the canteen for "a last stirrup-cup of tea (women's traditional panacea)" before they go out to their more tough-minded, masculine roles, "stomping out across the field in their great heavy sheepskin boots, lugging their forty pounds of equipment with them."[82]

War and Fashion   ■ While feature articles, commercial advertising, and government-created imagery promoted the idea that women could maintain their femininity even as they made the transition to employment as war workers, government campaigns also urged them to exchange their impractical and dangerous feminine clothes for the more practical wear needed for working with machines. In March 1943 Ann Rosener created an OWI photographic series called "Safe Clothes for Women War Workers," with captions telling stories of women being transformed from frivolously feminine job seekers to factory machine operators sensibly dressed.

Two photographs present images of Eunice Kimball. In one, she is being interviewed by a male employment manager at Bendix Aviation. Kimball, the caption warns, is dangerously dressed: she wears a fuzzy sweater, which could catch on machines; dangling jewelry, which could be a "constant menace" among moving machine parts; and open-toe shoes, which offer no protection from heavy falling tools. The second photograph reveals her wise transformation. Eunice is now "safely and becomingly dressed" in denim trousers, factory cap, and square-toed safety shoes as she takes on her new role operating a milling machine.

Images like these suggested the mixed signals women often received. Contributing to the war effort, the women's fashion industry in America and women's magazines also promoted images of women as both serious-minded and glamorous. Trying to make the factory environment a friendly one for female workers, clothing manufacturers sponsored fashion shows during work hours. In a story in its 1944 series "Her Job and How She Got It," *Harper's Bazaar* featured twenty-four-year-old Helen Weimer, who helped produce weekly fashion shows for the nine thousand women employed at the Air Service Command Headquarters, in Dayton, Ohio. Photographed on the airfield, Weimer sported a no-nonsense but pert blouse and a skirt of "gray men's wear flannel." Still, her businesslike guise was just a temporary one, for as the magazine added, Weimer wrote to her husband daily and "plans fervently for a home and family."[83]

While Weimer was imaged as serious-minded, governmental and

Left, Eunice Kimball, as the original caption reads, is "inappropriately dressed for a war plant job" when being interviewed for a job at Bendix Aviation, 1943. Office of War Information photograph by Ann Rosener. Library of Congress, FSA-OWI Collection.

Right, Eunice Kimball, now "safely and becomingly dressed" as she operates a horizontal milling machine at the Bendix Aviation plant, 1943. Office of War Information photograph by Ann Rosener. Library of Congress, FSA-OWI Collection.

magazine images also glamorized women in their war-related roles, even extending the glamour to female farmworkers. In her documentary images for the OWI, Ann Rosener photographed Mrs. William Wood, a farm wife who managed a 120-acre Michigan farm while her husband was at war. Wood appears hard-working yet also prettified in her shorts and flowered blouse as she drives a tractor in the fields. Farm women were made both mythic and glamorous in the *Harper's Bazaar* story "Give Your Vacation to the Land," describing the work of the Women's Land Army, made up of civilians recruited by the government's Agricultural Extension Service.[84] As photographed by Louise Dahl-Wolfe, one of these farmworkers, wearing a jaunty kerchief on her head and a bare-backed halter, sits half perched on the seat of her tractor. Photographed from a low angle, her strong back to the viewer, the woman is made heroic, her figure looming upward near a towering silo.

The intermingling of cosmetics, fashion, and femininity in the midst of war not only eased women's entry into jobs associated with a man's world but also helped reduce the stress of the transformation. In a peculiar inversion two central cultural signs of surface and superficiality — cos-

metics and fashion — were reconceived not as artifice but as substance, signs of virtue in a precarious world. But with cultural ambivalence, and reflecting the contingent nature of wartime, women's new defense jobs, their mechanical competence, and even their man-tailored work clothes were seen as an ephemeral mask, a temporary guise, a version of self that would be relinquished at war's end. Beneath the persona of the woman with tough-minded competence, beyond the wearing of no-nonsense overalls, or so the subtext implied, was a woman who would never forget her truer, domestic, quintessentially feminine self.

The reduction of technological seriousness to mere surface is evident in fashion photography of the era, which seized on mechanical imagery as fanciful props, continuing in the modernist tradition of technology chic. One of the more witty conflations of fashion and technology was seen in *Harper's Bazaar*'s photographic essay "The Blown-Glass Look" in 1945, with its photographs by the New York photographer Genevieve Naylor.[85] Naylor had worked as a photojournalist for *Fortune* and *Time* magazines and was one of the first women hired by the Associated Press.[86] Later becoming a fashion photographer, she created images for magazines. For "The Blown-Glass Look" she photographed models dressed in

Genevieve Naylor's photograph of a fashion model wearing an Adele Simpson suit, photographed through the transparent Du Pont Lucite of a B-29 turret manufactured at an aircraft plant on Long Island, New York. *Harper's Bazaar,* **September 1945. Courtesy of the Staley-Wise Gallery.**

high-fashion clothes. In one photograph, a woman wearing an Adele Simpson woolen suit "the color of steel wool" is photographed through a transparent bubble of the Du Pont Lucite of a B-29 gunner's turret, and other models are similarly photographed surrounded by the bubble-shaped transparent plastic surfaces.

(Twenty years later, during the Vietnam War years, the American pop artist James Rosenquist linked airplane nose cones and femininity in his painting *F-111*, in which a smiling young girl sits under a gleaming, metallic, cone-shaped hair dryer—a sardonic view of America's focus on cosmetic glamour, the war, and sophisticated military machines like the F-111, its newest fighter-bomber.)

Naylor's fashion photographs, which were not published until a few months after the war ended, reduced the plastic turret to an artful aesthetic shell, stripped of its military associations, the kind of associations seen in a recruitment poster issued by OWI in 1941 that proclaimed with patriotic fervor, "The more WOMEN at work the sooner we WIN!" (plate 41). The poster featured Alfred Palmer's photograph of a California woman dressed in red overalls standing inside the Plexiglas nose section of a B-17F Navy bomber to put the finishing touches on her work.

Naylor's *Harper's* photographs also made reference to technological modernity, yet decontextualized the seriousness of the role of science and technology in a world at war. Plastics such as Lucite, used during the

Women working on Plexiglas nose cones of A-20 Havloc light bombers in a Douglas Aircraft factory. Sparkling overhead lights are reflected in the plastic surfaces. Library of Congress, Washington, D.C.

war for the manufacture of aircraft nose cones and turrets, were associated with modernity, scientific progress, and America's war readiness. Yet in Naylor's fashion photographs the plastic airplane parts become signs of the modern without the associations of war. The image of the woman in her steel-colored suit is disjunctive, even surreal: with her gloved hands touching the Lucite panels, the woman stands in profile in the darkened transparent space, disconnected from an actual aircraft, from danger, from war itself.

Naylor's photographs contrasted sharply with documentary photographs of women manufacturing Plexiglas nose cones in factories, which served as dramatic signs of America's heated-up airplane production following the Japanese attack on Pearl Harbor on 7 December 1941. Yet even these factory images could be made glamorous. In a photograph of women producing A-20 nose cones in a Douglas Aircraft factory in California the radiating rows of nose cones, reflecting dots of light from the factory ceiling overhead, create an aura of celestial, starry glamour.

Yet none of these photographic images, whether fashion or industrial, suggested the actual danger associated with the acrylic nose cones and turrets. Gunners flying in long-range bombers such as B-17s and B-29s sat in ball turrets, round Plexiglas bubbles equipped with machine guns and often embedded in the belly of the plane. In actuality, the gunner's turret was a terrifying place to be. In the war poems of the American writer Randall Jarrell the turrets became the site of both isolation and vulnerability. In his poem "Siegfried," from his volume *Little Friend, Little Friend* (1945), Jarrell, who was trained as a pilot and became a control-tower operator at a B-29 base in Arizona during World War II, evoked the painful sense of disconnection experienced by a gunner flying over Japan in a long-range bomber. The poem's flier is encased in a ball turret, and he witnesses death through the bubble. Seated inside the machine, he feels invulnerable, protected by Plexiglas and steel, yet his sheltered space is fractured as the poem ends with a burst of fire spewing crystals of blood.[87]

Women High and Low ■ The glamorization of women in fashion and cosmetic images, a glamorization that often appropriated war imagery in a complex amalgam of patriotic surface and substance, also served to both heighten and diminish women's roles. While women were often portrayed as being lifted beyond the quotidian reality of factory life to movie-star glamour, the addition of elegance also reduced them to icon and image, mere conceptual beings.

More literal versions of heightening and lowering appeared in wartime factory photographs, such as those in Egbert P. Booth's *Women at War: Engineering*, a privately published booklet approved by Britain's Ministry of Supply that informed women about jobs in England's munitions factories, including the Royal Ordnance Factory. The book's text and photographs were clearly intended to demonstrate women's capacity to handle very large machines, for as Booth noted, "as some of the illustrations will show," the women were "in complete charge of these engineering leviathans." But Booth's text and images were also at war with each other. Rather than enhancing respect for the women's capabilities, the photographs, paradoxically, tended to diminish their role.[88]

Many of the images were designed to show the huge disparity between women and machine. The caption to a photograph showing a woman working in a turning bay (women turned gun barrels up to a 4-inch caliber) directs the reader to "notice the woman operator, the size of the machine and the gun barrel." The machine's hugeness is emphasized by photographs with exaggerated perspectives and angles of vision, distortions that tended to further diminish the size of the women. Using sharply receding diagonal lines, the photographer represented a woman rifling a huge gun barrel as a tiny figure at the far end of the machine (the editors noted that because the photographs illustrate the manufacture of "secret weapons," the factory backgrounds had been blanked out, which further emphasized the size of the machines). And while Booth praised the women's skills, he also represented these skills as an anomaly: the women not only "astonish business men and industrialists, as to their capabilities" but also "surprised themselves."[89]

Turning Bay, in Egbert P. Booth's *Women at War: Engineering,* 1943. The caption in the book reads, "Notice the woman operator, the size of the machine and the gun barrel."

Women and Expertise   ■ Through the rhetoric of surprise, books such as Booth's exalted women while also revealing implicit assumptions about their limitations. While government agencies, as well as magazines and advertisements, promoted the idea that women were capable workers, able to master technical skills, American manufacturing plants made alterations in production methods to accommodate women's needs. At aircraft-manufacturing plants counselors were sometimes employed to design jobs based on efficiency experts' assumptions about women's innate capabilities and proficiencies — their greater dexterity, their patience, their capacity for repetitive and detailed work. These assumptions helped determine, and often limit, the types of jobs and work assignments women were given.[90]

While women were said to have longer index fingers, presumably giving them greater dexterity, other physical characteristics were perceived as limitations. Women were assumed to be shorter, to weigh less, and to be weaker than men and even to tend to slow down when decisions were required. In war plants, jobs were adapted to accommodate these physical limitations and supposedly lesser skill aptitudes.[91] Complex jobs requiring a high level of skill were broken down into smaller tasks so they could be quickly learned, and at warplane-production plants production engineers set up operations for women "in detail so that the need for making decisions does not arise."[92]

In her study of job segregation during World War II, *Gender at Work*, Ruth Milkman documented how American industrial managers concentrated female workers in jobs deemed suitable for them, and excluded them from others, based on their rationalizations about women's inherent physical characteristics. Milkman cited an article in *Automotive War Production* that stated that women had been found to be quicker and more efficient than men at jobs requiring "high manipulative skill." The article argued that "engineering womanpower means realizing fully that women are not only different from men in such things as lifting power and arm reach — but in many other ways that pertain to their physiological and their social functions." Adding a disclaimer, it noted, "To understand these things does not mean to exclude women *from the jobs for which they are peculiarly adapted*, and where they can help to win the war. It merely means using them as women, and not as men." Citing specific differences, the article claimed that the average woman had 35 percent muscle mass, compared with the average man's 41 percent, and that the average woman's hand squeeze exerted 48 pounds of pressure, in contrast to the man's 81 pounds.[93]

Milkman contended that the alterations in production methods, ostensibly made to accommodate the special needs of women, actually helped both men and women and became a way to rationalize increased technological efficiency. Vultee Aircraft, for example, introduced simplified work methods and increased the use of mechanical aids such as conveyers, chain hoists, and lift loaders for female workers. Yet, as Milkman suggests, though management attributed its increased job simplification and new technology "to its desire to make jobs easier for women," an equally important factor was "the labor shortage and the opportunity to introduce new technology at government expense under war contracts."[94]

Women Instructing and Being Instructed for War

■ Wartime photographs of women being given job training by male instructors also glorified women yet implicitly kept them in their place. By 1942 American women were being trained in private schools, in colleges, in special federally funded defense-training classes organized through school systems, in industrial plants, and in training programs sponsored by the National Youth Administration. After the attack on Pearl Harbor in 1941 the number of women in training classes increased; in San Diego schools five hundred women received aircraft production training in 1942, with forty women trained as welders for one plant alone.[95]

In photographs taken during both world wars the instructors most often were men, though a small number of women also served as instructors during both wars. In England during World War I there were no technical schools open to women, but because of the need to train

A rare early image of a female driving instructor giving a steering lesson to a British soldier being trained for the Military Transport Division in World War I. *Sphere,* April 1916.

welders for aircraft manufacture, the Women's Service Bureau in 1915 created a small workshop, where a female metalworker served as the instructor.[96] In England and America female pilots, though barred from combat service, also worked as instructors teaching military men to fly. Hilda Hewlett, the first Englishwoman to receive a pilot's license, in 1911, ran a flying school and trained fighter pilots in World War I, and the American pilot Marjorie Stinson taught Canadian men preparing for the British Royal Flying Corps during the war.[97] Britain's *Sphere* magazine in 1916 published a relatively rare photograph, of a British soldier who was readying for the army's Military Transport Division being given a lesson in automotive steering by a female instructor in a school for women drivers.

Photographs of women being instructed often show them looking up respectfully at their male instructors, as in Aikins's photograph of a Gerrard Company worker being taught by a Marine. In another recurring pictorial convention, male instructors stand or hover over seated female students, as in Andreas Feininger's wartime photographs of women in Salt Lake City, Utah, being trained as bus drivers. In Feininger's photographs, taken for the Division of Information, War Production Board, the instructor frames and defines the image as he stands over the woman at the wheel, pointing to the gearshift or instructing her in the use of an oil gauge stick.

In other photographic images female students were also glamorized. A photograph taken by Russell Aikins for U.S. Steel shows women seated in a high-school classroom in Oil City, Pennsylvania, where they are being taught machine-shop work and tooling operations. The male instructor is clearly central, standing at the blackboard, beneath an American flag hanging on the wall. Aikins photographed the women facing the instructor with shafts of sunlight dramatically illuminating their backs. One of the students sits prettily posed on top of a desk, adding artfulness, if not cheesecake, to the instructional scene.

Contrasting with these images of women in training were those taken by the American photographer Gordon Parks for the OWI during World War II. Parks neither glamorized nor subordinated his subjects—young African American women learning new wartime skills—but focused on their luminous faces. During the war, the National Youth Administration's vocational training programs opened up job possibilities for young African American women aged 17–21, who learned skills for jobs in war industries, working eight hours a day and earning $40 a week.

In his photographs for the OWI Parks captured the beaming expressions of young African American women training for defense occupations at Bethune-Cookman College in Daytona Beach, Florida, where they received instruction in welding and other occupations. Rather than smiling self-consciously or looking suitably serious, the young women in Parks's photographs often have a wide, engaging smile, and in the heroizing idiom of the period they are photographed from a low angle, their bodies looming upward, invested with stature and dignity.

From Rosie the Riveter to Rosie the Housewife

■ But for all the stature given to women in wartime photography, there were plenty of reminders that women's wartime roles were only temporary.[98] During the course of the war, advertisements had told readers that women formerly engaged in housework were now capable of handling war production machines: as one advertisement proclaimed, "The Hand That Held the Hoover Works the Drill." But advertisers also began producing transitional ads that were reminders of eventual demobilization, a time when women would once again focus on household appliances as they returned to their primary role as housewives after the war.[99]

A young female welding student in the National Youth Administration School, Bethune-Cookman College, in Daytona Beach, Florida, in January 1943. Office of War Information photograph by Gordon Parks. Library of Congress, FSA-OWI Collection.

Picturing a stalwart Rosie in khaki work pants riveting on the job, Monsanto Chemicals in a 1943 advertisement asked, "What Has Rosie the Riveter to Do with Strawberries for Thanksgiving?" The answer was not just that Rosie, as the ad suggested, would be giving flippant "razz-berries" to the Japanese emperor but also that she, like the corporation itself, would have to radically change her direction after the war. While during wartime Rosie stored her aluminum rivets in a deepfreeze unit lined with Monsanto's insulating material Santocel, after the war, "When Rosie the Riveter becomes Rosie the Housewife," she would be storing her Thanksgiving berries in a new home refrigerator lined with the same Monsanto material.

A 1944 advertisement in the *Saturday Evening Post* for Adel Precision Products, maker of airplane hydraulic valves and other equipment, illustrated a factory-working mother in overalls, with one foot resting on her bicycle, talking to her similarly dressed young daughter. "Mother, when will you stay home again?" the daughter asks anxiously. The textual voice is reassuring: "Some jubilant day mother will stay home again, doing the job she likes best—making a home for you and daddy, when he gets back." Preparing for the company's own coming transition to domestic postwar production, the mother knows that "in her postwar home she'll want appliances with the same high degree of precision," which she will get when Adel "converts its famous Design Simplicity" to products for home and industry.[100]

Adel's reference to kitchen design—for example, in *Life* magazine's 1943 story "Kitchens of Tomorrow"—became a transitional discourse, hinting at women's redirected focus in the postwar world. The emphasis on "Design Simplicity" and women's preference for "precision" also allied Adel with contemporary notions of modernity and an engineering aesthetic, aimed at creating kitchens with clean lines and an efficient look, like the one in *Life*'s "Kitchen of Tomorrow," with its recessed appliances, built-ins (ovens that went into walls), sleek surfaces, and hidden hardware. Similarly, in 1944 the British appliance corporation Moffats advertised that it was preparing for the "Electrical Age" after the war and creating the efficiency of an all-electric kitchen with a "distinctive streamlined" electric stove.[101]

The prophecy of a postwar aesthetic of streamlined kitchens represented a continuation of a trend in appliance design that had been developing since the 1930s, in which designs were simplified and signs of mechanization were hidden under uncluttered surfaces. Adel's idea of "precision" also suggested that women were interested in engineering efficiency, an interest heightened by their experience working in factories during the war.

Also linking women's wartime factory experiences with domestic-product design, Britain's *Electrical Age for Women* reported in a 1944 story entitled "Mrs. America Looks Ahead" that America's Edison Electric Institute had predicted that after the war women would want to have better homes "so that they may have an opportunity to become better mothers, better companions and better citizens of the world." The institute had also predicted "that the job of running a home in post-war years would be 'streamlined' by introducing into the daily chores 'something akin to a production line,'" and that the 17.5 million American women fully employed in war jobs would "demand the same kind of efficiency in their homes as they have found in the factory or in business offices."[102] After the war, design simplicity did indeed continue as an appliance aesthetic for refrigerators and stoves, becoming the "sheer look" during the period 1954–64, when manufacturers in America produced square or boxy kitchen appliances, intended to look clean-lined and efficient.[103]

"A good washer is like a good man," advertisement for a Frigidaire washing machine, 1962. In postwar American advertising women appeared as happy homemakers in love with their household machines. Kettering / GMI Alumni Foundation's Collection of Industrial History.

A good washer is like a good man

As men returned from military service and production needs dropped sharply, there was a sharp reduction in the number of women employed in American and British war-related industries. During the peak employment of the wartime years women employed in Los Angeles aircraft industries constituted 40 percent of workers employed in the industry, but by 1946 the percentage had dropped to 18 percent, and by 1948 it had dropped to 11.9 percent.[104] Women's work roles after the war reverted back to prewar employment patterns, with the bulk of men and women remaining in prewar sex-typed job categories, women largely in clerical and domestic occupations and men in skilled jobs.[105]

The drop in the number of women in war industries buttressed the American government's own intentions. As many have argued, the government had not intended to make a permanent change in women's roles but had viewed women's employment as temporary.[106] The drop in numbers also seemed to verify the government's own (mistaken) assumptions that the bulk of women working in war industries were housewives with no previous work experience who would happily leave their jobs after the war. A Women's Bureau survey found that half of the women who had been housewives before the war wanted to continue working, and another contemporary survey found that women working in wartime wanted to keep their new employment: in 1944, 75–80 percent of women wanted to keep their jobs.[107]

In America's postwar period—with its economic boom, its large increases in productivity and in private home ownership, and expanding suburbs—advertising images of housewives happily using their new home appliances abounded. In these advertisements women's love for their fighting-soldier husbands was now redirected to their new kitchen appliances. The housewife in a Frigidaire advertisement from 1947 clasps her hands together as she gazes at a vision of her refrigerator surrounded by clouds and small red hearts and exclaims, "I just saw the new Frigidaire cold wall. It was love at first sight."[108]

In the early 1960s Frigidaire advertising copywriters more overtly anthropomorphized household appliances into male love objects for women. In an advertisement that ran in August 1962, a woman puts her hands to her cheeks as she gazes in admiration at a "sheer look" washing machine with a man's fedora hat perched on one corner. The ad proclaims: "A good washer is like a good man." In a talk to writers and editors that same year, Herman F. Lehman, a Frigidaire vice president, reported that according to Frigidaire's market researchers, the "overwhelming response" to this ad indicated that "women generally identify the washer as a masculine, sturdy, hard-working product."[109]

During the late 1960s the Whirlpool appliance company also portrayed its refrigerator as a prospective spouse. As its advertisement told women, "Whirlpool believes you should choose a refrigerator the way you'd choose a husband"; that is, they should choose a Whirlpool refrigerator, which was handsome, trustworthy, and "so nice to have around the house."

In postwar advertisements appliances were not only presented as masculine love objects for women but also personified as female figures — as maternal objects or beautiful and hard-working mechanical brides.[110] In an economic climate intent on promoting product consumption and keeping factories in high gear, manufacturers during the 1950s encouraged women to focus on product style and convenience and to see their appliances as extensions of themselves. Products like a Norge gas range were advertised as "glamourous" and "lovely to look at." Frigidaire's Herman Lehman reported that while women saw washers as masculine, refrigerators had a "feminine image" (a concept, apparently, not shared by Whirlpool), and Frigidaire also found that "the refrigerator bears a close alliance with the mother of the house" because it was " 'turned to' most often and enjoyed by everyone."[111]

In nineteenth-century America, women were encouraged by Harriet Beecher Stowe, among others, to become more efficient housekeepers, and during the early years of the twentieth century they were encouraged by women's magazines to view the home as a factory and themselves as

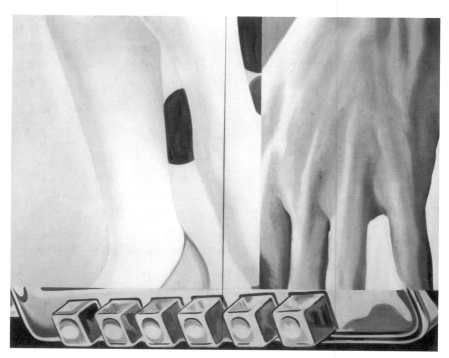

James Rosenquist, *Push Button* (1961). The Museum of Contemporary Art, Los Angeles, The Panza Collection, oil on canvas. © James Rosenquist / Licensed by VAGA, New York. Photo: Squidds and Nunns.

"household engineers."[112] But though kitchen appliances in the period following World War II may have been designed with engineering efficiency in mind, advertisers were apt to portray the women who used these new appliances as purveyors of emotional well-being rather than as household engineers. While images of women during wartime promoted the idea that they were competent yet still feminine, postwar appliance ads portrayed them as blissfully unconcerned with mechanical details. Their competence lay in making themselves good-looking and good housewives for their husbands in the home.

During the 1950s, women in home appliance advertisements were often portrayed as anything but machine-savvy factory workers. Instead they appeared as glamorous cooks working in improbably high heels and elegantly dressed. The "sheer look" in product design hid mechanisms that in an earlier era might have provided women with a technical knowledge and an indication of how appliances actually operated. As Thomas Hine suggested, though women in an earlier era might have been viewed as household engineers, in the electric push-button 1950s and 1960s they were assured that the only real knowledge they needed was how to turn on the machines (as well as their husbands).[113]

The disembodied woman's hand, which during World War II had come to signify women's mechanical expertise and femininity, became transformed, in a Plymouth automobile advertisement from 1955, into a hand gloved in sleek turquoise leather with one long finger pressing a transmission button. Updating Villiers de l'Isle-Adam's nineteenth-century fictional image of Hadaly as push-button lover, the American pop artist James Rosenquist, in his 1961 painting *Push Button*, wittily spoofed the erotic connections between women and the machine, juxtaposing automobile push buttons, a man's hand, and—this time—a pair of woman's legs, her feet in high heels. In the sardonic pop art world of the sixties, the self-sacrificing, machine-savvy Rosies of the war years seemed generations behind, as both women and men readily embraced the ease of the ever more automated age.

The ELECTRONIC EVE

and Late-Twentieth-Century Art

CODA IN OVID'S CLASSIC TALE of transformation the sculptor Pygmalion falls in love with his own artistic creation, a beautiful woman made of ivory, and Venus, acting as divine agency, brings the beautiful figure to life. As imagined by the French academic painter Jean-Léon Gérôme in *Pygmalion and Galatea* in 1890, Galatea emerges from the marble as a nude, sensuous lover, the object of the sculptor's deepest longing and desire (plate 42). But in other versions of transformation it is Venus herself who comes alive, this time through the wonders of science and technology. In Villiers de l'Isle-Adam's nineteenth-century novel *L'Ève future* the beautiful Alicia is likened to the Venus de Milo, and through the transforming power of electricity her female double, Hadaly, comes alive (see chapter 3). In the 1930s the photographic image of Venus in Man Ray's *Électricité* series is a charged, sizzling torso animated by electricity in the modern machine age.

But as the twentieth century drew to a close, female artists fashioned their own images of transformation, turning to electronic technologies to reconfigure and reanimate archetypal representations of the female in the late postmodern age.

In her 1990 computer animation *Venus & Milo* the artist Donna Cox created a witty spoof of technology, art-world icons, and computer visualizations themselves. Produced with supercomputer technology, Cox's Venus is a hot-pink plastic Venus who sits on a pedestal in a museum, where she is tended to by the cigar-smoking janitor Milo, a bug-eyed man with bushy eyebrows and few words (plate 43). On the museum's wall are three works with iconic images of the feminine: Botticelli's Renaissance painting *The Birth of Venus*, Seurat's painting *A Sunday on La Grande Jatte*, and an Andy Warhol silkscreen print of Marilyn Monroe. Nearby is the famed dada satire of the art world's worship of the precious object, Marcel Duchamp's bicycle wheel on a stool.

In Cox's animation Milo has his eyes not on Venus but on a box of chocolate-covered cherries, and he watches as the pink plastic Venus starts eating the candies. In a section sardonically labeled "Scientific Vi-sualization," Venus's body goes through contortions and changes shape as she tries to digest the candies. Finally she gives a big burp as the chocolates fly from her mouth, and Milo mutters, "What a mess," as he reaches for a vacuum cleaner. Vacuuming the candies that are resting under Venus's base, Milo accidentally sucks Venus into the machine. On an inspired whim, he plants the vacuum cleaner on the sculptural base, echoing Duchamp's mocking elevation of the quotidian found object. Delighted, he laughs and says, "I like it!" But at the end the truant vacuum cleaner, still running, sucks up Milo himself.

In Cox's comic send-up of the art world and technology, Venus, un-like the sculptural beauty in the Pygmalion myth, comes alive only to be devoured, not by the sculptor's adoring gaze, but by the household tool so often identified with women's household labors. Rather than being liber-ating, the vacuum cleaner—here a version of autonomous, Franken-steinian technology running out of control—turns on Venus as well as on the complacent Milo. With Cox's mocking nods to dada art—and one of its 1960s manifestations, pop art—the postclassical plastic Venus is sup-planted by the banal, the vacuum cleaner as art.

Venus & Milo was an outgrowth of Cox's earlier collaborative experi-ments creating computerized images of the sculptural Venus. During the late 1980s she worked with a team of scientists, computer scientists, and

The paleolithic Venus of Willendorf appears next to a digitally produced counterpart in Donna Cox's *Venus in Time*, silver print, 1987. Cover illustra-tion for *ACM Computer Graphics Quarterly*, January 1987. Courtesy of the artist.

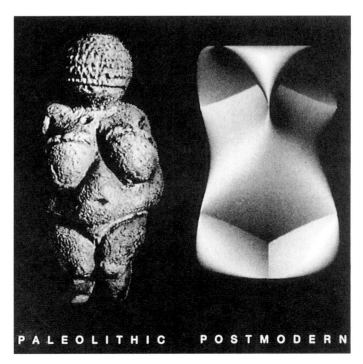

PALEOLITHIC   POSTMODERN

mathematicians using a supercomputer to visualize complex topological surfaces. Producing homotopies, or deformations, they "sculpted" and explored topological surfaces with interactive programs, producing abstracted, sculptural-looking images like the "Etruscan Venus," which appeared alongside one of the most ancient representations of the female figure, the prehistoric Venus of Willendorf, ca. 25,000–20,000 B.C., under the title *Venus in Time* on the cover of a computer-graphics journal.[1] Rather than being classically timeless, like the Hellenistic Venus de Milo, these homotopic images were plastic and changing, energized rather than classically still.

Harriet Casdin-Silver turned to another technology to create her 1991 *Venus of Willendorf.* Using lasers, she created a hologram of a nude, full-figured female, juxtaposing the image next to a photograph of the paleolithic fertility figure.[2]

During the late 1990s, digital artists created other archetypal images of female identity as they increasingly focused on reimaging the female body and concepts of the self. The New York artist Melanie Crean drew the title for her interactive kinetic sculpture *Golem (female),* of 1998, from the golem, an artificially created but not mechanical creature, generally a man, described in sixteenth-century Hebraic legends and legends inspired by Hasidic writings after the seventeenth century.[3]

In Crean's version, the menacing golem has become an animated,

Leaping golem with glowing hair, a digital image from Melanie Crean's interactive kinetic sculpture *Golem (female),* 1998. Courtesy of the artist.

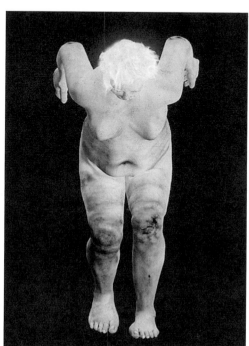

friendly female nude — an image created in part from a photograph of her ninety-one-year-old grandmother — who walks, jumps, and leaps. Using an old-fashioned mechanism, as well as electronic technology, Crean's installation asks the participant to turn the metal hand crank of an antique wine press and peer through an eyepiece into the press's wooden interior.

The image inside, fusing the face of the artist's grandmother with the bodies of the artist and her mother, becomes enlivened as the participant turns the crank. The faster the crank is turned, the more animated and manic the figure becomes, changing from a huddled figure rocking slowly back and forth to a woman with glowing gray hair who runs and jumps, turns handsprings, and even throws a karate kick.[4]

Updating late-nineteenth-century proto–movie machines — crank-handled mechanisms that whirled sequential still images around, creating the illusion of animation — Crean's installation, with its Macintosh software, produces an eerie black-and-white image that nods to cinematic roots while crossing into the cyber-realm of the synthetic. This is a grandmother who jumps jubilantly, but with the jerkiness and artifice of computer animation. Rather than being static, this whimsical grandmother moves in bold defiance of age.

Female artists of the 1980s and 1990s used electronic tools not only to fashion images of the female but to transform their own identity. In her songs, wry anecdotes, and electronic musical pieces the American performance artist Laurie Anderson used acoustic filters and harmonizers to alter her own voice while offering deadpan, acerbic views of America's late-twentieth-century technological culture.

In her four-part epic work *United States* (1979–83) Anderson used map imagery in her meditations on a media-driven electronic society. In 1921 Ruth Eastman's *Motor* magazine cover illustration of a woman standing near her automobile and reading a map suggested the new technological and cultural terrain opening up for women and women's capable mastery of modern machines (chapter 5). But in Anderson's early performance pieces maps were much more problematic. In part 1 of *United States* she describes driving in an automobile along a road at night, pondering America's signs as she repeatedly asks, "Can you tell me where I am?" In part 2 she stands before a huge projected image of a United States map and tells of a world that has lost its bearings in a cacophony of electronic language and advertising texts, a world in which one of the few defensive strategies — or perhaps ironic dictums — was to "Talk Normal."[5]

Revisiting airplane imagery, Anderson's song *From the Air*, from her album *Big Science* (1982), evokes a world of autonomous, out-of-control science and technology and insists, "Jump out of the plane. There is no pilot." Nearly twenty years later, in her work *Songs for A.E.* (2000), she paid tribute to Amelia Earhart, whose plane's radio technology tragically failed her on her effort at a round-the-world flight in 1937. But here Anderson, drawing on Earhart's flight notes, imaginatively focuses, not on technology's failings, but on a world in which Earhart ultimately affirms the feel and beauty of the machine: "My plane, her skin so smooth, it shines like a lucky dime."[6]

Anderson's performances also reflected a postmodern preoccupation with a world dominated by simulations and reproductions — the world of television, film, electronic images and sound. It is a world of disconnections, fragmentation, and reinventions of one's identity. Anderson herself often playfully used electronic equipment to change the timbre and pitch of her own voice, lowering it to a deep male voice imitating the standardized voice of broadcast journalism, raising it to a falsetto, multiplying it to create a choral effect. In a universe of shifting, synthetic images and suspect language the artist's own voice becomes both duplicitous and authentic, distorted yet also holding out the possibility of talking "normal."[7]

In the last decades of the twentieth century female artists like Jenny Holzer turned to electronic technologies for her wry reflections on societies inundated by competing signs. Starting in the late 1970s and continuing in the 1990s, Holzer's electronic signs flashed out taut and witty aphorisms spoofing the flood of texts and images generated by advertising media. The continuously running computer-programmed LED electronic display signs in her "Truisms" and "Living" series flashed such sardonic phrases as "LACK OF CHARISMA CAN BE FATAL," "PROTECT ME FROM WHAT I WANT," and "ACTION CAUSES MORE TROUBLE THAN THOUGHT." Her electronic messages — taping around the building at One Times Square in New York in 1982, flashing at Las Vegas in 1986, displayed at the 1990 Venice Biennale, where Holzer was America's representative artist at the U.S. pavilion — were prodding and provocative, appropriating the discourse of advertising for the artist's own reflections and exhortations.[8]

In the Orwellian year 1984, Holzer and the artist Barbara Kruger also directed a prophetic voice at the realm of big science and the dangers of nuclear disaster. A collaborative work aphoristically labeled by Holzer and spray-painted by the Brooklyn graffiti artist Lady Pink presented an irradiated human body overlaid with the ominous slogan in block letters "You Are Trapped on the Earth So You Will Explode." Kruger, whose montages fused oversized cropped photographs with printed texts, juxtaposed a photograph of an exploding nuclear bomb with the warning "Your Manias Become Science."[9]

Other artists in the eighties and nineties provided glimpses of imagery seen much earlier while also mapping out new terrain. June Leaf's playful revisiting of an unflattering paradigm — the female as automaton, rote-talker, and maniacal maenad — in the form of a sardonic sculptural woman's *Head* (1980) is made of painted aluminum and stainless steel and embedded with exposed gears and a hand crank that makes the woman's tongue move up and down. Turning the crank also makes a cone-tipped magneto behind her eyes eerily revolve.

Updating the dada and surrealist fascination with mannequins and dolls, the photographer Cindy Sherman, in her untitled series of 1992, presented grotesque views of segmented, blond-haired female manne-quins with arched bodies and legs spread wide, with an incongruous beautifying hairbrush on the floor nearby. By 1999 Sherman's color photographs of mannequins had transmuted into black-and-white pho-tographic images of grotesquely dismembered, mutilated, and recon-structed plastic toy dolls — mutant Barbie dolls and their counterpart Ken dolls engaging in violent sexual behavior, the idealized figures of Ameri-can mass marketing gone awry. While the eighteenth-century female automatons of Jacquet-Droz were technological wonders, and the fic-tional Hadaly embodied the cool perfection of artifice and scientific inge-nuity, Sherman's plastic dolls — dismembered and oozing bodily fluids — have the visceral messiness of the macabre.

Transporting the world of women and the automobile into a digital landscape, the American digital artist Adriene Jenik based her CD-ROM *Mauve Desert* (1997) on the 1987 novel *Le Désert mauve*, by the Qué-becois writer Nicole Brossard, and presents the teenage Mélanie, who steals her mother's Météor automobile and rides through the Arizona desert at night, her reflected face framed in the car's rear-view mirror. Going beyond the unified and contained conception of self framed by the automobile window in Tamara de Lempicka's *AutoPortrait*, Mélanie's in-teractive voyage negotiates its way through a layering of discourses, in-cluding an interview with the novelist Brossard, the personal diary of the actress who plays Mélanie, and the artist's own reflections on the creation of her work.[10]

Other artists continued to seize on technological imagery and media to probe political and social themes. In her installation *Amerika* (1990) the German artist Rebecca Horn startled viewers with surrealistic kinetic metal sculptures near the ceiling that created an atmosphere of menace, expelling flamelike electrical charges and sizzling sounds.[11]

But as the twentieth century drew to a close, and the world of cyber-space increasingly became a privileged arena for investigation, artists continued to be particularly haunted by the possibilities of constructing new bodies, new selves.[12] In her interactive website *Bodies©INCcorported* the California artist Victoria Vesna invited participants to create digital images of human bodies by choosing among selections of gender, size, texture, and materials.[13] The computer-generated photographs of the New York artist Nancy Burson expanded the possibilities of transformation and morphing, allowing participants to reconfigure their own faces and identity.

In Burson's *Aging Machine* (1990–92) courageous participants could view their own faces as they would look in the future, and in her *Composite Machine* (1988) participants could fuse scanned images of their own faces with the faces of celebrities, including film stars, art-world icons, political figures, and other popular figures, such as Marilyn Monroe, Barbara Bush, Cher, Marcel Duchamp, Andy Warhol, Mikhail Gorbachev, Elvis, Paul McCartney, and the television talk-show host Oprah Winfrey.[14]

In her untitled photographs merging the faces of young girls and dolls (see chapter 3), Burson used contemporary electronic-imaging techniques to present opposing images of women as artificial and real, mannequin and authentic self. In her *Composite Machine* she further illuminated the problematic nature of a fixed identity in the postmodern age. The digitally morphed photographs suggest the constructed nature of social identities and the blurring of boundaries between the artificial and real. Fusing their own faces with those of famous figures, participants in the *Composite Machine* showed a willingness to surrender themselves to a media-dominated society, to merge themselves with culturally defined images — and human icons — of success, beauty, glamour, and power.

Updating Ruth Eastman's image of the map, the scientist Donna Haraway in the age of electronic media and advancing biomedical research used it as metaphor for the dominant investigatory and cognitive discourses of the late twentieth century. In her book *ModestWitness@Second_Millennium. . .* , its title a parody of an e-mail address, she argued that "the chip, seed, or gene is simultaneously literal and figurative, we inhabit and are inhabited by such figures that map universes of knowledge, practice, and power." For Haraway, the task of the "mutated, modest witness" was to read these maps with attentive literacy, without framing them in terms of utopian "comedic resolution" or tragic, horrific outcomes, including "apocalyptic disasters."[15]

For Haraway and other observers of late-twentieth-century cyberculture, the new area for exploration remained the world of transgressed boundaries, the world of ambiguous and blurred distinctions between artifice and authenticity, the world of new fusions, new identities. Contemplating digital culture, some warned of the dangers of human dissolution — a loss of corporeality in which the human body would be supplanted by virtual bodies of cyberspace. With vatic vehemence, observers such as Vivian Sobchack warned that if we do not maintain a "subjective kind of bodily sense in mind as we negotiate our techno-culture, we may very well objectify ourselves to death."[16]

At the onset of the new millennium our exploration of still-transforming human bodies and human identities continues. The mysterious definitions of a newly embodied Eve, a new Venus, await our mapping.

. . . .

Notes

Introduction   1. British Museum Department of Prints and Drawings, *Catalogue of Political and Personal Satires*, vol. 11 (1829–32), ed. M. Dorothy George (London: British Museum, 1954), 343–44. The print originally appeared in the *Looking Glass* 1, no. 10 (1 October 1830). George notes that milling machines also appeared elsewhere in British satirical prints as an emblem of rejuvenation and transformation.

2. The idea that women were by nature more fastidious than men also had an impact on labor decisions in industry. Nineteenth-century female industrial workers were sometimes assigned duties meant to save them from dust and dirt, and fewer American women were employed in the woolen industry than in cotton mills because processes like carding were considered greasy, dirty, and wet (see Elizabeth Faulkner Baker, *Technology and Women's Work* [New York: Columbia Univ. Press, 1964], 18). At America's Centennial Exhibition in 1876, Emma Allison, of Grimsby, Iowa, who operated the 6-horsepower Baxter engine in the exhibition's Women's Pavilion, was praised by one writer for her mechanical ability yet also described as suitably feminine, for she offered "an example worth following to engineers of the male sex in the neatness of her dress and the perfection of cleanliness exhibited in both engine and engine-room" (James McCabe, *Illustrated History of the Centennial Exhibition* [Philadelphia: National Publishing Co., 1876], 591). In a later version of these gendered differences, Robert C. Post noted that among some male drag racers, intimacy with machines was linked to a willingness to be dirty (Robert C. Post, *High Performance: The Culture and Technology of Drag Racing, 1950–2000*, rev. ed. [Baltimore: Johns Hopkins Univ. Press, 2001], 277).

3. Shown in Joseph J. Corn et al., eds., *Yesterday's Tomorrows: Past Visions of the American Future* (1984; reprint, Baltimore: Johns Hopkins Univ. Press, 1996).

4. *New York Times*, 28 June 1992, Society section, 7.

5. See also Picabia's gouache railroad-machine diagram, *Fille née sand mère* (Girl born without a mother), ca. 1917, illustrated in K. G. Pontus-Hultén, *The Machine as Seen at the End of the Mechanical Age*, exh. cat. (New York: Museum of Modern Art, 1968), 83. The title was also used for his 1918 book of poems and drawings.

6. "Housewife" [Maud Lancaster], *Electric Cooking, Heating, Cleaning: A Manual of Electricity in the Service of the Home*, ed. E. W. Lancaster (London: Constable, 1914) 1–4, 24–27, cited in Gerrylynn K. Roberts, *Sources for the Study of Science, Technology, and Everyday Life, 1870–1950*, vol. 1, *A Primary Reader* (London: Hodder & Stoughton in association with the Open University, 1988), 16–17.

7. The ad appeared in the December 1942 issue of *Ladies' Home Journal*, p. 46. The airfield models, designed to help fool enemy bombardiers, were photographed as they would look from the air.

8. The Sunbeam advertisement showing the husband's face reflected in the shiny surface of the coffeemaker is illustrated in Ellen Lupton, *Mechanical Brides: Women and Machines from Home to Office*, exh. cat. (New York: Cooper-Hewitt National Museum of Design, Smithsonian Institution, 1993), 8; the version showing the housewife's face reflected appeared in the November 1950 issue of *Good Housekeeping* magazine.

9. For a thoughtful discussion of Russian female artists, see John E. Bowlf and Matthew Drutt, eds., *Amazons of the Avant-Garde: Alexandra Exter, Natalia Goncharova, Liubov Popova, Olga Rozanova, Varvara Stepanova, and Nedezhda Udaltsova* (New York: Guggenheim Museum, 2000). For an early overview of women industrial designers, see Isabelle Anscombe, *A Woman's Touch: Women in Design from 1860 to the Present Day* (New York: Viking Penguin, 1984).

1. Framing Images of Women and Machines

1. Dorothy Levitt, *The Woman and the Car*, ed. and intro. C. Byng-Hall (London: John Lane, 1909).

2. See Linda Nochlin's comments on mirroring, gender relations, and Diego Velázquez's painting *The Toilet of Venus (The Rokeby Venus)* in *Women, Art, and Power and Other Essays* (New York: Harper & Row, 1988), 26 ff. Londa Schiebinger discusses visual personifications of science as a woman, including Cesare Ripa's *Iconologia* (1618), in which a woman who holds a mirror in her hand "symbolizes the study of appearances leading to knowledge of essences" (*The Mind Has No Sex? Women in the Origins of Modern Science* [Cambridge: Harvard Univ. Press, 1989], 123).

3. The image is illustrated in Annette K. Baxter with Constance Jacobs, *To Be a Woman in America 1850–1930* (New York: Times Books, 1978), 197.

4. In the 1920s the role-reversal theme reappeared in an advertisement for a Maytag washing machine in which a rural woman driving a tractor smiles happily as her husband, wearing an apron, washes clothes (see Katherine Jellison, *Entitled to Power: Farm Women and Technology, 1913–1963* [Chapel Hill: Univ. of North Carolina Press, 1993], 53).

5. See Harvey Green, *The Light of the Home: An Intimate View of the Lives of Women in Victorian America* (New York: Pantheon, 1983), 162. Barbara Welter argued that in the nineteenth century "any form of social change was tantamount to an attack on woman's virtue, if only it was correctly understood" ("The Cult of True Womanhood: 1820–1860," *American Quarterly* 18 [summer 1966]: 157).

6. See Carolyn Marvin, *When Old Technologies Were New: Thinking about Electric Communication in the Late Nineteenth Century* (New York: Oxford Univ. Press, 1988), 70–74. Marvin notes that beginning in the late 1870s, work as a telegraph operator or at a telephone switchboard was considered respectable for middle-class women, but in stories in America's electrical journals these women were often portrayed as either virtuous or risqué (26–27).

7. See *Art and the Industrial Revolution*, exh. cat. (Manchester: Manchester City Art Gallery, 1968), 30.

8. Charles Dickens, *American Notes for General Circulation* (1842), in *New Oxford Illustrated Dickens* (London: Oxford Univ. Press, 1963), 53.

9. Samuel Breck, *Recollections of Samuel Breck with Passages from his Note-Books, 1771–1862*, ed. H. E. Scudder (Philadelphia: Porter & Coates, 1877), 276.

10. Helen Bullitt Lowry, "Woman's Place Is in the Tonneau," *Motor* 40, no. 6 (November 1923): 34.

11. There are extensive studies of the gendered nature of technology. See, for example, Cynthia Cockburn and Susan Ormrod, *Gender and Technology in the Making* (London: Sage, 1993); and idem, *Brothers: Male Dominance and Technological Change* (London: Pluto, 1983), 203. Cockburn ascribes these culturally defined gender characteristics to differences in role models, education, childhood exposure to technology and to segregation in the labor market. See also Judy Wajcman, *Feminism Confronts Technology* (University Park: Pennsylvania State Univ. Press, 1991), 155; and Ava Baron, ed., *Work Engendered* (Ithaca: Cornell Univ. Press, 1991).

12. [Mary Isabella Gascoigne], *The Handbook on Turning* (1842; reprint, London: Saunders & Otley, 1859), xiii. My thanks to Michael Wright, curator of mechanical engineering at the Science Museum, London, for information on this book. Gascoigne, whose book was reprinted under the title *The Turner's Companion* (Philadelphia: Henry Carey Baird, 1851), died in 1891; she was identified as Mary Isabella Gascoigne in a letter from Warren Ogden Jr. to the New York Public Library, 28 January 1953.

13. [Gascoigne], *Handbook on Turning*, xiii.

14. Women observing men at work in forges also appeared in the work of eighteenth-century Belgian painters like Léonard Defrance.

15. Judy Edgerton, *Wright of Derby* (London: Tate Gallery, 1990), 21, 58–59. Some commentators have attributed the painting's twisting body postures to the conventions of European mannerist art.

16. Illustrated in Katherine Baetjer and David James Draper, eds., *"Only the Best": Masterpieces of the Calouste Gulbenkian Museum*, exh. cat. (New York: Metropolitan Museum of Art, 1999), 36.

17. Delasalle was made a chevalier of the Legion of Honor in 1926.

18. Henri Frantz, *The Salon of 1900 and the Decennial Exhibition*, trans. Clarence Wason (Paris: Goupil, 1900), 36; Delasalle's painting appears opposite 34.

19. See Evelyn Fox Keller, *Reflections on Gender and Science* (New Haven: Yale Univ. Press, 1985), 76–77.

20. Pierre-Joseph Boudier de Villemert, *L'Ami des femmes* (1759), translated into English as *The Ladies' Friend* (Philadelphia: Matthew Carey, 1793), 13. Other English versions of this book, sometimes under the title *The Friend of Women*, were published in New Haven, Conn., and in London. In America, as Margaret Rossiter and others have documented, these stereotypes about women's limited scientific abilities continued into the nineteenth and twentieth centuries, particularly during the period 1820–1920, and helped perpetuate women's subordinate place in scientific enterprises: women were viewed as soft, delicate, and emotional, while science was considered rigorous, rational, impersonal, masculine, and unemotional. During the 1870s, in a backlash against women's entering the fields of medicine, science, and higher education, it was argued that women had small, light brains and that the study of science would impair them in mind and body (Margaret Rossiter, *Women Scientists in America* [Baltimore: Johns Hopkins Univ. Press, 1995], xv, 13).

21. Schiebinger, *The Mind Has No Sex?* 4, 36, 41, 44. This study focuses on the role of women in the rise of modern science in seventeenth- and eighteenth-century Europe.

22. Marilyn Bailey Ogilvie, *Women in Science: Antiquity through the Nineteenth Century* (Cambridge: MIT Press, 1986); for more theoretical issues, see also Sandra Harding, *Whose Science? Whose Knowledge? Thinking from Women's Lives* (Ithaca: Cornell Univ. Press, 1991). In the world of eighteenth-century astronomy, the famed astronomer Sir William Herschel's sister Caroline discovered three new nebulae in 1783 and eight comets during the period 1786–97, published a *Catalogue of Stars* in London in 1798, and became an honorary member of the Royal Society in 1835 (Ogilvie, *Women in Science*, 96–98). Schiebinger argues that Herschel's possibilities in science were limited and that she was largely confined to assisting her brother (*The Mind Has No Sex?* 261–64).

23. Abbé Jean-Antoine Nollet, *Essai sur l'électricité des corps* (Paris, 1746); the book went through several editions, the fifth published in 1771.

24. Mary Shelley, introduction to 1831 edition of *Frankenstein or The Modern Prometheus*, in *Frankenstein: Complete, Authoritative Text with Biographical and Historical Contexts, Critical History, and Essays from Five Contemporary Critical Perspectives*, ed. Johanna M. Smith (Boston: St. Martin's, 1992), 22. In the novel, however, Shelley does not state that the Monster was created through galvanism. See also Samuel Holmes Vasbinder, *Scientific Attitudes in Mary Shelley's "Frankenstein"* (Ann Arbor: UMI Research Press, 1984), 82.

25. See the eighteenth-century engraving *La Méchanique*, by the French artist J. Tavenet, in which a classicized female figure holds a square and is surrounded by small putti holding other emblems of architecture, mathematics, and science.

26. As a woman painter, Kauffmann was barred by social convention from the study of nude models, which was considered necessary for suitable academic training. Nevertheless, she received enough acclaim to become one of the founding members of Britain's Royal Academy of Art in 1768, and her neoclassical portraits and history paintings of classical subjects brought her praise and popularity. For more on Kauffmann, see Ann Sutherland Harris and Linda Nochlin, *Women Artists, 1550–1950* (Los Angeles: Los Angeles Museum of Art, 1976), 174; and Whitney Chadwick, *Women, Art, and Society* (New York: Thames & Hudson, 1990), 7, 142–47.

27. The tale was based in part on early biographies, which developed the probably apocryphal story that Watt was inspired while watching a Chinese tea kitchen, a type of covered samovar used as a steam boiler that had a fire box in the middle that was heated with charcoal. My thanks to Michael Wright, of the Science Museum, London, for this information.

28. The British engraver was James Scott. Dominique François Jean Arago's biography, *Eloge historique de James Watt*, was published in English as *Historical Eloge of James Watt by M. Arago*, trans. James Patrick Muirhead (London: James Murray, 1839), and included the apocryphal anecdote (6–7), said to have occurred in 1750. The translator's wording differs from that on Scott's engraving.

29. The "shavograph" is illustrated in Julie Wosk, *Breaking Frame: Technology and the Visual Arts in the Nineteenth Century* (New Brunswick: Rutgers Univ. Press, 1992), 96.

30. Henry Adams, "The Dynamo and the Virgin" (1900), in *The Education of Henry Adams* (1907; reprint, New York: Modern Library, 1931), 383. For further

discussion, see Leo Marx, *The Machine in the Garden: Technology and the Pastoral Ideal in America* (New York: Oxford Univ. Press, 1964), 345–50.

31. Agostino Ramelli, *Le Diverse et Artificiose Machine* (1588), translated into English by Martha Teach Gnudi as *The Various and Ingenious Machines of Agostino Ramelli* (Baltimore: Johns Hopkins Univ. Press, 1976).

32. Martha Banta, *Imaging American Women* (New York: Columbia Univ. Press, 1987), preface.

33. Serafina K. Bathrick, "The Female Colossus: The Body as Facade and Threshold," in *Fabrications: Costume and the Female Body*, ed. Jane M. Gaines and Charlotte Herzog (New York: Routledge, 1990), 79–80.

34. Abigail Solomon-Godeau offers an intriguing analysis in *Photography at the Dock: Essays on Photographic History, Institutions, and Practices* (Minneapolis: Univ. of Minnesota Press, 1991), 237.

35. In contrast, Elizabeth Beardsley Butler, in her reform-minded *Women and the Trades: Pittsburgh, 1907–1908* (1909; reprint, New York: Charities Publication Committee, 1911), 258–59, presented a much less sanguine view of box manufacturing, describing the hazards to women working on a Knowlton & Beach staying machine, which put glue-dipped clamps or manilla stays on the corners of box covers. Butler's monograph was part of *The Pittsburgh Survey*, ed. Paul Kellogg, 6 vols. (New York: Charities Publication Committee, 1909–14). The hands of female operatives were often dangerously near to the knives that came down to clip the stays. The situation was made even more dangerous by the women's practice of fastening back their protective knife guards so they would not interfere with their work. See the photograph titled *The Dangerous Staying Machine at Work*, a carefully ordered and balanced image that belies the danger of the work, adjacent to 258.

36. The engraving appears as a frontispiece in *Godey's Lady's Book and Magazine* (1860). The quotation is taken from the "Editors' Table," *Godey's Lady's Book and Magazine* 60 (January 1860): 74. As Barbara Welter wrote in "The Cult of True Womanhood," Americans identified the "True Woman" with "piety, purity, submissiveness and domesticity," and "domesticity was among the virtues most prized by the women's magazines" (152, 162).

37. Hearth light and oil lamps were often associated with the concept of home and appeared in more than half of all images of the home in popular-magazine illustrations in the 1890s, but increasingly they disappeared as symbols of the home as they were replaced by electric lights. By 1920 incandescent lights appeared in nearly half of these illustrations (see David Nye, *Electrifying America: Social Meanings of a New Technology* [Cambridge: MIT Press, 1990], 282; Bernard Finn, cited in ibid., 426; and idem, "The Incandescent Electric Light," *Bridge to the Future, Annals of the New York Academy of Sciences* 424 [1984]: 252).

38. Banta discusses the cliché of nymphs holding lamps, including the most notable example, the Statue of Liberty, in *Imaging American Women*, 530 n. 53, 532.

39. David Nye, *Image Worlds: Corporate Identities at General Electric, 1890–1930* (Cambridge: MIT Press, 1985), 116. Conflating the bad-girl look with domesticity, the German electrical company AEG featured a vamp with short black hair in its "Vampyr" vacuum cleaner advertisement in 1921, illustrated in Penny Sparke, *Electrical Appliances: Twentieth-Century Design* (New York: E. P. Dutton, 1987), 96. During the twenties and thirties the American illustrator

Maxfield Parrish was hired by General Electric to create calender art for Mazda lamps. The illustrations had suggestive titles like *Ecstasy*, in which transparently clad, wind-blown women stood with their arms raised amidst rocky landscapes and billowing clouds (see Coy Ludwig, *Maxfield Parrish* [New York: Watson-Guptill, 1973], 27).

40. David Nye, in *Image Worlds*, 121, suggests that the pose is derived from Jacques-Louis David's neoclassical oil painting *Madame Récamier* (1800).

41. The photograph created a sensation in Paris when it was published in the surrealist journal *Minotaure* in 1934 (Nancy Spector, "Meret Oppenheim: Performing Identities," in *Meret Oppenheim: Beyond the Teacup*, by Jacqueline Burckhardt, Bice Curiger, et al., exh. cat. [New York: Independent Curators, 1996], 35, 39–40). Spector notes that when the photograph was published in *Minotaure*, the phallic crank, suggesting a hermaphroditic image, was cropped from the picture, hence the idea "veiled erotic." The photograph is one of a series taken by Man Ray for the magazine in the studio of the cubist painter Louis Marcoussis (see also Jean-Hubert Martin, ed., *Man Ray: Photographs*, exh. cat. [New York: Thames & Hudson, 1982], 43, 69).

42. Roland Marchand, *Advertising the American Dream: Making Way for Modernity, 1920–1940* (Berkeley and Los Angeles: Univ. of California Press, 1985); Ellen Gruber Garvey, *The Adman in the Parlor: Magazines and the Gendering of Consumer Culture, 1880s to 1910s* (New York: Oxford Univ. Press, 1996).

43. See Donald Hoke, "The Woman and the Typewriter: A Case Study in Technological Innovation and Social Change," in *Business and Economic History*, 2nd ser., vol. 8, *Papers Presented at the Twenty-Fifth Annual Meeting of the Business History Conference, March 2–3, 1979*, 8:76–88. In a history of the typewriter published in New York's Herkimer County, where, in the town of Ilion, the Remington Rand Corporation was located until 1938, Christopher Latham Sholes is portrayed as a benefactor to women for helping them earn a livelihood. A composite photograph shows him envisioning a beatific lineup of women (Herkimer County Historical Society, *The Story of the Typewriter, 1873–1923* [Herkimer County, N.Y., 1923], 8, 142. See also Bruce Bliven Jr., *The Wonderful Writing Machine* [New York: Random House, 1954]).

44. Hoke, "Woman and the Typewriter," 82–83. The photograph of Lillian Sholes reappeared in Remington typewriter advertisements and in *A Brief History of the Typewriter* (New York: Remington Rand, 1955).

45. "An Improved Mangle," *Scientific American*, 10 May 1879, 294.

46. In 1879, the same year as the mangle story, the magazine began its American Industries series. The series, which ran for four years starting in 1879, included numerous images of women working in industries, but as David Hounshell has argued, the images were more promotional than descriptive. The stories, often written by the manufacturers, and the illustrations, at least some of which were done by the magazine's staff artists, were essentially front-page advertisements for the companies. With the help of the patent agent Munn & Company, companies could not only secure patents but also have their own illustrations and idealized stories placed in the magazine (Carroll W. Pursell Jr., "Testing a Carriage: The 'American Industry' Series of *Scientific American*," *Technology and Culture* 17 [January 1976]: 83; David A. Hounshell, "Public Relations or Public Understanding? The American Industries Series in *Scientific American*," ibid. 21 [October 1980]: 589–93).

47. "American Industries — No. 10, Sewing Machines," *Scientific American*, 3 May 1879.

48. "Electrical Invasion of the Home," ibid., 15 April 1911, 381.

49. See Nye, *Electrifying America*, 260. Appealing to wealthy consumers, a 1908 General Electric advertising photograph featured a woman wearing a dress with ornate lace sleeves who is holding an electric grill on a dining-room table, her expensive silver tea service prominently displayed.

50. "Electricity in the Household," *Scientific American*, 19 March 1904, 232. Electric irons were introduced in 1893.

51. "Cleaning with Electricity," ibid., 24 June 1911, 626.

52. Ruth Schwartz Cowan, *More Work for Mother: The Ironies of Household Technology from the Open Hearth to the Microwave* (New York: Basic Books, 1983). Early electric irons were also presented as labor-saving. While the artist Edgar Degas's painting *The Ironers* (1884) and Picasso's painting of a woman ironing from his blue period emphasized women's heavy labors as they worked bowed over their flatirons, the woman in *Scientific American*'s 1911 article "Electrical Invasion of the Home" contentedly uses her new home electric iron (381).

53. Brian Bowers, in *A History of Electric Light and Power* (London: Peter Peregrinus in association with the Science Museum, 1982) and *Lengthening the Day: A History of Lighting Technology* (Oxford: Oxford Univ. Press, 1998), discusses Britain's electrification rates; for more on early British electrical appliances, see Sparke, *Electrical Appliances*; and Rebecca Weaver and Rodney Dale, *Machines in the Home* (New York: Oxford Univ. Press, 1992). For comments on American electrification, see Nye, *Electrifying America*; and Susan Strasser, *Never Done: A History of American Housework* (New York: Pantheon, 1982), 80–81. The market for electrical appliances was also spurred on by a decrease in the number of servants available for middle-class households. Sparke argues that in America in the 1920s servants were available to only half the households that desired them in the 1870s, but she notes that the situation was not as extreme in Britain. Strasser, Cowan, and Nye, among others, discuss the links between the decline in the availability of servants and the promotion of appliances such as the vacuum cleaner. Sparke notes that after irons, vacuum cleaners were the best-selling domestic items during the first half of the twentieth century (*Electrical Appliances*, 10–11, 92).

54. Hoovers were available in Britain by 1912, and by 1919 Hoover had gained a national market in America, and British manufacturers, including Goblin, were soon marketing their own machines (Sigfried Giedion, *Mechanization Takes Command: A Contribution to an Anonymous History* [1948; reprint, New York: Norton, 1969]; Nye, *Electrifying America*, 18; Weaver and Dale, *Machines in the Home*, 54; Sparke, *Electrical Appliances*, 28, 45, 92–93; Strasser, *Never Done*, 78). Christina Hardyment argues that in Britain, where electric power became readily available less quickly, vacuum cleaners gained acceptance more slowly, also due to fears of shock (*From Mangle to Microwave: The Mechanization of Household Work* [Cambridge: Polity Press, 1988], 88).

55. Cowan, *More Work for Mother*, 199.

56. The Frantz ad is illustrated in Joseph E. Dispenza, *Advertising the American Woman* (Dayton, Ohio: Pflaum, 1975).

57. Grace Rogers Cooper, *The Sewing Machine: Its Invention and Early Development*, 2nd ed. (Washington, D.C.: Smithsonian Institution Press, 1976), 34–

35. Adrian Forty argues that Singer's first "Family" machine, introduced in 1858, was not a success because it was too expensive and because Americans needed to be persuaded that it was desirable to have what amounted to "a machine tool in the living room" (*Objects of Desire: Design and Society since 1750* [London: Thames & Hudson, 1986], 96–97). In England, Wheeler & Wilson also manufactured machines in London in 1859, and Singer opened a factory in Glasgow in 1867. British manufacturers began producing their own sewing machines, though American manufacturers remained a strong presence (Weaver and Dale, *Machines in the Home*, 48).

58. John Scott, *Genius Rewarded; or The Story of the Sewing Machine* (New York: J. J. Caulon for the Singer Sewing Machine Co., 1880).

59. Ibid., 7–8.

60. "Sewing Machine Clubs," *Godey's Lady's Book and Magazine* 61 (September 1860): 271.

61. Herbert Ladd Towle, "The Woman at the Wheel," *Scribner's* 57 (February 1915): 223.

62. *Sphere*, 4 May 1918. In a related phenomenon, sewing-machine companies later offered women another version of presumed emancipation: firms such as the Coventry Sewing Machine company in England began producing bicycles in 1868.

63. The advertising broadsheet appears in the Landauer Collection, New-York Historical Society.

64. Pamela Walker Laird, in "Progress in Separate Spheres: Selling Nineteenth-Century Technologies," *Knowledge and Society* 10 (1996): 19–49, also discusses a trade-card advertisement for a washing machine in which family order is restored.

65. David A. Hounshell, *From the American System to Mass Production, 1800–1932: The Development of Manufacturing Technology in the United States* (Baltimore: Johns Hopkins Univ. Press, 1984), 89.

66. "Improved Motion for Sewing Machines," *Scientific American*, 2 April 1866, 274; Philippe Perrot, *Fashioning the Bourgeoisie: A History of Clothing in the Nineteenth Century*, trans. Richard Bienvenu (Princeton: Princeton Univ. Press, 1994), 105; Karen Offen, "'Powered by a Woman's Foot': A Documentary Introduction to the Sexual Politics of the Sewing Machine in Nineteenth-Century France," *Women's Studies International Forum* 11, no. 2 (1988): 94.

67. For more on ornamented nineteenth-century machines, see Wosk, *Breaking Frame*, chs. 3, 5, and 6, as well as Julie Wosk, "Brunel Meets Brunelleschi," *American Heritage of Invention and Technology* 11 (summer 1995): 58–63.

68. My thanks to Beverly Brannan, curator of the Photography, Prints and Photographs Division, Library of Congress, Washington, D.C., for information on this image.

69. American women also worked in tenement apartments for contractors or subcontractors who hired small groups, often of fewer than ten people, who worked on individual tasks or at piecework (see Peter Liebhold and Harry R. Rubenstein, "Between a Rock and a Hard Place: The National Museum of American History's Exhibition on Sweatshops, 1820–Present," *Labor's Heritage* 9, no. 4 [1998]: 10, citing J. M. Fenster, "Seamstresses," *Invention and Technology*, winter 1994, 41–44).

70. *Harper's Bazaar*, 18 February 1871; *Frank Leslie's Illustrated Newspaper*,

3 November 1888. See Harry R. Rubenstein, "Symbols and Images of American Labor: Dinner Pails and Hard Hats," *Labor's Heritage* 1, no. 3 (1989): 41. In nineteenth-century England, as Lisa Tickner has argued, images in paintings and prints presented comparable, if sentimentalized views, of the oppressed "sweated seamstress" (*The Spectacle of Women: Imagery of the Suffrage Campaign, 1907–1914* [London: Chatto & Windus, 1987], 51).

71. Butler, *Women and the Trades*, 10–11, 16. For more on Hine, see Judith Mara Gutman, *Lewis W. Hine, 1874–1940: Two Perspectives* (New York: Grossman, 1974); and *America and Lewis Hine: Photographs 1904–1940*, exh. cat. (New York: Aperture, 1977). The photographs, among the thousands made by Hine over a long career of documenting labor and American social conditions, appeared in publications like the *Survey*, which documented New York labor conditions, and in Butler, *Women and the Trades*, 103–4, 108–9. In 1930 the Berlin-born surrealist Alice Lex-Nerlinger created a modernist reconfiguring of a woman and a sewing machine in her photograph *Seamstress*, a montage juxtaposing a woman at her sewing machine with a larger image of a woman's face gazing intently at the viewer. The photograph appears in Naomi Rosenblum, *A History of Woman Photographers*, 2nd ed. (New York: Abbeville, 2000), 133.

72. Shelley, *Frankenstein*, quotation on 142. For further discussion, see George Levine's trenchant essay "The Ambiguous Heritage of *Frankenstein*," in *The Endurance of Frankenstein: Essays on Mary Shelley's Novel*, ed. George Levine and U. C. Knoepflmacher (Berkeley: Univ. of California Press, 1979), 16, where he discusses Shelley's theme that "the self expressed in technology can only be . . . monstrous." Chris Baldick, however, strenuously objected to what he called the "technological reduction" of the 1970s and 1980s, which saw the story "chiefly as an uncanny prophecy of dangerous scientific inventions" (*In Frankenstein's Shadow: Myth, Monstrosity, and Nineteenth-Century Writing* [Oxford: Clarendon; New York: Oxford Univ. Press, 1987], 7).

73. Judith A. McGaw, *Most Wonderful Machine: Mechanization and Social Change in Berkshire Paper Making, 1801–1885* (Princeton: Princeton Univ. Press, 1987), 340, 344, 368, notes that though paper making may have seemed safe for women, it was unhealthful: women developed respiratory problems because of the particles in the rag rooms.

74. *The Penny Magazine of the Society for the Diffusion of Useful Knowledge* (London), monthly supplement, 28 September 1833, 284.

75. Herman Melville, *The Paradise of Bachelors and The Tartarus of Maids*, first published in *Harper's Monthly*, 1855, reprinted in *The Writings of Herman Melville*, ed. Harrison Hayford et al., vol. 9 (Evanston: Northwestern Univ. Press; Chicago: Newberry, 1987). All page references are to this edition.

76. Melville, *Paradise of Bachelors*, in ibid., 322.

77. McGaw notes that in 1851 Melville himself made a journey from his home in Pittsfield to nearby Dalton to purchase paper. His portrayal of the mill and its workers is essentially accurate, she notes though he failed to mention the male workers (*Most Wonderful Machine*, 335–36).

78. Melville, *Paradise of Bachelors*, 326–27.

79. Ibid., 328, 333.

80. Ibid., 330, 334.

81. Ibid., 334.

82. Powered by water wheels, some of these spinning frames could spin

forty-eight lengths of yarn at one time (Brooke Hindle and Steven Lubar, *Engines of Change: The American Industrial Revolution, 1790–1860* [Washington, D.C.: Smithsonian Institution Press, 1986], 16, 19, 59 ff.).

83. John Dyer, *The Fleece: A Poem in Four Books* (1757; reprint, ed. Edward Thomas, London: T. Fisher Unwin, 1903), pt. 3, pp. 86, 89–90, 98. For a thoughtful discussion of the impact of the mechanization of Britain's textile industry on women's work, see Deborah Valenze, *The First Industrial Woman* (New York: Oxford Univ. Press, 1995); on the impact of the spinning jenny, see p. 83.

84. In 1827, 90 percent of the work force at the Merrimack Manufacturing Company in Lowell were women and girls, and in 1855 the Lowell mills employed 8,800 women and 4,400 men, though the number of women would decrease by 1870 as looms and weaving processes became faster and more men and boys were employed (see Thomas Dublin, *Transforming Women's Work: New England Lives in the Industrial Revolution* [Ithaca: Cornell Univ. Press, 1994], 110–18; Hindle and Lubar, *Engines of Change*, 188–202; and Baker, *Technology and Women's Work*, 10–11. Baker draws heavily on Edith Abbott's *Women in Industry: A Study in American Economic History* [1910; reprint, New York: Arno, 1969], which argues that American textile mills had a larger proportion of female workers than did British mills [90]. See also Helena Wright, "The Uncommon Mill Girls at Lowell," *History Today*, January 1973, 10–19). As Deborah Valenze has noted about female workers in Britain, the revolution in textile production and the mechanization of spinning "gave another occasion for the revaluation of women's work as inferior." British women were excluded from unions, and when power-loom weaving allowed women into factories in the 1830s, their claim to skill came under siege (Valenze, *First Industrial Woman*, 183–84).

85. Job-typing and assumptions about women's capabilities continued into the twentieth century. The Alexander Hamilton Institute in New York, a business-management training institute staffed by male businessmen, described women as being "most successful in jobs that are light, unskilled, routine and repetitive" for "they are patient, painstaking" (Alexander Hamilton Institute, *Women in Industry* [Astor Place, N.Y., 1918], 12, 16). A large body of literature has long argued that the sex-typing of jobs was due less to inherent biological differences and differing natures than to differing socialization experiences: women came to factories having had previous experience in cleaning, laundering, and, later, sewing machines, all of which entailed repetitive work that fostered precise movements and visual discrimination (see Cowan, *More Work for Mother*; and McGaw, *Most Wonderful Machine*, 353–54). And as Deborah G. Douglas has pointed out, women in the American aviation industry during World War II were routinely assigned such jobs as inspection, requiring fine, precise work, which was, however, considered to have lesser status than jobs requiring physical strength (*United States Women in Aviation, 1940–1985*, Smithsonian Studies in Air and Space, 7 [Washington, D.C.: Smithsonian Institution Press, 1990], 24). Cynthia Cockburn, in *Machinery of Dominance: Women, Men, and Technical Know-How* (London: Pluto, 1985), discusses studies that validated the premise that even when industries became less arduous and mechanization of printing made linotype work less heavy, women were still assigned to less skilled jobs.

86. Women also operated roving and spinning frames at mills in Lowell and Waltham, Massachusetts. While almost all the weavers in the Waltham mills in 1821 were women, in England it was men who worked as weavers. In two typical

mills in Massachusetts 40 percent of the workers were adult women, who ran power looms and spinning machines, 25 percent were children, and one-third of the jobs were held by men working as supervisors, mule spinners, hand-loom weavers, and machinists, among other jobs (see Baker, *Technology and Women's Work*, 16; Ruth Schwartz Cowan, "From Virginia Dare to Virginia Slims: Women and Technology in American Life," *Technology and Culture* 20 [January 1979]: 55; Abbott, *Women in Industry*, 94; and Hindle and Lubar, *Engines of Change*, 192).

87. Edward Baines Jr., *History of the Cotton Manufacture in Great Britain* (London: H. Fisher, R. Fisher, & P. Jackson, 1835), 456. Francis Klingender noted that Baines, son of a reformer, was an economist and sociologist who took a benign view of issues such as child labor (*Art and the Industrial Revolution*, rev. ed. Arthur Elton [1968; reprint, New York: Schocken, 1970], 212). Before the federal government took control of currency in 1863, state and local banks commissioned artists and engravers to create images for bank notes. In 1836 a local bank in Coventry, Rhode Island, appropriated one of Allom's images of well-dressed young female textile workers for the vignette on engraved antebellum bank notes (see Francine Tyler, "The Angel in the Factory: Images of Women Workers Engraved on Ante-bellum Bank Notes," *Imprint* 19, no. 1 [spring 1994]: 3).

88. While Thomas Allom's engraving shows a girl awkwardly crouching in the British mill as the supervisor looked down at her work, the American photographer Lewis W. Hine, in his photograph *The Boss Teaches a Young Spinner in a North Carolina Mill* (1908), presented a benign view of a factory supervisor instructing a young girl at a spinning machine. Though known for his arresting, documentary images charting the plight of child labor, Hine here casts the supervisor in a kindly, paternalistic role.

89. James Geldard, *Hand-Book on Cotton Manufacture, or a Guide to Machine-Building, Spinning, and Weaving* (New York: John Wiley, 1867), 106.

90. William Cullen Bryant, *The Song of the Sower* (New York: Appleton, 1871), 29.

91. Wright, "Uncommon Mill Girls at Lowell," 17.

92. Harriet Farley, letter 3, in "Letters from Susan," *Lowell Offering* 4 (1844): 237–40, reprinted in *The Lowell Offering: Writings by New England Mill Women (1840–1845)*, ed. Benita Eisler (New York: Harper Colophone, 1977), 57.

93. Sarah G. Bagley, *Pleasures of Factory Life*, in *Lowell Offering* 1 (1840): 25–26, reprinted in ibid., 64.

94. *A Week in the Mill*, in *Lowell Offering* 5 (1845): 217–18, reprinted in ibid., 75.

95. Josephine L. Baker, *A Second Peep at Factory Life*, in *Lowell Offering* 5 (1845): 97–100, reprinted in ibid., 78.

96. In one of the earliest photographs of a woman at a factory loom, a daguerreotype taken in America about 1850, a female weaver in New England works at her machine, yet the photograph may have been an illustration of a new or improved model of the machine rather than a portrait. The journal *SIAN* noted in November 1978 that the fact that the woman appears alone with her loom suggests that she was photographed, not in a factory weaving room, but in the factory's machine shop using a new or improved model and that the lighting requirements for a daguerreotype may have necessitated the machine to the shop. The photograph appears in Helen Wright, "Machine Portrait," *History of Pho-*

tography 3, no. 2 (April 1979): 156, and in Hindle and Lubar, *Engines of Change*, 198.

97. Richard Oestreicher, "From Artisan to Consumer: Images of Workers, 1840–1920," *Journal of American Culture* 4, no. 1 (1981): 49, 51; Harry R. Rubenstein, "With Hammer in Hand: Working-Class Occupational Portraits," in *American Artisans: Crafting Social Identity, 1750–1850*, ed. Howard R. Rock et al. (Baltimore: Johns Hopkins Univ. Press, 1995), 140. Rubenstein argues that the poses were influenced by nineteenth-century graphic reorientations of workers seen in prints, images found in encyclopedias on trades, and labor-association graphics (190).

98. The painting reflects the Italian and Russian futurist artists' and writers' celebration of "dynamism" and the machine. In a later view, the Russian designer Valentina N. Kulagina—reflecting the ethos of the Soviet Union of the 1930s dedicated to promoting economic development, national duty, and idealized images of labor—created her own forceful image of a strong-armed female textile worker standing at her loom in a poster celebrating International Women Workers Day. With her chiseled, abstracted features, this mammoth woman becomes a powerful emblem of workers'—and women's—strength. During the same period, the American photographer Margaret Bourke-White photographed women working in Russian textile factories. In her photograph *The Woman Who Wept For Joy* she created a timeless, abstracted image of a woman stooped beside the frame of bobbins examining threads for breaks as they pass through the riders and the warpers. An interpreter told Bourke-White that the woman was crying with happiness because the photographer had chosen her "instead of any of the pretty young girls on the floor" (see Margaret Bourke-White, *Eyes on Russia* [New York: Simon & Schuster, 1931], 22–23, 55–57).

99. Illustration for Mary E. Wilkins's *Giles Corey, Yeoman*, in *Harper's New Monthly Magazine* 86 (December 1892), 23. The drama appears on pp. 20–40.

2. Wired for Fashion

1. In 1856 the *New Monthly Belle Assemblée* noted that some women were wearing sixteen petticoats for evening attire (Alison Carter, *Underwear: The Fashion History* [London: Batsford; New York: Drama Book, 1992], 45).

2. James Laver, *Taste and Fashion: From the French Revolution to the Present Day*, rev. ed. (London: Harrap, 1945), 49.

3. Mary Hillier, *Automata and Mechanical Toys: An Illustrated History* (1976; reprint, London: Bloomsbury, 1988), 91.

4. E. T. A. Hoffmann, *The Sandman*, in *Nachtstücken*, in *Eight Tales of Hoffmann*, trans. J. M. Cohen (London: Pan Books, 1952). Freud cites the German psychologist Ernst Jentsch, who gives as examples of the "uncanny" "doubts whether an apparently animate being is really alive" and "uncertainty whether a particular figure in the story is a human being or an automaton" (quoted in Sigmund Freud, *The Uncanny*, in *The Standard Edition of the Complete Psychological Works of Sigmund Freud*, ed. and trans. James Strachey, vol. 17 [London: Hogarth Press, 1955], 226–27). Freud, however, discounts this uncertainty as the central issue in the tale, focusing instead on images of eyes and castration themes.

5. Henri Bergson, *Le Rire: Essai sur la signification du comique* (1900), quoted in *Comedy: An Essay*, ed. Wylie Sypher (New York: Doubleday, 1956), 218.

6. C. Willett Cunnington, *Feminine Attitudes in the Nineteenth Century* (London: Heinemann, 1935), 233.

7. Green, *Light of the Home*, 122.

8. See ibid., 130, where Green writes that for unmarried women, corsets suggested their potential for childbearing; and Thorstein Veblen, "The Economic Theory of Women's Dress," *Popular Science Monthly* Nov. 1894: 198–203.

9. C. Willett Cunnington and Phyllis Cunnington, *The History of Underclothes*, rev. ed. (London: Faber & Faber, 1981), 98–99, 105.

10. James Laver, *Costume and Fashion: A Concise History* (rev. ed., 1985; reprint, New York: Thames & Hudson, 1988), 95–97. First published in the United States in 1969.

11. James Laver, *Clothes* (London: Burke, 1952), ch. 5.

12. *Ladies' Companion*, April 1856; Alison Gernsheim, *Fashion and Reality (1840–1914)* (London: Faber & Faber, 1963), reprinted as *Victorian and Edwardian Fashion: A Photographic Survey* (New York: Dover, 1981), 44.

13. Alison Gernsheim cites Milliet as the inventor of the cage crinoline in *Victorian and Edwardian Fashion*, 45.

14. Elizabeth Ewing, *Dress and Undress: A History of Women's Underwear* (New York: Drama Book Specialists, 1978); Cunnington and Cunnington, *History of Underclothes*, 104–5. In 1868 America's *Demorest's Monthly Magazine* reported that another contemporary invention, the sewing machine, was being used to sew tapes to wires, eliminating the need for bothersome metal fasteners, including clasps and buttons (*Demorest's Monthly Magazine* 14 [March 1868]: 175; Gernsheim, *Victorian and Edwardian Fashion*).

15. Laver, *Clothes*, 111; Carter, *Underwear*, 152.

16. "Douglas & Sherwood's Hoop Skirt Factory," *Harper's Weekly*, 29 January 1859, 68; "Thomson's Crown-Skirt Factory," ibid., 19 February 1859, 125. For similar figures on Thomson's London factory, see W. Born, "Crinoline and Bustle," *CIBA Review (Corset and Underwear Review)*, no. 46 (1943).

17. John Leander Bishop, *A History of American Manufactures from 1608–1860*, 3rd rev. and enl. ed. (Philadelphia: E. Young, 1868), 211.

18. Cunnington and Cunnington, *History of Underclothes*, 104, 112; Anne Buck, *Victorian Costume and Costume Accessories*, rev. 2nd ed. (Bedford: R. Bean, 1984), 47.

19. Laver, *Taste and Fashion*, 51–52, 98; Laver, *Costume and Fashion*, 179.

20. Edward Philpott, *Crinoline in Our Parks and Promenades from 1710 to 1864* (London, [1864]), cited in Cunnington and Cunnington, *History of Underclothes*, 105; Gernsheim, *Victorian and Edwardian Fashion*, 46.

21. In her memoirs, Lady Dorothy Nevill wrote about "that Monstrosity 'The Crinoline' which once came near to costing my life." As she stood near a fireplace, her skirt was set ablaze, but none of the women present could help her because their "enormous crinolines rendered them almost completely impotent to deal with the fire." She herself finally subdued the fire (*The Reminiscences of Lady Dorothy Nevill*, ed. Ralph Nevill [London: Thomas Nelson, 1906], 96).

22. Charles Eastlake, *Hints on Household Taste in Furniture, Upholstery, and Other Detail*, 2nd rev. ed. (London: Longmans, Green, 1869), 232–33.

23. *Punch*, 21 August 1858.

24. *Guardian*, 22 July 1713, cited in William B. Lord, *The Corset and the Crinoline: A Book of Modes and Costumes* (London: Ward, Lock, & Tyler, 1868), 110.

25. "The Farthingdale Reviewed; or More Work for the Cooper. A pan-

egyric on the late but most admirable invention of the hooped petticoat," cited in ibid., 218–19.

26. "The Rise and Fall of the Crinoline," *Harper's Bazaar*, 15 February 1868, 256, and 21 March 1868, 336.

27. George Meredith, *Ipswich Journal*, quoted in Duncan Crow, *The Victorian Woman* (New York: Stein & Day, 1971), 124.

28. Montaigne, *Essays*, ch. 15, bk. 2, "That Our Desires Are Augmented by Difficulty," in *The Complete Works of Montaigne: Essays, Travel Journal, Letters*, trans. Donald M. Frame (Stanford: Stanford Univ. Press, 1957), 465. The link between Montaigne's quotation and hooped skirts is cited in David Kunzle, *Fashion and Fetishism: A Social History of the Corset* (Totowa, N.J.: Rowman & Littlefield, 1982), 74.

29. *Punch*, 29 August 1857, 94.

30. Account from an unnamed newspaper ca. 1860, reporting on an article in the *Sentinelle* of Toulon, quoted in Laver, *Clothes*, 116.

31. Newspaper report ca. 1860 quoted in ibid., 116–17.

32. Gernsheim, *Victorian and Edwardian Fashion*, 44–45; Sarah Leavitt, *Victorians Unbutton'd: Registered Designs for Clothing, and Their Makers and Wearers, 1839–1900* (London: George Allen Unwin, 1986), 38–39. *Punch* cartoons satirizing the inflated-tube crinolines appeared in 1849 and continued in 1856 and 1857. A stereograph of a young woman in her air-tube crinoline appears in Carter, *Underwear*, 59.

33. Joseph Addison, *The Spectator*, 5 January 1709, quoted in Laver, *Clothes*, 108–9.

34. *Harper's Weekly*, 3 April 1858, 221.

35. Kunzle, *Fashion and Fetishism*, 70. Laver, *Clothes*, also comments on fertility and the French Revolution; and see Perrot, *Fashioning the Bourgeoisie*, 72, 110.

36. Bathrick, "Female Colossus," 80–85.

37. J. Carl Flügel, *The Psychology of Clothes* (1930; reprint, London: Hogarth Press and the Institute of Psycho-Analysis, 1966), 47.

38. Kunzle, *Fashion and Fetishism*, 77; Carter, *Underwear*, 20.

39. Carter, *Underwear*, 20, 27.

40. Ibid., 34, 36, 64; Green, *Light of the Home*, 120.

41. An advertisement for Izod's patent steam-molded corsets is illustrated in Carter, *Underwear*, 62. For more on Hogarth and his discussion of beautifully designed stays, see David Kunzle, "The Corset as Erotic Alchemy: From Rococo Galanterie to Montaut's Physiologies," in *Woman as Sex Object: Studies in Erotic Art, 1730–1970*, ed. Thomas B. Hess, and Linda Nochlin, in *Art News Annual* 38 (October 1972): 99, 101.

42. Valerie Steele, *Fashion and Eroticism* (New York: Oxford Univ. Press, 1985), 224, 161. Steele cautions, however, against stereotyping the social meanings of corsets, since women have worn them since the Renaissance.

43. "An Article on Corsets," *Godey's Lady's Book and Magazine* 68 (June 1864): 529.

44. Steele, *Fashion and Eroticism*, 174–75.

45. Kunzle, "Corset as Erotic Alchemy," 105.

46. Ibid., 105 ff. Kunzle dates Basset's print to the early 1790s. This and other examples of what he calls the "windlass theme" appear in ibid.

47. Lowell's illustration appeared in *Judge*, May 1917.

48. The phrase "bum rolls," from Ben Johnson's *The Poetaster*, is quoted in Laver, *Costume and Fashion*, 99. Walpole's comment appears in his letter of 1783 to the countess of Ossory, cited in Cunnington and Cunnington, *History of Underclothes*, 61.

49. Buck, *Victorian Costume and Costume Accessories*, 62, 87–88.

50. The 1888 advertisement, from *Young Ladies' Journal*, 1888, is illustrated in Carter, *Underwear*, 60. An advertisement for Health Braided Wire Bustles, manufactured by Weston & Wells in Philadelphia, is in the Warshaw Collection, Archives Center, National Museum of American History, Smithsonian Institution, Washington, D.C.

51. Laver, *Costume and Fashion*, 198.

52. Oskar Fischel and Max von Boehn, *Manners and Morals of the Nineteenth Century*, 4 vols., rev. and enl. ed. (1927; reprint, New York: Benjamin Blom, 1970), 2:161.

53. This painting is illustrated in Marie Simon, *Fashion in Art: The Second Empire and Impressionism*, trans. Edmund Jephcott (London: Zwemmer, 1995), 64.

54. Linda Nochlin, *The Politics of Vision: Essays on Nineteenth-Century Art and Society* (1989; reprint, New York: Harper-Collins, 1991), 173.

55. Ibid., 180.

56. Meyer Shapiro, "Seurat and 'La Grande Jatte,'" *Columbia Review* 17 (1935): 14–15.

57. *Journal des dames et des modes*, December 1811, quoted in Nora Waugh, *Corsets and Crinolines* (London: Batsford, 1964), 100.

3. The Electric Eve   1. Georg Matthias Bose, *L'Électricité, son origine et ses progrès: Poème traduit de l'Allemande . . .* (Leipzig: Chez les Heritiers Lankisch, 1754), thought to have been translated by Bose himself; originally published as *Die Electricität nach ihrer Entdeckung, und Fortgang, mit poetischer Feder entworffen, von George Mathias Bose* (Wittenberg: J. J. Uhlfelden, 1744). The French version is quoted and translated in J. L. Heilbron, *Electricity in the Seventeenth and Eighteenth Centuries: A Study of Early Modern Physics* (Berkeley and Los Angeles: Univ. of California Press, 1979), 267. Over a hundred years later, the New York journal *Electrical World* wryly reported that the king and queen of Spain were sending kisses to each other by telegraph and mocked this "dry method of transmitting kisses" (*Electrical World* 1 [23 July 1883]: 468).

2. Geoffrey V. Sutton, *Science for a Polite Society: Gender, Culture, and the Demonstration of Enlightenment* (Boulder, Colo.: Westview, 1995), 314–15; Sutton notes that Jean-Antoine Nollet in a treatise on electricity in 1753 attributed the electrified kiss to Bose (371 n. 21). See also Stig Ekelöf, ed., *Catalogue of Books and Papers in Electricity and Magnetism Belonging to the Institute for the History of Electricity* (Göteborg, Sweden, 1991), 85. My thanks to Dr. Ellen Kuhfeld, curator, and Elizabeth Ihrig, librarian, of the Bakken Museum and Library, Minneapolis, for their help in locating and annotating these materials.

3. Heilbron, *Electricity in the Seventeenth and Eighteenth Centuries*, 263, 267.

4. Sutton, *Science for a Polite Society*, 223–25.

5. Eusebio Sguario, *Dell'elettricismo . . .* (Naples: G. Ponzelli, 1747); Peter

Johann Windler, *Tentamina de causa electricitatis* (Naples: Ex Regia Typographia S. Porsile, 1747). Sguario's book was probably the first book on electricity written in Italian.

6. See also works by the eighteenth-century French artist Hubert Francois Gravelot, illustrated in Jacques Monnier-Raball et al., *Autour de l'électricité: Un siècle d'affiche et de design* (Lausanne: Editions de la Tour, 1990), 86–87. For another perspective on French art, see Rosi Huhn, "Kunst und Elektrizität," in Staatliche Kunsthalle Berlin, Neue Gesellschaft für Bildende Kunst (NGBK), *Absolut modern sein: zwischen Fahrrad und Fliessband: culture technique in Frankreich, 1889–1937*..., exh. cat. (Berlin: Elefanten, 1986), 329–40.

7. Albert Robida's *La Vie électrique* was first published in 1890 with *Le Vingt-ième Siècle*, for which his drawing of the woman was the frontispiece. The drawing appeared again titled *L'Électricité (la grande esclave)* in *Le Vingtième Siècle: La Vie électrique* (Paris: Librairie Illustrée, Montgredien, 1893), pt. 3.

8. His invited guests were his old classmates, members of the electrical engineering Society of Seventy-Seven in Newark (*Electrical Diabolerie*, undated pamphlet published by Hammer Archives, IEE [Institution of Electrical Engineers] Society, London).

9. As reported in ibid.; *New York World*, 3 January 1885; *Newark (N.J.) Daily Advertiser and Journal*, 3 January 1885; and *St. Augustine News*, 1 April 1888, 8, in Hammer Archives, National Museum of American History.

10. David E. Nye, "Electrifying Expositions, 1880–1939," in *Fair Representations: World's Fairs and the Modern World*, ed. Robert W. Rydell and Nancy E. Gwinn (Amsterdam: VU Univ. Press, 1994), 143.

11. In a more somber image of the mythic goddess, the Internationale Elektrotechnische Ausstellung (International Electromechanical Exhibition) in Frankfurt am Main in 1891 featured a performance of the ballet *Pandora oder Götter-Funken* (Pandora or the spark of the gods), with a libretto by William Hock, in which a female as an emblem of culture stood on a flower-decorated centerpiece bearing the names of Galvani and Volta. In the ballet, Prometheus brings intellectual force and fire to humanity, but Pandora, as the archetypal seducing female, tries to prevent men from using their force and intellectual power and opens a box that unleashes plague and pestilence. In the second act the battery is invented, and in the final scene, "The Victory of Light," the electric light itself is used to symbolize the reconciliation of man and woman, rationality and passion. The ballet is illustrated and discussed in *Eine Neue Zeit!* exh. cat. (Frankfurt am Main: Historisches Museum, 1991), 289–91.

12. In its story about the performance the journal speculated that the fifty lights in Beyval's costume were activated when the dancer stood on metal plates and that her manager-electrician operated electrical switches in the wings ("Une Chanteuse Electrique," *Electricity*, 20 July 1892, 9).

13. Thomas Edison, "Electricity Man's Slave," originally published in *New York Tribune*, reprinted in *Electrical World* 5 (24 January 1885): 8–9. See also Marvin, *When Old Technologies Were New*.

14. In the eighteenth century, Pierre Jean Etienne Mauduyt de la Varenne described a procedure for using electricity to help women start their suppressed menstrual periods that involved applying a metal rod to their bodies both over and under their clothes (*Mémoire sur les différentes manières d'administrer l'électricité* ... [Paris: Impr. Royal, 1786], 80, 83, illustrated near 301; Francis

Lowndes, *Observations on Medical Electricity* [London: D. Stuart, 1787], 40. For a broader study, see Margaret Rowbottom and Charles Suskind, *Electricity and Medicine* [San Francisco: San Francisco Press, 1984]).

15. Alonzo Rockwell and George Beard, *Medical Uses of Electricity* (1871), discussed in Green, *Light of the Home*, 137–41; Rachel P. Maines, *The Technology of Orgasm: "Hysteria," the Vibrator, and Women's Sexual Satisfaction* (Baltimore: Johns Hopkins Univ. Press, 1999).

16. "Electric Jewelry," *Scientific American*, 25 October 1879, 263.

17. "Luminous Jewels," *Electrical Review*, 24 January 1884, 3.

18. "Electric Jewelry," *Electrical World* 1 (4 August 1883): 488–89.

19. Auguste Villiers de l'Isle-Adam's *L'Ève future* has been edited and translated as *Tomorrow's Eve/Villiers de l'Isle-Adam* by Robert M. Adams (Urbana: Univ. of Illinois Press, 1982). All page references are to this English edition.

20. Ibid., 43, 29.

21. Ibid., 37, 43.

22. Ibid., 60–61.

23. See "The Traumas of Transport," ch. 1 of Wosk, *Breaking Frame*, on images of railroad and steamship disasters.

24. Villiers de l'Isle-Adam, *Tomorrow's Eve*, 64–65.

25. *Scientific American*, 26 April 1890, cover story; Hillier, *Automata and Mechanical Toys*, 93; Charles Bartholomew, *Mechanical Toys* (London: Hamlyn, 1979), 85–86; Walter L. Welch and Leah Brodbeck Stenzel Burt, *From Tinfoil to Stereo: The Acoustic Years of the Recording Industry, 1877–1929* (Gainesville: Univ. Press of Florida, 1994), 47–51.

26. Years later, improvements were made in French versions of the phonographic doll, which had a jeweled stylus instead of Edison's metal one, and during the 1920s celluloid rather than brown wax was used for the records, which made them less subject to wear (Neil Maken, *Hand-Cranked Phonographs: It All Started with Edison . . .* [Huntington Beach, Calif.: Promar, 1993], 6). Felicia Miller-Frank discusses contemporary attitudes towards Edison's phonograph and the mechanical capturing of the human voice in *The Mechanical Song: Women, Voice, and the Artificial in Nineteenth-Century French Narrative* (Palo Alto, Calif.: Stanford Univ. Press, 1995) and in "Edison's Recorded Angel," in *Jeering Dreamers: Villiers de l'Isle-Adam's L'Eve future at Our Fin de Siècle: A Collection of Essays*, ed. John Anzalone (Amsterdam: Rodopi, 1996). She suggests that Edison himself may have been the source of Villiers's idea for an android because in 1877 Edison constructed a voice-activated mechanical doll for his daughter, which created a furor in France when Edison reportedly got the idea by talking through his hat (Miller-Frank, "Edison's Recorded Angel," 149, citing Jacques Périault, *Mémoires de l'ombre et du son: Une archéologie de l'audio-visuel* [Paris: Flammarion, 1981], 203).

27. Quoted in Hillier, *Automata and Mechanical Toys*, 94.

28. In his *Discourse on Method* and *Philosophical Letters* Descartes had argued that human beings, except for having a mind or soul, were essentially machines analogous to the popular eighteenth-century automatons. La Mettrie, in his essay *L'Homme machine* (1748), extended the metaphor of humanity as a well-tuned mechanism akin to a self-functioning timepiece, writing, "The human body is a machine which winds its own springs. It is the living image of perpetual movement" (René Descartes, *Discourse on Method*, trans. F. E. Sutcliffe [Balti-

more: Penguin, 1968], 73–74; idem, *Philosophical Letters*, trans. Anthony Kenny [Oxford: Clarendon, 1970]; Julien Offray de La Mettrie, *Man a Machine*, trans. Gertrude Carman Bussy [La Salle, Ill.: Open Court, 1912], 135, 93).

29. The lady musician is described in Alfred Chapuis and Edmond Droz, *Automata: A Historical and Technological Study*, trans. Alec Reid (Neuchâtel, Switzerland: Éditions du Griffon; London: Batsford, 1958), 279–83. For more on automatons, see chap. 2 in Wosk, *Breaking Frame*.

30. Villiers de l'Isle-Adam, *Tomorrow's Eve*, 61.

31. For extensive discussions of nineteenth-century electroplating and electrotyping, see Wosk, *Breaking Frame*, chs. 3–5.

32. For comments by critics, see ibid.

33. *Art Journal*, American ed., 1876, 304.

34. Villiers de l'Isle-Adam, *Tomorrow's Eve*, 154.

35. Ibid., 54. The concept of the copy displacing the original, indeed becoming "better than real" and superior to the original, was a recurring theme in the late-twentieth-century discourse on electronic reproductions.

36. Nina Auerbach, *Woman and the Demon: The Life of a Victorian Myth* (Cambridge: Harvard Univ. Press, 1982), 64, 74, 107.

37. Villiers de l'Isle-Adam, *Tomorrow's Eve*, 86, 199.

38. Ibid., 64, 122–23.

39. Ibid., 68, 81, 133. This vision of Hadaly as slave or servant suggests nineteenth-century sculptural images of the woman enslaved, such as the popular marble sculpture *The Greek Slave* (1853), by the American artist Hiram Powers, which was mechanically duplicated. As Joy Kasson noted, idealized nineteenth-century figures of women often portrayed them as slaves or victims, figures that "affirmed a sentimental view of woman's vulnerability by their emphasis on captivity subjects." The image of women under duress softened the idea of female strength. Women's power, in the nineteenth-century view, lay in acquiescence and abnegation. Kasson adds, however, that there are also nineteenth-century fantasies of overthrow and reversal in Victorian literature in which the victim woman gains empowerment (*Marble Queens and Captives: Women in Nineteenth-Century American Sculpture* [New Haven: Yale Univ. Press, 1990], 242, 147).

40. See Kasson, *Marble Queens and Captives*, 243.

41. Col. Rookes Evelyn Crompton, *Reminiscences* (London: Constable, 1928), 109, quoted in Caroline Davidson, *A Woman's Work Is Never Done: A History of Housework in the British Isles, 1650–1950* (London: Chatto & Windus, 1982), 39–40.

42. The Electrical Association for Women, founded in 1924, lobbied for improvements in electrical appliances (Caroline Haslett, ed., *The Electrical Handbook for Women* [London: Electrical Association for Women and Hodder & Stoughton, 1934], 369). American utility companies, like Denver Gas and Electric from 1907 to 1909, also sent female home-service agents to help demonstrate the uses of electrical irons and stoves in the home (Mark H. Rose, *Cities of Light and Heat: Domesticating Gas and Electricity in Urban America* [University Park: Pennsylvania State Univ. Press, 1995], 73, 83–87, where Rose writes that gas and electric appliances were often defined as appropriate for women's use alone; Nye, *Electrifying America*, 265–76; James C. Williams, "Getting Housewives the Electric Message: Gender and Energy Marketing in the Early Twentieth Century," in *His*

and Hers: Gender, Consumption, and Technology, ed. Roger Horowitz and Arwen Mohun [Charlottesville: Univ. of Virginia Press, 1998], 93–113; Davidson, *A Woman's Work Is Never Done*, 34–36; Bowers, *History of Electric Light and Power*, 232–36. For a more recent study, see Bowers, *Lengthening the Day*).

43. Steven Zdatny, "Fashion, Technology, and Gender" (paper presented at the annual conference of the Society for the History of Technology, Pasadena, Calif., 1997), 4. This paper includes a bibliography.

44. Mary Chamot, *Goncharova: State Designs and Paintings* (London: Orekco, 1979), 54.

45. Sonia Delaunay, *Nous irons jusqu'au soleil* (Paris: Editions Robert Laffont, 1978), original French on 43. Delaunay's *Zig-Zag*, a two-dimensional neon-light relief sculpture, was awarded a first prize by the Société d'Électricité de France in the publicity mural competition. See also *Sonia Delaunay: A Retrospective*, exh. cat. (Buffalo, N.Y.: Albright-Knox Gallery, 1980), 82–83; and *Absolut modern sein*, 339.

46. Delaunay is said to have done many sketches for her 1914 *Electric Prisms* in the prewar years, and it was also one of several paintings done in 1913 and 1914 that she called by the same name (Stanley Baron with Jacques Damase, *Sonia Delaunay: The Life of an Artist* [London: Thames & Hudson, 1995], 43, 52–53). During the 1930s she produced advertising posters for Mica-tube, a light bulb and neon sign manufacturer, and her design *Zig-Zag* was also used as an advertising poster (*Sonia Delaunay: A Retrospective*, 82–83).

47. The poem is dedicated to W.S.M., her brother William Monroe, who managed a power plant. It appeared in Harriet Monroe, *The New Poetry: An Anthology* (1917), and was reprinted in later editions, including Harriet Monroe and Alice Corbin Henderson, eds., *The New Poetry: An Anthology of Twentieth-Century Verse in English* (New York: Macmillan, 1930), 351–54. Rather than being intimidated by technology, Monroe in *A Poet's Life* described how at age 16 she went with her family to visit the Centennial Exhibition in Philadelphia in 1876 and was enthralled by "the Corliss engine, turning its great wheels massively," a machine that "impressed me more vividly than the art gallery" (Harriet Monroe, *A Poet's Life: Seventy Years in a Changing World* [New York: Macmillan, 1938], 34).

48. The drawing appeared on the cover of the avant-garde journal *291* in June 1915. Picabia's mechanical "object-portraits" were inspired by both Apollinaire and Duchamp, as well as by illustrations in popular magazines. A similar image of this spark plug appeared, for example, as an advertisement for a "Red-Head" priming plug manufactured by the Emil Grossman Manufacturing Company in Brooklyn and advertised in *The Motor*, December 1914, 97 (see William A. Canfield, "The Mechanistic Style of Francis Picabia," *Art Bulletin* 48 [September–December 1966]; and William Innes Homer, "Picabia's *Jeune fille américaine dans l'état de nudité* and Her Friends," ibid. 57 [March 1975]: 110–15). Homer (115) noted that the word *forever* appeared in the text of the Grossman ad, which read, "Every part so good we can guarantee them forever." Debates continue over the meanings of the spark plug, which was variously viewed as an image of American culture dominated by technical values (by Picabia's wife), as an eroticized portrait of American womanhood, or as a reference to Agnes Meyer, friend of the Stieglitz group in New York and a patron of modern art (Homer, "Picabia's *Jeune fille américaine*," 110, 115). Caroline A. Jones, in her 1998 essay on

Picabia, drew links between the artist's hermaphroditic imagery, seen in the phallic spark plug, and his own struggles with neurasthenia ("The Sex of the Machine: Mechanomorphic Art, New Women, and Francis Picabia's Neurasthenic Cure," in *Picturing Science, Producing Art*, ed. Caroline A. Jones and Peter Galison [New York: Routledge, 1998], 145–80). Alfred Jarry, in his surreal novel *Le Surmâle (The Supermale)* (1902), like the dada artists, spoofs mechanized sex. The character compliantly agrees to participate in an experiment in which a version of the electric chair—which gives sudden shocks—enables sex to become mechanized and robotic. Picabia and Jarry are briefly discussed in National Museum of American Art, *Perpetual Motif: The Art of Man Ray*, exh. cat. (New York: Abbeville, 1988), 145, 147.

49. For more on Höch's *Mädchen*, see Nochlin, *Women, Art, and Power*, 28–29.

50. See Man Ray, *Électricité: Dix rayogrammes de Man Ray et un text de Pierre Bost* (Paris: La Compagnie Parisienne de Distribution d'Électricité, 1931), a series of ten photogravures of original *rayogrammes* by Man Ray.

51. Commissioned for distribution to executives and special customers, the portfolio was reproduced as gravures in an edition of five hundred (*Man Ray: Photographs from the J. Paul Getty Museum* [Los Angeles: J. Paul Getty Museum, 1998], 62; Foris M. Neususs and Renate Heyne, "The Rayographs," in *Man Ray: Photography and Its Double*, ed. Emmanuelle de l'Ecotais and Alain Sayag, exh. cat. [Corte Medera, Calif.: Gingko Press, 1998], 188).

52. In his text for Man Ray's *Électricité* Bost also argued that this push-button electrical society itself would become passé in the future.

53. Miller, who was introduced to Ray by a letter from photographer Edward Steichen, was also the model for Ray's photograph of an armless, classical Greek nude female torso in 1930 (National Museum of American Art, *Perpetual Motif*, 208, 210–11; see also Ecotais and Sayag, *Man Ray: Photography and Its Double*).

54. For these solarized images, electric light was turned on during the processing, resulting in images in which part appeared as a positive, part as a negative (National Museum of American Art, *Perpetual Motif*, 208–9, 212, 237).

55. Ibid., 211–12.

56. "La Fée électricité," *L'Art vivant*, no. 228 (special issue, 1937): 70–71. See also *Absolut modern sein*, 331.

57. By 1984, though, thoughts of taming technology were far from celebrations of electricity and electronics in art, as in the massive exhibition *Electra: L'Électricité et l'électronique dans l'art au XXe siècle*, held in 1983 at the Musée d'art moderne de la Ville de Paris, where the female goddess served as muse. Katherine Dieckmann, in her review essay, complained about the utopian celebration of electronic technologies and the lack of probing critique ("Electra Myths: Video, Modernism, Postmodernism," *Art in America* 73 [fall 1985]: 195–204).

58. In a 1999 interview for this book Burson stated that she was pregnant when she created the photographs and that the images reflect her own concerns about her unborn child. For more on Burson's art, see Nancy Burson with Richard Carling and David Kramling, *Composites: Computer-Generated Portraits* (New York: William Morrow, 1986).

59. Donna J. Haraway, *Simians, Cyborgs, and Women: The Reinvention of Nature* (London: Free Association Books, 1991), 152, 154, 157.

4. Women and the Bicycle

1. *Wheelwoman*, 23 May 1896, quoted in Andrew Ritchie, *King of the Road* (London: Wildwood House; Berkeley: Ten Speed Press, 1975), 160.

2. Pryor Dodge, *The Bicycle* (Paris: Flammarion, 1996), 20.

3. Ibid., 18.

4. Ritchie, *King of the Road*, 146. Ritchie provides a useful overview of responses to the bicycle in nineteenth-century magazines.

5. In contemporary accounts, male riders told of the exertion required to learn to ride a velocipede and of the bruises and scratches they had received in doing so (see John Woodforde, *The Story of the Bicycle* [London: Routledge & Kegan Paul, 1977], 17–34). Pickering & Davis, in New York, manufactured a lady's model with a slightly dropped frame that made riding somewhat easier (see Ritchie, *King of the Road*, 150; Robert A. Smith, *A Social History of the Bicycle: Its Early Life and Times in America* [New York: American Heritage Press, 1972], 5, 7; and Dodge, *Bicycle*, 40).

6. Ritchie, *King of the Road*, 149.

7. Dodge, *Bicycle*, 4, 149; James McGurn, *On Your Bicycle: An Illustrated History* (New York: Facts on File, 1987), 41–42; *Harper's Weekly*, 13 February 1869.

8. According to Pryor Dodge, between 1868 and 1870 there were three women's races in Belgium and twenty in France (*Bicycle*, 50, 136).

9. *Le Monde illustré*, 21 November 1868; "Nouveau Steeple Chase," *Harper's Weekly*, 19 December 1868, 12; while not citing *Harper's* specifically, McGurn points out the discrepancy (*On Your Bicycle*).

10. Joseph Firth Bottomley, *The Velocipede: Its Past, Its Present and Its Future* (London: Simpkins, Marshall, 1869), 73–77.

11. Ibid., 80–81.

12. Dodge writes that d'Antigny was also the inspiration for Zola's character Nana (*Bicycle*, 38).

13. Ritchie, *King of the Road*, 151. The waning popularity of these machines in America was due in part to their being banned from sidewalks in many cities and to the rough condition of streets. My thanks to Roger White for this information.

14. A four-wheeled version was also introduced in 1877. Though there had been earlier versions of tricycles, technical developments in the 1870s helped make the machine popular (Dodge, *Bicycle*, 76).

15. Wiebe E. Bijker, *Of Bicycles, Bakelites, and Bulbs: Toward a Theory of Sociotechnical Change* (Cambridge: MIT Press, 1995), 59; Dodge, *Bicycle*, 73.

16. *Harper's Weekly*, 17 July 1886.

17. The company promoted this twelve-color lithograph as a work of art, "the most attractive bicycling picture ever published," and offered to send a version suitable for hanging, mounted on cloth and with decorated brass ferrules on the top and bottom, for five two-cent stamps. The poster with its text is illustrated in Ross D. Petty, "Peddling the Bicycle in the 1890s: Mass Marketing Shifts into High Gear," *Journal of Macromarketing* 15, no. 1 (1995): 35.

18. David V. Herlihy, "The Bicycle Story," *American Heritage of Invention and Technology* 7 (spring 1992): 55–58; Ritchie, *King of the Road*, 155; Smith, *Social History of the Bicycle*, 13–14; Dodge, *Bicycle*, 96–110.

19. "The World Awheel," *Munsey's*, May 1896, 131, 157. Ellen Gruber Gar-

vey attributed these favorable stories about bicycling for women in part to the fact that approximately 10 percent of advertisers in American middle-class magazines were related to bicycle manufacture. Mindful of this fact, fictional stories in special bicycling issues often helped configure bicycling as socially accepted and approved behavior (Garvey, *Adman in the Parlor*, 107; idem, "Reframing the Bicycle: Advertising-Supported Magazines and Scorching Women," *American Quarterly* 47 [March 1995]: 82).

20. "World Awheel," 159.

21. Anne O'Hagen, "The Athletic Girl," *Munsey's*, August 1901, 737. Among sports practiced by women — golf, tennis, fencing, basketball, horseback riding — the magazine cited the drop-frame bicycle for doing "more than anything else to admit women to the world of outdoor sport."

22. Green, *The Light of the Home*, 137 (quotation); George Beard, *American Nervousness: Its Causes and Consequences* (New York: G. P. Putnam's Sons, 1881); John S. Haller Jr., "Neurasthenia: The Medical Profession and the 'New Woman' of the Late Nineteenth Century," *New York State Journal of Medicine* 71 (15 February 1971): 474. Haller, neurasthenic symptoms, and the sexual etiology of neurasthenia are discussed in Rachel P. Maines, *The Technology of Orgasm: "Hysteria," the Vibrator, and Women's Sexual Satisfaction* (Baltimore: Johns Hopkins Univ. Press, 1999), 34–35.

23. Lillias Campbell Davidson, *Handbook for Lady Cyclists* (London: H. Nisbet, 1896).

24. Ibid., 86–87.

25. Ibid., 119, 86–87.

26. Ibid., 87, 90.

27. "World Awheel," 159.

28. Ibid., 157–58.

29. On the chaperone issue, see Garvey, *Adman in the Parlor*, 124.

30. A Chaperone Cyclists' Association ad in *The Queen* in 1896 stated that "many ladies have a great objection to their daughters cycling without a proper and efficient escort" (Dodge, *Bicycle*, 133. See also Davidson, *Handbook for Lady Cyclists*, 36, 75, 76, 120).

31. Lillias Campbell Davidson, "Cycling for Ladies," in *Cycles and Cycling*, by H. Hewitt Griffin, 2nd ed. (London: George Bell, 1893), 119.

32. Ibid.

33. *Outing*, June 1896; *Harper's Weekly*, 21 May 1898, 484.

34. "World Awheel," 159.

35. *Cycling*, 8 October 1892.

36. Davidson, *Handbook for Lady Cyclists*.

37. See a Remington bicycle advertisement published in *Harper's Weekly*, 13 June 1896, illustrated and discussed in Garvey, "Reframing the Bicycle," 75, 77; see also Garvey, *Adman in the Parlor*, 116.

38. *Cycling*, 7 October 1893.

39. Garvey, *Adman in the Parlor*, 117–20; Garvey, "Reframing the Bicycle," 74–78. Garvey notes that the medical discourse against women's riding bicycles appeared largely in the medical press.

40. Humber Bicycle advertisement, 1895, in Jack Rennert, *One Hundred Years of Bicycle Posters* (New York: Harper & Row, 1973): 40. See also Nick Sanders, *Bicycle: The Image and the Dream* (Derbyshire: Red Bus, 1991), 129, which

makes the comment about riding one-handed; Sanders cited Petty, "Peddling the Bicycle in the 1890s," 38.

41. Green, *Light of the Home*, 162.

42. *Cycling*, 8 March 1905, illustrated in *The George Moore Collection*, vol. 6 (London: Beekay, 1985), 62.

43. For a discussion of the gigantic woman and bicycles, see Patricia Marks, *Bicycles, Bangs, and Bloomers: The New Woman in the Popular Press* (Lexington: Univ. Press of Kentucky, 1990), 197.

44. *Puck*, 4 December 1895, German-language edition.

45. Marks, *Bicycles, Bangs, and Bloomers*, 174–78, 197.

46. "World Awheel," 158.

47. Reportedly the first reference to the "New Woman" was made by the radical novelist Sarah Grand in "The New Aspect of the Woman Question," *North American Review*, March 1894, and the concept was widely satirized in plays by Sidney Grundy, George Bernard Shaw, and others. For further discussions of the "New Woman" in literature, see Jean Chothia, introduction to *The New Woman and Other Emancipated Woman Plays*, ed. Chothia (Oxford: Oxford Univ. Press, 1998); and introduction by Viv Gardner to *The New Woman and Her Sisters: Feminism and Theatre, 1850–1914*, ed. Viv Gardner and S. Rutherford (New York: Harvester Wheatsheaf, 1992).

48. Davidson, *Handbook for Lady Cyclists*, 27–28.

49. Ritchie, *King of the Road*, 151.

50. See Bijker, *Of Bicycles, Bakelites, and Bulbs*, 2–3, 277, for a discussion of the interrelations between bicycle design and women's clothing.

51. Marguerite Merington, "Woman and the Bicycle," *Scribner's* 17 (June 1895): 703; idem, "Woman and the Bicycle," in *Athletic Sports*, ed. D. A. Sargent et al. (New York: Charles Scribner's Sons, Out of Door Library, 1897), 211–12.

52. *Life*, June 1897, 512, quoted in Marks, *Bicycles, Bangs, and Bloomers*, 193.

53. The quotation from *Cycling* magazine appears in Smith, *Social History of the Bicycle*, 100; see also "World Awheel," 159.

54. "World Awheel," 159.

55. *Life*, 29 August 1895.

56. Dodge, *Bicycle*, 128; "How Cyclists Used to Dress," *Boneshaker* 5 (spring 1968): 192–93. My thanks to Pryor Dodge for drawing my attention to this article.

57. Miss F. J. [Fannie] Erskine, *Lady Cycling* (London: Walter Scott, 1897), 19–23, 77–78.

58. Davidson, "Cycling for Ladies," 116.

59. Ibid., 114–16.

60. Merington, "Woman and the Bicycle," *Scribner's* 17 (June 1895): 703.

61. Ibid.

62. Davidson, "Cycling for Ladies," 112.

63. Ibid., 112–13.

64. Davidson, *Handbook for Lady Cyclists*, 115.

65. Davidson, "Cycling for Ladies," 114.

66. Davidson, *Handbook for Lady Cyclists*, 94–95.

67. Erskine, *Lady Cycling*, 115, 126.

68. Maria Ward, *Bicycling for Ladies* (New York: Brentano's, 1896).

69. Ibid., ix–x.

70. Ibid., 112.

71. Frances E. Willard, *A Wheel within a Wheel: How I Learned to Ride the Bicycle, With Some Reflections By the Way* (Chicago: Women's Temperance Publishing Association, 1895), revised and retitled *How I Learned to Ride the Bicycle: Reflections of an Influential Nineteenth Century Woman*, ed. Carol O'Hare (Sunnyvale, Calif.: Fair Oaks, 1991), 74–75. All references are to this edition.

72. Willard, *A Wheel within a Wheel*, 24. Dodge notes that women were velocipede instructors in Boston schools (Dodge, *Bicycle*, 44).

73. Willard, *A Wheel within a Wheel*, 76. In this final photograph Willard's male riding assistant appears as a small figure receding in the distance.

74. Susan B. Anthony, quoted in Dodge, *Bicycle*, 130; Frances E. Willard, quoted in Garvey, *Adman in the Parlor*, 112–13 (see also her n. 14).

5. Women and the Automobile

1. Gilles Néret, *Tamara de Lempicka, 1898–1980* (Cologne: Benedikt Taschen, 1993), 7. A more recent study clarifies that the artist was born in Moscow, not Warsaw as she claimed (see Laura Claridge, *Tamara de Lempicka: A Life of Deco and Decadence* [New York: Clarkson Potter, 1999]).

2. In America, the first automobile licensing took place at different times in the varying cities and states. For comments on early registration and licensing statistics in Los Angeles and Tucson, see Virginia Scharff, *Taking the Wheel: Women and the Coming of the Motor Age* (New York: Free Press, 1991), 25–26. Clay McShane provides early statistics on automobiles registered in women's names in New Hampshire and Maryland (which included the District of Columbia) and notes that in Maryland in 1905 only 1.8 percent of automobiles were registered to women (*Down the Asphalt Path: The Automobile and the American City* [New York: Columbia Univ. Press, 1994], 259).

3. Patrick Robertson, *The Shell Book of Firsts* (London: Ebury, 1974), 205.

4. *Automobile Club Journal* (London) 8 (11 August 1904).

5. Motor Vehicle Manufacturers' Association, *Automobiles of America* (Detroit: Wayne State Univ. Press, 1974), 21, cited in Scharff, *Taking the Wheel*, 25. For statistics on Chicago women, see *Automobile Topics* (New York), 1, no. 17 (9 February 1901): 614; and *Automobile Topics*, ca. 1902, reprinted in *Car Illustrated* (London), 25 June 1902, 172.

6. Milton Lehman, "The First Woman Driver," *Life*, 8 September 1952, 83–92.

7. Ibid.

8. "A Lady's Easter Tour," *Autocar* 6 (27 April 1901): 390; "A Lady's Tour of 1,500 Miles," ibid. 6 (11 May 1901): 446. A large photograph of Butler in her Renault in England appears in Butler's own story, "Cars and How to Drive Them. No. X—The Renault," *Car Illustrated*, 3 September 1902, 57–58; the story reveals her substantial automotive knowledge.

9. "An Expert Lady Driver," *Car Illustrated*, 18 June 1902, 123.

10. Cecelia, "My Motoring Diary," ibid., 27 August 1902, 13.

11. "Women Autoists Skillful Drivers. Not Content With Electrics, Now Using High-Powered Gasoline Cars," *New York Times*, 29 September 1907, sec. 4, 3.

12. Ibid.

13. *Automobile Review*, November 1899; *New York Herald Tribune*, 2 July

1899, cited in Rudolph E. Anderson, *The Story of the American Automobile* ([Washington, D.C.]: Public Affairs Press, 1950), 192, 193, 197.

14. Edwin Wildman, "The City of the Automobile," *Munsey's*, February 1900, 704–12.

15. New York's *Automobile Topics* chose the same image for the cover of its first issue. For comments on the Automobile Club of America, see Scharff, *Taking the Wheel*, 68. On American women's efforts to form their own club, see "Women Plan Auto Club To Be Patterned After Ladies' Automobile Club of London," *New York Times*, 10 November 1903, 3.

16. See "Cars at Stafford House," *Car Illustrated*, 25 June 1902, 165, describing the duchess of Sutherland.

17. "An Expert Lady Driver," 123.

18. See Maurice Hamel, *Les Salons de 1901* (Paris: Goupil, 1901); the painting is also illustrated in Gilles Néret and Hervé Poulain, *L'art, la femme, et l'automobile* (Paris: Editions E.P.A., 1989), 170.

19. "The Automobile in Painting," *Automobile Topics* 2, no. 3 (4 May 1901): 78. The title of the painting is unknown.

20. Patricia K. Webster, quoted in "Ready for War Call: Motor Schools and Classes Prepare — Feminine Auxiliaries to Aid U.S.," *Motor Age*, 19 April 1917, 38–39. Though Webster counters the idea of feminine fragility, she also makes reference to a stereotypical feminine ailment, writing that the automobile "cannot help but have a beneficial effect on the nerves."

21. *Ladies' Home Journal*, August 1924, cited in Jennifer Scanlon, *Inarticulate Longings: The Ladies' Home Journal, Gender, and the Promises of Consumer Culture* (New York: Routledge, 1995), 48.

22. Mrs. Edward Kennard, "Motor Bicycling for Ladies," *Car Illustrated*, 27 August 1902, 16; Lord Montagu of Beaulieu, *Lost Causes of Motoring* (London: Cassell, 1969), 203; Mrs. Edward Kennard, *Car Illustrated*, 25 June 1902, 172.

23. Michael Brian Schiffer, *Taking Charge: The Electric Automobile in America* (Washington, D.C.: Smithsonian Institution Press, 1994), 122.

24. Scharff, *Taking the Wheel*, 52–55. See also Ronald Kline and Trevor Pinch, "Users as Agents of Technology Change: The Social Construction of the Automobile in the Rural United States," *Technology and Culture* 37 (October 1996): 779; and McShane, *Down the Asphalt Path*, 163.

25. Julie Lynn Wilchins, "The Squeaky Wheel Gets the Grease: Women, the Automobile, and the Fight for Suffrage, 1900–1920" (senior honors thesis, Harvard University, 1994), n. 1.

26. "Women Autoists Skillful Drivers." See also *New York Times*, 12 July 1905, cited in Scharff, *Taking the Wheel*, 73.

27. Ronald Kline and Trevor Pinch survey the contradictory evidence about the role of automobiles in furthering the mobility of rural women in America. Some commentators argue that women used their autos for visiting, loosening their ties to the farmstead. Others argue that by providing women with an easier means, for example, to shop for domestic goods, the automobile simply reinforced conventional gender roles for both rural and suburban women (see Kline and Pinch, "Users as Agents," 780–81, citing Michael L. Berger, *The Devil Wagon in God's Country: The Automobile and Social Change in Rural America, 1893–1929* [Hamden, Ct.: Archon Books, 1979], ch. 2; Scharff, *Taking the Wheel*, 142–45;

Jellison, *Entitled to Power*, 122–24; and Mary Neth, *Preserving the Family Farm: Women, Community, and the Foundations of Agribusiness in the Midwest, 1900–1940* [Baltimore: Johns Hopkins Univ. Press, 1995], 246–47).

28. See Virginia Scharff, "Of Parking Spaces and Women's Places: The Los Angeles Parking Ban of 1920," *N[ational] W[omen's] S[tudies] A[ssociation] Journal* 1, no. 1 (1988): 50; and idem, *Taking the Wheel*, 142–45, 150.

29. Levitt, *Woman and the Car*, 86–87.

30. "Women Auto Drivers in Endurance Run," *New York Times*, 31 October 1909, sec. 4, 4; Scharff, *Taking the Wheel*, 76–77.

31. Claudia M. Oakes discusses the controversy over the title of first American female aviator and the role of Bessica Faith Raiche, another contender, in *United States Women in Aviation through World War I*, Smithsonian Studies in Air and Space, 2 (Washington, D.C.: Smithsonian Institution Press, 1978), 17–19, 27.

32. "Cross-Country Run Ends," *Motor Age*, 4 August 1910, 25.

33. "Woman to Drive Auto to Frisco," *New York Times*, 15 March 1910, sec. 4, 4.

34. Ibid.

35. British women over age thirty gained suffrage in 1918, followed by those over twenty in 1921. They were less apt than American suffragists to use automobiles in publicity campaigns.

36. Scharff, *Taking the Wheel*, 79.

37. Schiffer, *Taking Charge*, 122.

38. "Suffragists Storm National Capital," *New York Times*, 19 April 1910, 1. For useful discussions of the automobile and the American women's suffrage movement, see Scharff, *Taking the Wheel*; and Wilchins, "Squeaky Wheel Gets the Grease," ch. 3, "The Road to the Polls: The Automobile and the Fight for Female Suffrage." Both Scharff and Wilchins provide useful references to *New York Times* coverage.

39. "Big Auto Suffrage Parade," *New York Times*, 30 July 1914, 9.

40. "Truck for Suffragists: Big Motor Will Carry Leaders to Meetings in City Squares," ibid., 20 July 1914, 7.

41. "Suffrage Auto Campaign," ibid., 2 April 1916, sec. 1, 21.

42. Ibid.; "Suffrage Autoists Motor 10,700 Miles: Mrs. Burke and Miss Richardson Back in City after a Tour to Pacific Coast," *New York Times*, 1 October 1916, sec. 1, 19.

43. Ida Husted Harper, ed., *History of Woman Suffrage*, vol. 5 (New York: National American Woman Suffrage Association, 1922): 481–82.

44. Michael L. Berger, "Woman Drivers! The Emergence of Folklore and Stereotypic Opinions concerning Feminine Automotive Behavior," *Women's Studies International Forum* 9, no. 3 (1986): 261, 258. Berger argued, however, that electric automobiles were not considered a threat to men since the electric, being clean and easy to operate, was considered a woman's car.

45. *Automobile Topics*, 1902, reprinted in *Car Illustrated*, 25 June 1902, 172.

46. "Women Automobile Drivers in Endurance Run," 4.

47. Scharff, *Taking the Wheel*, 75.

48. Winifred M. Pink, "Motor Racing for Women," *Woman Engineer* (London) 2 (December 1928): 325.

49. "A Speed Demon Reappears, a Ghost Now at the Wheel," *New York*

Times, 27 April 1997, sec. H, 39. Nice's Grand Prix career was short, however: while driving an Alfa Romeo Monza in a Grand Prix in São Paolo, Brazil, in July 1936, she swerved to avoid a collision, lost control, and killed six spectators. Although she continued racing until 1951, she did not race professionally again (see *Classic and Sports Car*, June 1997, 78–81).

50. Berger, "Woman Drivers!" 259–60 ff.

51. "Beauty at the Helm: Mrs. Harold Harmsworth," *Car Illustrated*, 2 July 1902, 191.

52. "Another Lady Expert: Mrs. S. F. Edge," ibid., 9 July 1902, 224.

53. "Women and Their Cars," *Vogue*, 15 October 1949, 93–94, 96–97, quotations on 93.

54. Towle, "Woman at the Wheel," 214.

55. Ibid., 214, 217–20, 223.

56. Mrs. Edward Kennard, "Motor Bicycling for Ladies. — III," *Car Illustrated*, 29 October 1902, 314.

57. Anderson, *Story of the American Automobile*, 203, cites an advertisement for the Oldsmobile in the *Ladies' Home Journal* in 1903 that portrays it as the first one with a special appeal for women, promoting it as an "ideal vehicle for shopping and calling."

58. Scharff, *Taking the Wheel*, 36, 42.

59. Chauffeurs often were given the task of charging and maintaining the automobile batteries (General Electric, *Charging the "Electric" at Home with the Mercury Arc Rectifier* [Schenectady, N.Y., March 1911], 3).

60. Scharff, *Taking the Wheel*, 150. Gerald Silk, however, argued that the causal relationship between the development of the electric self-starters and a wish to appeal to a female market is difficult to substantiate (see *Automobile and Culture* [New York: Harry Abrams; Los Angeles: Museum of Contemporary Art, 1984], 216).

61. Marchand, *Advertising the American Dream*, 185. Erving Goffman, in *Gender Advertisements* (Cambridge: Harvard Univ. Press, 1979), commented on these awkward poses as suggestive of women's social subordination.

62. "Forces That Buy," *Automobile* 30 (5 February 1914): 382, quoted in James J. Flink, *The Car Culture* (Cambridge: MIT Press, 1975), 145.

63. Cynthia Cockburn and Susan Ormrod, in *Gender and Technology in the Making*, 6–7, make a useful distinction between "projected identities" (like that of the housewife portrayed and constructed by appliance designers and advertisers) and "subjective identities" (the gendered sense of self created and experienced by the individual).

64. Mrs. A. Sherman Hitchcock, "Woman at the Motor Wheel," *American Homes and Gardens* 10 (suppl. 6, April 1913).

65. Levitt, *Woman and the Car*, 31.

66. S. C. H. Davis also admired Levitt for managing to look dirt-free (*Atalanta: Women as Racing Drivers* [London: G. T. Foulis, (195–)], 32–33, 40).

67. Levitt, quoted by C. Byng-Hall in "Dorothy Levitt: A Personal Sketch," introduction to Levitt, *Woman and the Car*, 9.

68. Byng-Hall, "Dorothy Levitt," 4, 7.

69. Ibid., 4.

70. Davis, *Atalanta*, 32–33, 40.

71. Levitt, *Woman and the Car*, 25–26.

72. Ibid., 86–87.

73. Ibid., 41.

74. Ibid., 51–52, 87.

75. For more on Nicholls, see chapter 7.

76. "Little Things about a Car That Every Woman Who Means to Drive One Ought to Know," *Ladies' Home Journal*, March 1917, 32.

77. Ann Murdock, "The Girl Who Drives a Car," ibid., July 1915, 11.

78. Ibid.

79. Scharff, *Taking the Wheel*, 91–92.

80. Diane Atkinson, *Suffragettes in Pictures* (Stroud, Gloucestershire: Sutton; London: Museum of London, 1996), 178.

81. Diana Condell and Jean Liddiard, comps., *Working for Victory? Images of Women in the First World War, 1914–18* (London: Routledge & Kegan Paul, 1987), 7, 100.

82. Mrs. H. M. Usborne, ed., *Women's Work in Wartime: A Handbook of Employments* (London: T. Werner Laurie, 1917), 81.

83. *Sphere*, 4 May 1918, 87.

84. Jellison, *Entitled to Power*, chs. 5 and 6; Mabel E. Deutrich and Virginia C. Purdy, eds., *Clio Was a Woman: Studies in the History of American Women* (Washington, D.C.: Howard Univ. Press, 1980), 136–46.

85. Scharff, *Taking the Wheel*, 92–97; Dorothy Schneider and Carl J. Schneider, *Into the Breach: American Women Overseas in World War I* (New York: Viking, 1991), 99. The Schneiders cite as a notable example Helen Douglas Mankin, who drove an ambulance in France and later spent three years in the United States Congress (101).

86. Oakes, *United States Women in Aviation through World War I*, 33–36.

87. Scharff, *Taking the Wheel*, 96–97; Gertrude Stein, *The Autobiography of Alice B. Toklas* (New York: Harcourt Brace, 1933), 214–35. "Auntie" was named after Stein's Aunt Pauline, and Stein called the next car "Godiva" because, as she explained, the car had "come naked into the world" and friends had given them something with which to bedeck it (235).

88. Scharff, *Taking the Wheel*, 100; "Ready for War Call," 38–39.

89. "Woman's Work in the World War," *Motor Age* 31 (26 April 1917): 32.

90. "Ready for War Call," 38–39. See also "The American Doughgirls," *New York Times*, 1 October 1918, 12.

91. As noted by Jellison, *Entitled to Power*, and others, after the war driving a tractor was still considered man's work.

92. Laura Breckinridge McClintock, "Why Take a Man Along?" *Motor* 40 (July 1923): 25, 92, 94.

93. Lowry, "Woman's Place Is in the Tonneau," 34.

94. Marie Russell Ullman, "Of Course It's a Woman!" *Motor* 41 (December 1923–March 1924): 27.

95. *Saturday Evening Post*, 22 March 1966, inside front cover.

96. *Women and Automobiles: A Study of Dealers and Consumers in the Car Purchasing Process* (n.p.: A Joint Project of Woman's Day Magazine, National Automobile Dealers Association and Audits & Surveys, Inc., 1979), cited in Charles L. Sanford, "'Woman's Place' in American Car Culture," in *The Automobile and American Culture*, ed. David L. Lewis and Laurence Goldstein (Ann Arbor: Univ.

of Michigan Press, 1983), 150. Sanford's essay includes an examination of sex-role typing and stereotyping in American car advertisements during the 1970s.

97. Richard Langworth, *Illustrated Dodge Buyer's Guide* (Osceola, Wis.: Motorbooks International, 1995), 53–57; Ron Kawalke, ed., *Standard Catalog of American Cars, 1946–1975*, 4th ed. (Iola, Wis.: Krause, 1997), 334–36; Chrysler Corporation, 1955 sales brochure for the Dodge La Femme.

98. *Sonia Delaunay: A Retrospective*, 79, 87.

6. Women and Aviation

1. Louise McPhetridge Thaden, *High, Wide, and Frightened* (New York: Stackpole Sons, 1938), 138.

2. Claudia M. Oakes, *United States Women in Aviation, 1930–1939* (Washington, D.C.: Smithsonian Institution Press, 1991), 1, 4.

3. Marjorie Stinson, "The Diary of a Country Girl at Flying School," *Aero Digest* 12 (February 1928): 168.

4. Ibid.

5. Oakes, *United States Women in Aviation, 1930–1939*, 24.

6. The theme was a common one in contemporary automobile advertisements. Early European posters advertising aviation meets depict airplanes flying over emblems of aesthetic and technological achievement; for example, one shows the French pilot Leon Morane flying over the Rouen Cathedral, the airplane becoming the capstone in the history of creative invention.

7. See *Sonia Delaunay: A Retrospective*, 85. For photographs of Delaunay's murals on site, see Jacques Damase, *Sonia Delaunay: Rhythms and Colors* (Greenwich, Conn.: New York Graphic Society; London: Thames & Hudson, 1972), 247.

8. For an overview of artists using aerial imagery, including Sonia Delaunay and Elsie Driggs, see Julie Wosk, "The Aeroplane in Art," *Art and Artists* (London), no. 219 (December 1984): 23–28.

9. My thanks to John A. Bagley, former curator of aviation at the Science Museum, London, for first alerting me to the probable identity of "Gamy." In his study *Art and the Automobile*, D. B. Tubbs notes that Ernest Montaut set up his own firm in Paris, and that after his death in 1909 the firm hired a number of artists, including one named "Gamy," which may be an anagram of Magy, for Marguerite Montaut (New York: Grosset & Dunlap, 1978), 37, 39.

10. A similar image, Gamy's lithograph *Voyage de Buc à Chartres*, shows a peasant girl in wooden shoes kneeling by a stream as she pauses in her task of washing clothes to gaze up at an aircraft flying over Gaillardon.

11. *Hannah Höch*, exh. cat. (Paris: Musée d'art moderne de la Ville de Paris, 1976), 11, cited in Dawn Ades, *Photomontage*, rev. ed. (London: Thames & Hudson, 1986), 20. Maud Lavin, in *Cut with the Kitchen Knife: The Weimar Photomontage of Hannah Höch* (New Haven: Yale Univ. Press, 1993, 37), suggests that in Weimar photocollages, images associated with aerial photography often read as a "tribute to technology and corresponding new ways of seeing."

12. Grahame-White wrote in a 1911 newspaper article that he regretted teaching women to fly because he feared that in an emergency they would panic and lose control, perhaps leading to their death (Oakes, *United States Women in Aviation through World War I*, 1).

13. Oakes, *United States Women in Aviation, 1930–1939*, 11, 13.

14. Mrs. Elliott-Lynn [Lady Sophie Heath], "Engineering and Aviation," *Woman Engineer* 2 (March 1927): 198–202. Women of the period were, however, admitted to the Royal Aeronautical Society.

15. Amelia Earhart, *The Fun of It: Random Records of My Own Flying and of Women in Aviation* (New York: Harcourt Brace, 1932), 179.

16. Bruce Gould, "Milady Takes the Air," *North American Review*, December 1929, 691–92.

17. Fred L. Hattoom, "Teaching Women to Fly," *Aero Digest* 16 (March 1930): 63.

18. Ibid., 64.

19. Ibid.

20. In her 1914 diary Marjorie Stinson recorded her wishes to fly solo. About her instructor Howard Rinehart she wrote, "Rinehart is simply glued to the plane. I don't need him at all, but it isn't flying etiquette to tell him so." And she added, "I think that if I could persuade this man Rinehart to get out of the plane, I could fly it and land it myself" (Stinson, "Diary of a Country Girl at Flying School," 296, 169). Louise McPhetridge Thaden, "Training Women Pilots," *Western Flying*, 9 (February 1931): 22, cited in Oakes, *United States Women in Aviation, 1930–1939*, 7.

21. Hattoom, "Teaching Women to Fly," 64.

22. Oakes, *United States Women in Aviation, 1930–1939*, 30–31.

23. Blanche Stuart Scott, quoted in Valerie Moolman, *Women Aloft* (Alexandria, Va.: Time-Life, 1981), 18, 44. Moolman writes that Scott's plane lifted 40 feet off the ground, but there is some controversy about whether this was deliberate or was due to a gust of wind. Raiche, however, after her first solo flight of a few feet in 1910, received recognition from America's Aeronautical Society as the "First Woman Aviator of America."

24. C. Griff, "Aviation Notes," *Woman Engineer* 2 (December 1927): 264. Griff, whose first name remains a mystery, was a member of the British Aeronautical Engineering Society and had worked for Vickers Aviation in World War I.

25. Ibid. 2 (September 1926): 174; Sicéle O'Brien, "Flying as a Career for Women," ibid. 2 (June 1928): 285.

26. Moolman, *Women Aloft*, 29.

27. On limitations placed on early female fliers, see Joseph J. Corn, *The Winged Gospel: America's Romance with Aviation, 1900–1950* (New York: Oxford Univ. Press, 1983), ch. 4.

28. "Youngest Flier in America a San Antonio Girl," *Aerial Age Weekly*, 17 April 1916, 149.

29. Michael Balint, *Thrills and Regressions* (New York: International Universities Press, 1959), 83, 85.

30. Betty D. Thornley, "Madame, the Aeroplane Waits," *Vogue*, 15 June 1920, 35–36, 110.

31. Ibid., 35, 108.

32. "Plastics in 1940," *Fortune*, October 1940, 89–96, 106–8. The magazine also noted that America was also working on what was considered a prime emblem of military progress, experimental all-plywood and plastic flivver planes.

33. Jeffrey Meikle, *American Plastic: A Cultural History* (New Brunswick, N.J.: Rutgers Univ. Press, 1995), 66–68.

34. "Plastics in 1940," 89.

35. Charles Gibbs-Smith, *Aviation: An Historical Survey from Its Origins to the End of World War II* (London: His Majesty's Stationery Office, 1970), 133; Oakes, *United States Women in Aviation through World War I.*

36. Ruth Law Oliver, oral history interview. Columbia University Oral History Research Office, published for the first time in *The American Heritage History of Flight* (New York: American Heritage, 1962), 157. Law's records included flying at 14,700 feet in 1917.

37. Amelia Earhart, "Plane Clothes," reprinted in *Harper's Bazaar: 100 Years of the American Female*, ed. Jane Trahey (New York: Random House, 1967), 40–41.

38. The photograph appears in Moolman, *Women Aloft*, 24.

39. Jacqueline Cochran, *The Stars at Noon* (Boston: Little, Brown, 1954), 29, 33, 56.

40. Earhart, "Plane Clothes." A photograph of Lady Heath dressed in helmet, boots, and beaver fur appears in Trahey, *Harper's Bazaar*, 42. See also Lady Sophie Mary Heath and Stella Wolfe Murray, *Woman and Flying* (London: J. Long, 1929).

41. The advertisement appeared in *Harper's Bazaar*, October 1944, 125.

42. Earhart, "Plane Clothes." In her article "Women's Influence on Aviation" Earhart suggested that women rally for better airport facilities for women (*Sportsman Pilot*, 3 [April 1930]: 341).

43. Douglas, *United States Women in Aviation*, 19–21; Christine Kleinegger, "The Janes Who Made the Planes: Grumman in World War II" (1998), unpublished paper, citing *Aviation Facts and Figures* (Washington, D.C.: American Aviation, 1959), 74. My thanks to Kleinegger for sharing Grumman information with me. For an early assessment of women's labor in World War II aviation industries, see also U.S. Department of Labor, Women's Bureau, *Women's Factory Employment in an Expanding Aircraft Production Program*, bulletin 189-1 (Washington, D.C.: U.S. Government Printing Office, 1942).

44. Kleinegger, "Janes Who Made the Planes"; Sherna Berger Gluck, *Rosie the Riveter Revisited: Women, the War, and Social Change* (Boston: Twayne, 1987): 17, xi–xii.

45. Moolman, *Women Aloft*, 135–39; Roy Nesbit, "What *Did* Happen to Amy Johnson?" *Aeroplane Monthly*, January 1988, 6–9.

46. Moolman, *Women Aloft*, 143–51.

47. "Avenger's WASPs," *Pegasus* 2 (November 1943): 5–6.

48. Ibid., 5.

49. Ibid., 13.

50. See Dickey Chapelle, *What's a Woman Doing Here? A Reporter's Report on Herself* (New York: William Morrow, 1961). After writing eight books on aviation during the war, Chapelle was at Iwo Jima as a war correspondent, and she was in Saigon, South Vietnam, at the time she was writing this book.

51. Dickey Meyer [Dickey Chapelle], *Girls at Work in Aviation* (Garden City, N.Y.: Doubleday, Doran, 1943), 36, 40–44. Although she wrote about women who had successfully challenged gender stereotypes, Chapelle seems to have shared some conventional ideas about women's inherent abilities and strengths. In her autobiography, *What's a Woman Doing Here?* she praised the abilities of female airplane mechanics, noting that two-thirds of the operations involved in

servicing airplanes were within their capabilities, the rest requiring strength for heavy lifting, but also wrote that "much of the work is of the precise, detailed type in which women excel" (75–76).

In its bulletin *Women's Factory Employment in an Expanding Aircraft Production Program* the Women's Bureau of the U.S. Department of Labor cited women's ability to operate a variety of machines, including punch presses, drilling machines, milling machines, lathes, and grinders, but ruled out their ability to operate engine lathes, shapers, large automatic screw machines, and to lift heavy parts, in part because of the need for lengthy experience and training. It also cited statistics on women's strength: the average woman's strength was said to be only 57 percent of a man's.

52. Meyer, *Girls at Work in Aviation*, 42.

53. Ibid., 74.

54. Arthur Wauters, *Eve in Overalls* (n.d.; reprint, London: Imperial War Museum, 1995), 13–14. Anthony James, in *Informing the People* (London: His Majesty's Stationery Office, 1996), 113, wrote that the undated book was probably published in 1941 or 1942 and that although many of the photographs are official government photos, it may not have been published by the Ministry of Information or His Majesty's Stationery Office.

55. Wauters, *Eve in Overalls*, 13, 17.

56. Ibid., 14, 17.

57. Muriel Rukeyser, *Theory of Flight*, in *Theory of Flight* (New Haven: Yale Univ. Press, 1935), 33–60.

58. Mrs. Elliott-Lynn, "Engineering and Aviation," 201.

59. Gertrude Stein, *Everybody's Autobiography* (New York: Random House, 1937), 191–92. In an unpublished manuscript written in 1933 Stein explored the interrelationship between Wilbur Wright, actors, and painting and mused about what type of painter Wright might have been. The thought was occasioned by seeing a monument dedicated to him in France (see idem, *Four in America*, intro. Thornton Wilder [New Haven: Yale Univ. Press, 1947]).

60. Gertrude Stein, *Picasso* (1938; reprint, Boston: Beacon, 1959), 49–50. Years later United Airlines, drawing on old stereotypes of the timid and fearful woman, suggested that rather than delighting in an aerial view, women needed to readjust their vision to feel comfortable at all. In a company advertisement from 1944 titled "First Flight" a woman in mink holds the hand of her daughter as they approach a plane with trepidation, but after twenty seconds in the air, "perspective lines vanish so not a trace of the dizziness or dread of high places she may have anticipated is present" (*Harper's Bazaar*, December 1944, 10).

The allure of aerial perspectives is seen also in the work of the American photographer Berenice Abbott, whose 1932 photograph *Nightview* displays the jutting verticalities of Manhattan skyscrapers with their pinpoints of lighted windows seen at night, and in the American painter Yvonne Jacquette's 1990s aerial views of American buildings at night.

61. In a 1986 interview with the author in New York, Driggs said she did not know that other artists were working on industrial subjects. See also Thomas C. Folk, *Elsie Driggs: A Woman of Genius*, exh. cat. (Trenton, N.J.: New Jersey State Museum, 1990); the cover illustration is Driggs's painting *Aeroplane*.

62. Fernand Léger, "The Machine Aesthetic, The Manufactured Object,

The Artisan and the Artist," *Bulletin de l'effort moderne,* nos. 1 and 2 (January–February 1924).

63. Sheldon Cheney and Martha Cheney, *Art and the Machine* (New York: Whittlesey House, McGraw-Hill, 1936), 97. See also Donald J. Bush, *The Streamlined Decade* (New York: Braziller, 1975); and Jeffrey L. Meikle, *Twentieth Century Modern: Industrial Design in America, 1925–1939* (Philadelphia: Temple Univ. Press, 1979).

64. Charles Sheeler's 1939 painting *Yankee Clipper,* which focuses on the jutting diagonal of the plane's nacelle, more clearly reveals the precisionists' admiration for the spare geometries of machine forms in the 1930s.

65. Folk, *Elsie Driggs,* 18. For critics' comments about Delasalle, see chapter 1.

66. For an earlier discussion, see Julie Wosk, "The Distancing Effect of Technology in Twentieth Century Poetry and Painting," *San José Studies* 11 (spring 1985): 22–41.

67. Stella Wolfe Murray, ed., *The Poetry of Flight: Anthology* (London: Heath Crantton, 1925), 12.

68. Astra, "The Gift of Flight," in Murray, *Poetry of Flight,* 135–39.

69. Janet Dunbar, *Laura Knight* (London: Collins, 1975), 163–64.

70. Margaret Bourke-White, quoted in Vicki Goldberg, *Margaret Bourke-White: A Biography* (New York: Harper & Row, 1986), 145.

71. Jonathan Silverman, ed. and comp., *For the World to See: The Life of Margaret Bourke-White* (New York: Viking, 1983), 73–75.

72. "Life's Bourke-White Goes Bombing," *Life,* 1 March 1943, 17.

73. Margaret Bourke-White, *Portrait of Myself* (New York: Simon & Schuster, 1963), 224, 232. The photograph of Bourke-White in her flying suit was also a frontispiece in this volume.

74. Ibid., 230.

75. Silverman, *For the World to See,* 115–16.

76. Bourke-White, *Portrait of Myself,* 226; Silverman, *For the World to See,* 195; Goldberg, *Margaret Bourke-White,* 268.

77. Silverman, *For the World to See,* 195.

78. "A New Way to Look at the U.S." *Life,* 14 April 1952, 128–39; see also Silverman, *For the World to See,* 191. Bourke-White's interest in aerial abstraction would be reflected again in her color photographs of American farmlands published in *Life* in 1955. These photographs, taken in 1954, appeared in "Base of Abundance—The Land," *Life,* 3 January 1955, 4–14.

79. "Strategic Air Command," ibid., 27 August 1951, 86–100; Silverman, *For the World to See,* 190.

80. Judith Tannenbaum has written that the isolation of objects in Celmins's paintings during the 1960s may also have reflected the artist's own feeling of solitude while living in California, separated from her family for the first time (*Vija Celmins* [Philadelphia: Institute of Contemporary Art, University of Pennsylvania, 1992], 12).

81. Audrey Flack, "Conversation with Audrey Flack," interview by Cindy Nemser, *Arts Magazine* 48, no. 5 (1974): 34–37.

82. Lucy Lippard, *A Different War: Vietnam in Art* (Bellingham, Wash.: Whatcom Museum of History and Art; Seattle: Real Comet Press, 1990), 42.

83. Nancy Spero, "Defying the Death Machine," interview by Nicole Jolicoeur and Nell Tenhaaf, *Parachute* 39 (June–August 1985): 50.

84. Nancy Spero, interview by Jon Bird, 1986, in *Nancy Spero*, exh. cat. (London: Institute of Contemporary Arts, 1987), 34.

85. The painting's title refers to a terse military phrase used during the war: *search and destroy* signified maneuvers in which soldiers searched villages for people believed to be Vietcong combatants and then destroyed them, their villages, and their arms caches. General William Depuy, one of the military's main architects of the search-and-destroy policy, was quoted as saying, "The solution in Vietnam is more bombs, more shells, more napalm . . . till the other side cracks and gives up," a policy that failed (William Depuy, quoted in Daniel Ellsberg, *Papers on the War* [New York: Simon & Schuster, 1972], 234).

86. Spero's titles take a sardonic view of the ways rhetoric employed by the military attempted to create verbal distancing, using euphemisms to soften and mask the deadly impact of war. *Pacification* refers to one of the military's euphemisms for the bombing of Vietnamese. After the Vietnam War, female writers also explored issues of rhetorical and aerial distancing as they considered gender, language, and military policy (see Carol Cohn, "Sex and Death in the Rational World of Defense Intellectuals," *Signs* 12, no. 4 [summer 1987], 687–718, discussed at length in Jane Roland Martin, "Aerial Distance, Esoterism, and Other Closely Related Traps," ibid. 21, no. 3 [spring 1996]: 584–614).

87. Shannon Lucid, quoted in "Marathon Woman," *Time*, 30 September 1996, 63. For comments on the encircling image in art and technology, see Wosk, *Breaking Frame*, 100–104.

7. Women in Wartime

1. "Women's Work in War Time," *Sphere*, 4 May 1918, 76.

2. Harriet Sisson Gillespie, "Where Girls Are Really Doing Men's Jobs," *Ladies' Home Journal*, November 1917, 83.

3. F. A. Stanley, "The United States Arsenal at Frankford—III," *American Machinist*, 4 January 1906, 1–6; "Women Machine-Tool Operators and Bench Workers in a French Munition Plant," ibid., 2 March 1916, 388.

4. *Scientific American*, 14 December 1918, 481; Mary Brush Williams, "Industrial Amazons," *Saturday Evening Post*, 14 November 1917, 28–34.

5. Williams, "Industrial Amazons," 28.

6. Ibid., 30, 33.

7. Ibid., 34.

8. Helen Fraser, *Women and War Work* (New York: G. Arnold Shaw, 1918), 102.

9. See also Perry Barlow's *New Yorker* cartoons during World War II.

10. Elizabeth Ewing, *Women in Uniform: Through the Centuries* (Totowa, N.J.: Rowan & Littlefield, 1975), ch. 8.

11. Ezra Bowen, *Knights of the Air* (Alexandria, Va.: Time-Life, 1980), 151.

12. "An Exhibition of Women's Work in the Engineering Industry: Noteworthy Achievements in Internal Combustion Engine Production," *Automobile Engineer* (London), April 1917, 96–98. In 1916, 300,000 British women and girls were employed in the metal and chemical trades, including those working at ordnance factories, and by early 1917 the number of women working in munitions factories—producing machine guns, howitzers, or explosives—had reached

691,000 (statistics from Member of Parliament Mr. Kellaway in his speech opening the British exhibit of women's work); see also I. William Chubb, "Women and Machine Tools," *American Machinist*, 22 June 1916, 1057.

13. Chubb, "Women and Machine Tools," 1062. The photographs for Chubb's article were specially taken and are not among the official Ministry of Munitions photographs.

14. Ibid., 1057.

15. Ibid.

16. Ibid.

17. Hall Caine, *Our Girls, Their Work for the War* (London: Hutchinson, 1916), 20–21.

18. Ibid., 24–25.

19. Ibid., 51–52, 28–29, 38, 60.

20. British munitions worker, 1975, quoted in Margaret Brooks, "Women in Munitions, 1914–1918: The Oral Record," in *Imperial War Museum Review*, no. 5 (1990): 4.

21. Brooks, "Women in Munitions," 9.

22. Joan Williams, *A Munition Worker's Career at Messrs Gwynnes, Ltd., Chiswick, 1915–1917*, a contemporary account in the collection of the Imperial War Museum's Women's Work Collection, cited in Mary Wilkinson, "Patriotism and Duty: The Women's Work Collection at the Imperial War Museum," *Imperial War Museum Review*, no. 6 (1991): 35.

23. Gail Braybon, *Women Workers in the First World War: The British Experience* (London: Croom Helm; Totowa, N.J.: Barnes & Noble, 1981), 157–70.

24. Caine, *Our Girls*, 49.

25. Brooks, "Women in Munitions," 6–8, which includes citations to specific audiotaped interview reels in the collection of the Imperial War Museum; Caine, *Our Girls*, 49, 53, 72.

26. L. K. Yates, *The Woman's Part: A Record of Munitions Work* (New York: George H. Doran; London: Hodder & Stoughton, 1918), 12–13, 20–21.

27. "On Board a War Ship: Fitting Electrical Wires," *Sphere*, 4 May 1918, 83.

28. "Women at Work in a South London Gasworks," ibid., 88–89.

29. Quoted in Braybon, *Women Workers in the First World War*, 160. See "Women Engineers," *Daily Mail*, 30 March 1916.

30. "Woman's Sphere in War Time," *Sphere*, 4 August 1917, iv.

31. Condell and Liddiard, *Working for Victory?* 4; Gareth Griffiths, *Women's Factory Work in World War I* (Stroud, Gloucestershire: Alan Sutton, 1991), 4. Deborah Thom argued in a study published after this chapter was written that the Ministry of Information photographs often focused on women in novel occupations, implying that their work itself was a novelty due to war and therefore reversible after the war. The photographs also gave the erroneous impression that many of the women were novices and that a larger number of women were new to the work force, replacing men, than was the actual case (*Nice Girls and Rude Girls: Women Workers in World War I* [London: I. B. Tauris, 1998], 87–90).

32. Jane Carmichael, "Home Front, 1914–1918: The Photographs of G. P. Lewis and Horace Nicholls," *Creative Camera* 247–48 (July–August 1985): 62–63. See also Francis Pugh's essay in *The British Worker: Photographs of Working Life*

1839–1939 (London: Arts Council of Britain, 1981), 23; *Report of the Imperial War Museum, 1918–1919* (London: His Majesty's Stationery Office, 1919), 9–10; and Wilkinson, "Patriotism and Duty," 32.

33. Carmichael, "Home Front, 1914–1918," 58, 63. See also Jane Carmichael, *First World War Photographers* (London: Routledge, 1989).

34. Nicholls's photograph appears in *British Industrial Photography, 1843–1986* (London: Photographers Gallery, 1986).

35. For an overview of smithing and forge scenes starting in the sixteenth century, see Klingender, *Art and the Industrial Revolution*, 59–61.

36. Condell and Liddiard, *Working for Victory?* 124–28, 174.

37. The Office of War Information photographs are in the Prints and Photographs Division, Library of Congress, Washington, D.C. The captions, based on photographers' notes, were later edited and sometimes rewritten by typists. In a parallel sign of transformation, the captions also noted manufacturing changes, for example, "Midwest vacuum cleaner plant converted to gas mask production."

38. "Women in Steel," *Life*, 9 August 1943, 75.

39. Sources differ about the number of women employed in industry. Sherna Gluck, in her study *Rosie the Riveter Revisited*, 7, 10, argued that 11.5 million American women were employed in 1940 and that by 1944, 16 percent of employed women were working in war industries.

40. "Women in Steel," 75.

41. "Speaking of Pictures . . . Hollywood Girls Are All Things to Servicemen," *Life*, 9 August 1943, 8–9, 11.

42. "Kitchens of Tomorrow May Look Like This," ibid., 53–56.

43. My thanks to Peter Liebhold, at the National Museum of American History, for information on the Aikins collection.

44. Sir Cecil Walter Hardy Beaton, *Air of Glory: A Wartime Scrapbook* (London: His Majesty's Stationery Office, 1941), 7, 66, 74. The book was issued by the Ministry of Information.

45. Carl Fleischhauer and Beverly W. Brannan, eds., *Documenting America, 1935–1943* (Berkeley and Los Angeles: Univ. of California Press, 1988); Maren Stange, " 'The Record Itself': Farm Security Administration Photography and the Transformation of Rural Life," in *Official Images: New Deal Photography*, by Pete Daniel et al. (Washington, D.C.: Smithsonian Institution Press, 1987). For a focused study of female FSA and OWI photographers, see Andrea Fisher, *Let Us Now Praise Famous Women: Women Photographers for the U.S. Government, 1935–1944: Esther Bubley, Marjorie Collins, Pauline Ehrlich, Dorothea Lange, Martha McMillan Roberts, Marion Post Wolcott, Ann Rosener, Louise Rosskam* (London: Pandora, 1987).

46. My thanks to Beverly Brannan, curator of photography, Prints and Photographs Division, Library of Congress, for the information about the photographs' captions.

47. "Women Make Tools of War," *Parade*, 29 March 1942, 13, illustrated in Daniel et al., *Official Images*, 136.

48. "Ann Rosener," *Minicam Photography* 6 (January 1943): 58–59.

49. Mary Elizabeth Pidgeon, *Your Questions as to Women in War Industries*, U.S. Department of Labor, Women's Bureau, bulletin 194 (Washington, D.C.: U.S. Government Printing Office, 1942), 9; Maureen Honey, *Creating Rosie the*

Riveter: Class, Gender, and Propaganda during World War II (Amherst: Univ. of Mass. Press, 1984), 21.

50. Honey, *Creating Rosie the Riveter*, 47–49, 32, 34.

51. Leila J. Rupp, *Mobilizing Women for War: German and American Propaganda, 1939–1945* (Princeton: Princeton Univ. Press, 1978), 155–56; Miriam Frank, Marilyn Ziebarth, and Connie Field, *The Life and Times of Rosie the Riveter: The Story of Three Million Working Women during World War II*, study guide for the documentary film produced and directed by Connie Field (Emeryville, Calif.: Clarity Educational Productions, 1982), 92.

52. Gluck, *Rosie the Riveter Revisited*, 13; Rupp, *Mobilizing Women for War*, 142–43; Honey, *Creating Rosie the Riveter*, 9.

53. Honey, *Creating Rosie the Riveter*, 20.

54. U.S. Department of Labor, Women's Bureau, *Women at Work: A Century of Industrial Change*, bulletin 161 (Washington, D.C.: U.S. Government Printing Office, 1939), 59–61. A similar text appeared earlier, in bulletin 115 (1934).

55. Honey, *Creating Rosie the Riveter*, 54; Melissa Dabakis, "Norman Rockwell's *Rosie the Riveter* and the Discourses of Wartime Womanhood," in *Gender and American History Since 1890*, ed. Barbara Melosh (New York: Routlege, 1993), 186; Frank, Ziebarth, and Field, *Life and Times of Rosie the Riveter*, 16; Douglas, *United States Women in Aviation*, 21.

56. Honey, *Creating Rosie the Riveter*, 50–51.

57. See Douglas Alan Fisher, *Steel in the War* (New York: U.S. Steel, 1946), 146.

58. For comments on segregated housing, see Martin Wachs and Margaret Crawford, eds., *The Car and the City: The Automobile, the Built Environment, and Daily Urban Life* (Ann Arbor: Univ. of Michigan Press, 1992).

59. *Harper's Bazaar*, October 1943, 141, and September 1943, 149.

60. Frank, Ziebarth, and Field, *Life and Times of Rosie the Riveter*, 98.

61. Augusta H. Clawson, *Shipyard Diary of a Woman Welder* (New York: Penguin, 1944), cited in *America's Working Women: A Documentary History, 1600 to the Present*, ed. Rosalyn Baxandall and Linda Gordon with Susan Reverby, rev. ed. (New York: W. W. Norton, 1995), 251–53. The editors note that until World War II women were not accepted in shipbuilding and that by January 1944 they made up 9.5 percent of workers in the industry, though women had difficult physical and psychological adjustments to make because this industry was unprepared for female workers. They add that Clawson's purpose in going to the Oregon shipyard was to "write up her own experiences and encourage other women (251).

62. Dunbar, *Laura Knight*, 156–62; Corinne Miller et al., *Images of Women* (Leeds: Leeds Art Gallery, 1989), 9.

63. Camel Cigarettes advertisement, *Harper's Bazaar*, December 1944, back cover.

64. *Saturday Evening Post*, 29 May 1943. Rose Monroe, a real-life Rosie the Riveter working at the Willow Run Aircraft Factory in Ypsilanti, Michigan, appeared in a promotional film for war bonds with Walter Pidgeon. Kay Kyser's song *Rosie the Riveter* was said to have been inspired by a New Yorker, Rosalind P. Walter (Tony Marcano, "Famed Riveter in War Effort, Rose Monroe Dies at 77," *New York Times*, 2 June 1997, B13).

65. Pidgeon, *Your Questions as to Women in War Industries*, 1.

66. Janet Owen, "Victory in the Hands of Women," in *American Women at War* (New York: National Association of Manufacturers, 1942), 24–27, quotation on 24.

67. Ibid., 24, 26.

68. Mary Hornaday, "Factory Housekeeping," in *American Women at War*, 38.

69. Pete Martin, "Right Face," *Saturday Evening Post*, 13 March 1943, 21.

70. *Harper's Bazaar*, September 1943, and May 1944, 4. For a more general overview of the social history of cosmetics, see Kathy Peiss, *Hope in a Jar: The Making of America's Beauty Culture* (New York: Holt, 1998), which was published after this chapter was written.

71. "She's Engaged!" *Harper's Bazaar*, February 1943, 91. The ad, which went to press after Huntington got married, included a newspaper wedding announcement.

72. Moolman, *Women Aloft*, 115–16.

73. Martin, "Right Face," 48; the first quotation is from Tania Long, the second is from Martin.

74. *Drugs and Cosmetics Review*, [1940s], Picture Collection, Mid-Manhattan Library, New York Public Library.

75. Bourke-White, *Portrait of Myself*, 213.

76. See Camel Cigarettes, "Women in the War" series, *Harper's Bazaar*, January 1944, back cover.

77. "Hands on the Job," ibid., February 1943, 53.

78. Ibid., 52–53.

79. Ibid., 53.

80. Lesley Blanch, "Pilots in Nailpolish," *Pegasus* 2 (October 1943): 13–16. The Air Transport Auxiliary employed mostly British women but also included Polish women, American women, women from other Allied nations, as well as men considered unfit for service in the Royal Air Force.

81. Ibid., 12.

82. Ibid., 13, 16.

83. "Her Job and How She Got It," *Harper's Bazaar*, January 1944, 82. See also ibid., February 1944, 124, showing a female private in the army, daughter of Mrs. Walter Lippman and H. Fish Armstrong.

84. "Give Your Vacation to the Land," ibid., June 1944, 30.

85. "The Blown-Glass Look," ibid., September 1945, 112–15.

86. See Rosenblum, *History of Women Photographers*, 232, 235.

87. Randall Jarrell, "Siegfried," in *Little Friend, Little Friend* (1945), reprinted in Randall Jarrell, *Selected Poems* (New York: Farrar, Straus, & Giroux, 1969), 146.

88. Egbert Booth, *Women at War: Engineering* (London: Bognor Regis; Sussex: J. Crowther, 1943), 10.

89. Ibid., 5–7, 10, 11, 17.

90. Robert Ferguson, "The Gender Shift: Adapting to Rosie the Riveter in the Southern California Aircraft Factories during World War II" (paper delivered at the August 1996 symposium of the International Committee for the History of Technology). Ferguson cites D'Ann Campbell, *Women at War with America: Private Lives in a Patriotic Era* (Cambridge: Harvard Univ. Press, 1984), 115.

91. W. Gerald Tuttle, "The Influence of Women on Aircraft Production Methods," *Aero Digest* 42 (March 1943): 143; "Dexterity Tests for Women Workers," *American Machinist*, 12 November 1942, 1240; W. Gerald Tuttle, "Women Who Work for Victory," *Mechanical Engineer* 65, 9 (September 1943): 657.

92. Tuttle, "Women Who Work for Victory," 657.

93. Ruth Milkman, *Gender at Work: The Dynamics of Job Segregation by Sex during World War II* (Chicago: Univ. of Illinois Press, 1987), 59. Milkman cites "Engineers of Womanpower," *Automotive War Production* 2 (October 1943): 4–5.

94. Milkman, *Gender at Work*, 60.

95. Pidgeon, *Your Questions as to Women in War Industries*, 9.

96. Fraser, *Women and War Work*, 111; Barbara McLaren, *Women of the War* (New York: George H. Doran, 1918), 136.

97. Moolman, *Women Aloft*, 29.

98. On the *Magazine War Guide*, created by the OWI in 1943, see Honey, *Creating Rosie the Riveter*, 55; and Rupp, *Mobilizing Women for War*.

99. Hoover advertisement, in *Electrical Age for Women* 4 (April 1944): 194.

100. *Saturday Evening Post*, 6 May 1944, illustrated in Honey, *Creating Rosie the Riveter*, 125.

101. *Electrical Age for Women* 4 (April 1944): 190.

102. "Mrs. America Looks Ahead," ibid. 4 (July 1944): 216, 220.

103. Thomas Hine, *Populuxe* (New York: Knopf, 1986), 4, 10, 64.

104. Gluck, *Rosie the Riveter Revisited*, 17.

105. Honey, *Creating Rosie the Riveter*, 22.

106. See, e.g., Rupp, *Mobilizing Women for War*, 138, 166.

107. See the Tobias and Anderson survey described in ibid., 161–62; and Honey, *Creating Rosie the Riveter*, 23.

108. *Good Housekeeping*, May 1947, 109. After World War I, as several historians have argued, advertisers often promoted the idea that women's domestic chores, including laundering, were an expression of love (see Strasser, *Never Done*, 78; Ruth Schwartz Cowan, "The 'Industrial Revolution' in the Home," *Technology and Culture* 17 [January 1976]: 192–93; Joann Vanek, "Household Technology and Social Status," ibid. 19 [July 1978]: 361–75; *Scientific American*, November 1974, 116–19; and Wajcman, *Feminism Confronts Technology*, ch. 4).

109. Herman F. Lehman, "The Mystery of the Missing Million," talk delivered to writers and editors, 25 October 1962, Frigidaire Collection, GMI Alumni Historical Collection, Flint, Michigan folder 79-10.7-59, [5]. My thanks to Shelley Kaplan Nickles for this information. Before running the ad, the company, which had found that women were most interested in product dependability, tested masculine-sounding words like *husky*, *faithful*, and *rugged* and ruled out *husky* because most people identified it with a dog. The most popular word was *sturdy*.

110. Lupton, *Mechanical Brides*. Joy Parr, in her study of the Canadian postwar washing-machine industry, noted that Beatty, a Canadian manufacturer of wringer washing machines, promoted a new model as feminine, "An Old Friend in New Dress" complete with a choice of two-tone colors ("What Makes Washday Less Blue? Gender, Nation, and Technology Choice in Postwar Canada," *Technology and Culture* 38 [January 1997]: 161).

111. Lehman, "Mystery of the Missing Million," [5].

112. See Cecelia Tichi, *Shifting Gears: Technology, Literature, Culture in Mod-*

ernist America (Chapel Hill: Univ. of North Carolina Press, 1987): 19–22. As Tichi argues "progressive housekeeping" often meant the willingness and ability to use new "labor saving" machines as well as efficient, nonwasteful housekeeping.

113. Hine, *Populuxe*, 64, 124.

Coda.
The Electronic Eve
and Late-Twentieth-
Century Art

1. See Donna J. Cox, "Using the Supercomputer to Visual High Dimension: An Artist's Contribution to Scientific Visualization," *Leonardo* 21, no. 3 (1988): 233–42. *Venus in Time*, juxtaposing the prehistoric sculpture Venus of Willendorf with a digital image by George Francis, Ray Idaszak, and Donna Cox, appeared on the cover of *ACM Computer Graphics Quarterly*, January 1987. The designation "Etruscan Venus" for the torsolike form reflects the artists' notion that the homotopy used is a simpler, or more primitive, version of the original homotopy worked on by Cox and her colleagues (see Ivars Peterson, "Twists of Space," *Science News*, 24 October 1987, 265).

2. Harriet Casdin-Silver, cover illustration, *Sculpture* 10 (special issue on art and technology, May–June 1991).

3. Chajim Bloch, *The Golem: Legends of the Ghetto of Prague, by Chayim Bloch*, trans. Harry Schneiderman (Blauvelt, N.Y.: Rudolf Steiner, 1972); Moshe Idel, *Golem: Magical and Mystical Traditions on the Artificial Arthropoid* (Albany: State Univ. of New York Press, 1990). For more on golem legends and artificial humans, see Wosk, *Breaking Frame*, 85.

4. For a discussion of this work, see Kirsten Solberg, "The Extra-sensitized Environment," *Leonardo* 31, no. 5 (special issue, sixth annual New York Digital Salon, 1998): 408.

5. *United States I–IV* was released in 1984 as a Warner Brothers record set and later as a CD. *Talk Normal* reappeared on her recording *Home of the Brave*, Warner Bros., 1986. See Craig Owens, "Amplifications: Laurie Anderson," *Art in America* 69 (March 1981): 120–23; Howard Smagula, *Currents: Contemporary Directions in the Visual Arts* (New York: Prentice-Hall, 1983), 243–59; and Roselee Goldberg, *Laurie Anderson* (New York: Harry Abrams, 2000).

6. Laurie Anderson, *Songs for A.E.*, first performed at Carnegie Hall, 27 February 2000. In her earlier song *O Superman* (1981) she intoned, "Here come the planes," and mocked modern technologies, including the surrogate speech of telephone-answering machines.

7. In the 1990s, in her CD and CD-ROM productions — including her *Nerve Bible Tour* and her interactive CD-ROM *Puppet Motel* (1995) — Anderson would become more technically sophisticated with 3-D digital graphics and animations, viewer interactivity, and an Internet website even as she continued with her familiar tools and images.

8. Grace Glueck, "And Now, a Few Words from Jenny Holzer," *New York Times Magazine*, 3 December 1989, 42, 108; David Joselit, "Holzer: Speaking of Power," *Art in America* 78 (October 1990): 155–57.

9. Both images appear in Nina Felshin, comp., *Disarming Images: Art for Nuclear Disarmament*, exh. cat. (New York: Adama Books, 1984), 48–49.

10. Adriene Jenik, *Mauve Desert: A CD-ROM Translation*, Shifting Horizons Productions, 1997. See also Holly Willis, "Cyberwomen: Feminist Strategies in Multimedia," *Artweek* 28 (February 1997): 14–15.

11. Brooks Adams, "Rebecca Horn at Marian Goodman," *Art in America* 79

(April 1991): 161. In 1997–99 Horn's installations continued to include mechanized, kinetic sculptures, but she increasingly endowed her work with political themes such as migration and refugee displacement. Also in the late 1990s, the New York painter Ellen K. Levy took an intriguing look at the role of display and representation in science and technology exhibits at American and European museums, seen in her painting *Damien's Gliders / Agassiz's Chart* (1999).

12. In his review essay "Modest Reviewer Goes on Virtual Voyage," Geoffrey C. Bowker discussed books focusing on the implications of new ontologies and the creation of new human entities in cyberspace (Bowker, *Technology and Culture* 39 [July 1998]: 499–511). The uses of morphing software programs to create new selves continued in the 1990s: in contrast to *L'Ève future*, a front cover of *Time* magazine in 1993 featured what Donna Haraway called a "homogenized SimEve," a morphed compilation of multiethnic and multiracial faces created through a computer-graphics program and suggesting a citizen of the future melting-pot America (Donna J. Haraway, *ModestWitness@Second_Millennium. FemaleMan©_Meets_OncoMouse™: Feminism and Technoscience* [New York: Routledge, 1997], 262; cover, *Time*, special issue, fall 1993).

13. Promoting fantasies, Vesna also asked participants to decide whether their computer-constructed figures were sex partners, significant others, or alter egos. See also Glenn A. Kurtz, "Victoria Vesna at the San Francisco Art Institute," *Artweek* 28 (April 1997): 20. In another website version of designer bodies, scanned images of the organs of members of Fakeshop in a 1999 performance were bought and sold by network participants, who redesigned the performers' bodies on screen (see Catherine Bernard, "Bodies and Digital Utopias," *Art Journal* 59 [winter 2000]: 28–29).

14. Nancy Burson, *Faces* (Houston: Contemporary Arts Museum, 1992), 13. The technology used for Burson's *Aging Machine*, which she developed with the computer scientist David Kramlich, was also used by America's FBI to help locate missing children by helping to project what they would look like in the future. In 2000 Burson developed a *Human Race Machine*, in which participants could merge their own faces with those of members of other races, and earlier she created an *Androgeny Machine*, which would allow them to create images of themselves based on varying percentages of male and female.

15. Haraway, *Modest-Witness*, 11.

16. Vivian Sobchack, "Beating the Meat / Surviving the Text; or How to Get Out of This Century Alive," in *Cyberspace/Cyberbodies/Cyberpunk: Cultures of Technological Embodiment*, ed. Mike Featherstone and Roger Burrows (London: Sage, 1995), 213.

Index

An Abbé Lacing a Lady's Stays (Monsiau; engraving), 63

Abbott, Berenice, 270n60

"Abominations of Modern Science" (*Punch*; cartoon), 13–14, 15

accelerator. *See under* bicycle

Adams, Henry, 18

Addison, Joseph, 57–58

advertising: of appliances to women, xvii, 26, 226–30, 245n49, 256n42, 277nn109,110; and art, 235; of automobiles, 120, 127, 133, 145, 147, 265n57; of automobiles to women, 120, 133–36; and aviation, 158–59, 161–62, 207, 269n60; of bicycles, 20, 89–90, 102, 103–4, 260n19; of cosmetics, 210–16; and decorative images of women, xii, xiii, xviii, xix, 18, 83, 84, 145, 240n7; and demobilization, 225–30; and electricity, 83–85; factory-machine, 20, 24, 25; and fashion, xvi–xvii, 135, 147, 216, 239n4; to gender machines, xviii, 1, 23–24; images of female pilots in, 207; and images of housewives, 228–30, 277n108; of incandescent lights, 20, 21, 243n37; of machines, xii–xiii, 20, 24, 25, 27–33; and new role of women, 6; and photography, 120, 127, 133–36, 145, 147, 218–20, 265n57; postwar, xvii, 228–30; in *Scientific American*, 25, 26, 31, 244n46; of sewing machines, 27–33; and sexuality, xii, xiii, 17–23, 31, 243n39; and transition to postwar, 225–26; of typewriters, 23–24; of vacuum cleaners, 24, 26–27; for women war workers, 206–10

Aero Digest (magazine), 155

Aeroplane (Driggs; painting), 170, 171–72

Aeroplane over Train (Goncharova; painting), 152

African-American women, xvi, 205, 224–25

The Age of Iron: Man As He Expects to Be (Currier & Ives; lithograph), 3–5, 6

Aging Machine (Burson; computer-generated photograph), 237

Agricultural Extension Service, 217

Aide-toi, le ciel t'aidera (Tito; print), 133

Aikins, Russell, 199, 200, 205, 206, 224

Air Minded (illustration), 160

airplane mechanics, 155, 168, 187, 196, 207, 269n51

airplanes, 148, 220; and artists, 148, 152–54, 161, 169–81, 207, 219, 234–35, 267nn8,9, 271nn66,80; and decorative images of women, 149, 156, 159, 160–63, 168–69; jet engine, xii, xviii; manufacture of, 185, 196, 209; views from, 170, 176, 269n60, 270n60; as war machines, 172–73, 176, 178, 179–80. *See also* aviation

Air Service Command Headquarters, 216

Air Transport Auxiliary (ATA), 164, 207, 215, 276n80

Albertina Rasch's Celebrated Dancers, 122, 123

Alken, Henry, xi

Allom, Thomas, 39

amazons, xiii, 19, 20, 60, 105, 137, 184

American Airlines, 158

American Automobile Association, 116, 129

American Braided Wire Company, 65

An American Dream of Venus (photomontage), 161, 162

American Machinist (magazine), 183–84, 187–88

American Medical Monthly (journal), 28–29

American Nervousness: Its Causes and Conse-quences (Beard), 98
American Plastic (Meikle), 161
The American Scene (James), 170
American Women at War (National Associa-tion of Manufacturers; anthology), 209, 210
Amerika (Horn; sculptures), 236
L'Ami des femmes (Villemert), 12
Analysis of Beauty (Hogarth), 62
Anderson, Laurie, 234–35, 278nn6,7
Anthony, Susan B., 114
appliances, electrical, xvi, xviii, 1, 87, 206; advertising of, xvii, 26, 226–30, 245n49, 256n42, 277nn109,110; design of, 30, 31, 226–30; irons, 26, 245n52, 256n42; as labor-saving, 25, 26, 27–28, 198, 245n52, 278n112; and photography, 86; in postwar homes, 226–30; women as customers for, 83–84, 225; women as demonstrators of, 256n42
archetypes, female, xviii, 3, 17–23, 254nn11,12; and electricity, 69–83, 87–88; in industry, 32, 38; in wartime, 198
Arden, Elizabeth, 210
Arkwright, Sir Richard, 38
Army Expeditionary Force, 196
artists: and advertising, 235; in antiwar movement, 178–79; and automobiles, x, 115, 119–20, 122, 123, 133, 147–48, 152, 236; and aviation, 148, 152–54, 161, 169–81, 207, 234–35, 267nn8,9, 271nn66,80; digital, 86–88, 231–38, 258n58, 279nn12,13,14; eighteenth-century, 13, 69, 242n26; and electric irons, 245n52; and electricity, 84–85, 86–87, 258n50; and female identity, x, xvii, 6, 115, 170–71, 176–78, 192, 231–38; and forges, 10–11, 172, 192, 241n14; and French salon, 11, 119, 171, 172; and machines, xiv, 153, 171, 220–21; medieval, 11; nineteenth-century, 11, 65–66, 67, 119, 171–72, 192, 231; Pop-art, xiv, 219, 229–30, 231, 237; pre-cisionist, 170–72, 271n64; Renaissance, 6, 78, 231; surrealist and dada, xiv, 22, 23, 85, 86, 153–54, 161, 236, 247n71, 257n48, 278n11; and weaving, 43, 84, 152, 250n98. *See also* images
Art Journal (magazine), 8, 79
Associated Press, 218
Astor, Mrs. John Jacob, 118
astronomy, 12–13

ATA. *See* Air Transport Auxiliary
Auerbach, Nina, 79
Austen, Alice, 111–13
automatons, xiv, 46, 75–78, 80, 87–88, 236, 255n28; in *Metropolis*, xv, 74, 81, 82–83
Automobile Club of America, 118
Automobile Club of France, 20
automobiles, 115–48; advertising of, 120, 127, 133, 145, 147, 265n57; advertising to women of, 120, 133–36; ambivalence about driving, 140–41, 144; and art by women, x, 115, 122, 123, 133, 147–48, 152, 236; cross-country tours of, 124–27; designs for, 148; electric, 2, 83, 115, 117, 122, 131, 133, 134, 135, 264n44; and fashion, 135; and femininity, 116, 136; and gender stereotypes, xii, 115, 127–33, 134, 137, 139; and grounded women, 132; handbooks for drivers of, 136–41; images of women with, xii, 115, 118–23, 126, 128, 131, 132, 136, 139, 142, 144, 145, 147; and independence, 115, 120, 122–23, 125, 135, 143–44; licensing of, 115–16, 262n2; and mobil-ity, 263n27; and modernity, 115, 122, 140, 152; and photography, 118–23, 139–40; racing in, 129–30, 137, 265n49; repairing, 2, 9, 115, 124–25, 130–33, 137–44, 194–95; representations of, by women, 147–48; and rural women, 122–23, 263n27; self-starters in, 134, 265n60; and speed trials, 137; steam, 2, 115, 122; and suffragists, 125–27, 141, 264nn35,38; and understanding of maps, 123, 138; women drivers of, xix, 3, 9, 115, 116–20, 121, 127–29, 132–35; and women's identity, ix–x, xii, xiii, xvi, xviii, 3, 9, 28, 67, 114, 115, 128–29
Automotive War Production (magazine), 222
AutoPortrait (DeLempicka; self-portrait), x, 115, 152, 236
aviation, 149–81; and advertising, 158–59, 161–62, 207, 269n60; ambivalence about, 158, 160, 169; and competence, 168–69; and emotional detachment, 170–81, 272n86; and female identity, 152, 153; and femininity, 162, 164; and gender differences, 155; gender stereo-types in, 154–57, 167–68, 191, 269n51; imagery of, 153–54, 159, 169–70, 176–81; job assignments in, 183, 248n85; and

modernity, 152, 154, 267nn6,10,11; and photographers, 146, 147, 150–53, 156, 165–67, 173–76, 177, 187–88, 197, 212, 250n98, 271n78; poems about, 169–70, 172, 220; and women's freedom, 149, 154, 158, 163, 181; in World War II, 164–70, 207, 215, 276n80

Aviation Research Instructor School for Women, 168

aviators, x, 3, 152, 224; clothing for, x, 160–63; and cosmetics, 163, 211, 213; exhibition, 124, 142, 150–51, 155, 157, 159–60; first women, 124, 151, 157, 264n31; stereoypes of, 154–57, 191; test pilots, 164, 165, 207, 215; and women's freedom, 169; and women's grounding, 158–59, 166–69; in World War I, 187. *See also* aviation

Bagley, Sarah G., 41

La Baignoire, au théatre des Variétés (Béraud; painting), 65

Baines, Edward, Jr., 39

Baker, Josephine L., 41, 43

Balint, Michael, 160

ballooning, 56, 57, 168–69

A Balloon Site, Coventry (Knight; painting), 168

Bara, Theda, 22

Barnes, Florence Lowe "Pancho," 157

Bathrick, Serafina, 19, 60

Beard, George, 98

Beaton, Cecil, 201

Bel Geddes, Norman, 171

The Bell-Tower (Melville), 35

Bendix Prize, 149, 163

Béraud, Jean, 65

Berg, Mrs. Hart O., 162

Berger, Michael L., 127, 128

Bergson, Henri, 46

Berlin, Irving, 158

Betinet, Maurice, 95

Beyval, Nada, 71

bicycles, ix, xii–xiv, xviii, 3, 17, 67, 89–114, 246n62; accelerator, 89–91; advertising of, 20, 89–90, 102, 103–4, 260n19; ambivalent attitudes toward, 90, 97, 104, 105, 113; in art, 95, 100, 103, 112, 259n12; and chaperones, 100, 260n30; and clothing, x, xvi, 106–11; and decorum, 100–101, 102, 107–8, 110–11; in 1870s, 259n13; 1890s craze for, 6, 89, 97; and freedom, 99–100, 101, 105, 114;

and gender stereotypes, 89, 106, 114; handbooks for, 109–12; and health, 29, 98–100, 102–3, 260n39; high-wheeler (ordinary), 95–96, 97; and independence, x, 6, 94, 111; and mobility, 96–97, 100; motor, xvi, 132; photographs with, 105–6, 111–14; racing on, 92, 102, 259n8; repair of, 101, 111, 113; safety of, x, 2, 89, 97–102, 106, 110; satires of, 89, 90, 95, 104, 105, 106, 107; as social threat, 6–7, 101–3, 105–6; and suffragists, 113–14; tandem, 94, 96

Bicycling for Ladies (Ward), 111–13

Big Science (Anderson; music album), 234

The Birth of Venus (Botticelli; painting), 231

Bishop, John Leander, 51

The Blood of a Poet (Cocteau; film), 87

Bloomer, Amelia Jenkins, 107

bloomers, 107, 108, 143

"The Blown-Glass Look" (Naylor; photo essay), 218

Bodies©INCcorported (Vesna; interactive website), 236–37

Bomb Bay of a B-36 Bomber (Bourke-White; photograph), 177

Booth, Egbert P., 220–22

Bose, Georg Matthias, 68

Bost, Pierre, 87, 258nn50,52

Botticelli, Sandro, 231

Bottomley, Joseph Firth, 94, 99

Bourke-White, Margaret, 146, 147, 173–76, 177, 197, 212, 250n98, 271n78

Braybon, Gail, 190, 192

Breadline during the Louisville Flood (Bourke-White; photograph), 146, 147

Breck, Samuel, 8

Bride Stripped Bare by Her Bachelors, Even (The Large Glass) (Duchamp; art work), xiv

British Royal Flying Corps, 224

Brossard, Nicole, 236

Bryant, William Cullen, 40, 41

Burke, Alice, 126–27

Burson, Nancy, 87, 88, 237, 258n58, 279n14

Bush, Barbara, 237

Buss, R. W., 14, 15

bustles, xiv, 46, 48, 50, 51, 64–67

Butler, Elizabeth Beardsley, 34

Butler, Vera Hedges, 116–17

Byng-Hall, C., 137

Byron, Lord, 12

Caine, Hal, 188–90, 191

Car Illustrated (magazine), 117, 118, 119, 129, 130, 132

Casidin-Silver, Harriet, 233

Cassier's (magazine), 13, 14

Catherine the Great (Russia), 10

Cellini, Benvenuto, 78

Celmins, Vija, 176–77, 179, 271n80

Centennial Exhibition (Philadelphia; 1876), 19, 20, 60, 239n2, 257n47

Chapelle, Dickey (Dickey Meyer), 167–68, 269nn50,51

Chaperone Cyclists' Association, 100

Charging the "Electric" at Home (booklet), 134

Le Charivari (newspaper), 54, 55, 56

Chauffeur: Trouble on the Road (Nicholls; photograph), 195

Cheney, Martha, 171

Cheney, Sheldon, 171

Cher, 237

Christy, Howard Chandler, 145

Chubb, I. William, 187–88

Churchill, Lady Randolph, 84

"Circuit Européen" (M. Montaut; poster), 153

Clark, Benton, 208

Clawson, Augusta, 206

clothing: and advertising, xvi–xvii, 135, 147, 216, 239n4, 240n7; of automobilists, x, 118, 138; of aviators, x, 160–63; of bicyclists, x, 106–11; changing clothes, ix, x, xv, xvi, xvii, 216–17; factory, x, xvi, 189, 193; and freedom, x, xiv; and technology, 45–67; of war workers, 216–18

Coalbrookdale by Night (Loutherbourg; painting), 192

Cochran, Jacqueline, 163, 164, 166, 211

Cocteau, Jean, 87

Coleman, Bessie, 154

Collins, Marjorie, 203, 207

Columbian Exposition (Chicago; 1893), 19, 60, 257n47

Companie Parisienne de Distribution d'Électricité, 86

Composite Machine (Burson; computer-generated photograph), 237

Conant, Cornelia W., 33

corsets, 48, 59, 61–64, 251n8, 252n42; abandonment of, 67; steam-molded, 62, 252n41; steel, 46, 61; whalebone, 61, 108; and windlass devices, 63–64, 252n46

cosmetics: and aviators, 163, 211, 213; and femininity, 210–16; and patriotism, 210–13, 218

Courbet, Gustave, 95

Cowan, Ruth Schwartz, 25, 26

Cox, Donna, 231–33

Crawford, Lady Gertrude, 10

Crean, Melanie, 233–34

crinolines, 49–61; abandonment of, 67; in art, 46, 65; and class, 48–49, 59; and crinolettes, 51, 52; as flying machines, 56–57; and gigantism, 57–58; hazards of, 52, 53, 251n21; inflated-tube, 57, 252n32; men's views of, 53–56; pannier, 49, 64; satires of, 47–48, 53–54, 58; steel-cage, xiv, 45–57, 59, 61, 251nn13,14; types of, 50–52

"La Crinolomanie" (Daumier; lithographs), 54

Cruikshank, Robert, 90–91

Crystal Palace Exhibition (New York; 1853–54), xiii, 19, 20, 60, 105, 137, 184

CTC. *See* Cyclists' Touring Club

Cuneo, Joan Newton (Mrs. Andrew), 122, 129

Currier & Ives, 3–5, 6

Curtiss, Glenn, 154

Cycling (magazine), 102–3, 104, 108

"Cycling for Ladies" (Davidson; article), 101

Cyclists' Touring Club (CTC), 98, 108

Cyclops Steel and Iron Works, 193

dada art, xiv, 22, 23, 85, 86, 153–54, 161, 236, 247n71, 257n48, 278n11

Dahl-Wolfe, Louise, 217

Die Dame (magazine), 115

Les Dames Goldsmith au bois de Boulogne (*En Promenade*; Stewart; painting), 119

d'Antigny, Blanche, 95, 259n12

Daughter Come Down (song), 158

Daumier, Honoré, 54, 55, 56, 168

Davidson, Lillias Campbell, 98–101, 102, 107–11

Davis, Bette, 130

Davis, Sydney "Sammy" Charles, 137–38

Dawn of the Century (Paull; musical piece), 19

Delangle, Hélène (Hellé Nice), 129–30, 265n49

Delasalle, Angèle, 11, 172

Delaunay, Robert, 148, 152

Delaunay, Sonia, 85, 147, 148, 152,
257n46
DeLempicka, Tamara, x, 115, 135, 152,
236, 262n1
Dell'elettricismo . . . (Sguario; treatise), 69
Descartes, René, 77, 255n28
Le Désert mauve (Brossard), 236
Diaghilev, Sergei Pavlovich, 148
Dickens, Charles, 8
Dietrich, Marlene, 120
digital art, 86–88, 231–38, 258n58,
279nn12,13,14
Le Diverse et Artificiose Machine (Ramelli;
treatise), 18
dolls: in art, 236; mechanical, 46–47,
255n26; phonographic, 76–77, 78,
255n26; in photographs, 87, 88, 237,
258n58, 279n14; women as, 87–88. *See
also* automatons
Doner, H. Creston, 198
Don't Get the Clothes Too Blue! (Kilburn;
stereograph), 7
Doolittle, James, 173
Douglas, Deborah G., 248n85
draisine. *See* bicycle: accelerator
Dress and the Lady (illustration), 47
Driggs, Elsie, 170–72
Duchamp, Marcel, xiv, 86, 231–32, 237,
257n48
Duke, Mrs. Angier Biddle, 130
Duncan, Isadora, 122
Dyer, John, 38
"The Dynamo and the Virgin" (Adams),
18
Dynamo Machine (Goncharova; painting),
84

Earhart, Amelia, 149, 151–52, 153, 155,
162–63, 235, 269n42
Eastlake, Sir Charles, 52
Eastman, Ruth, 123, 144, 145, 234, 237
EAW. *See* Electrical Association for
Women
Eddy, Henry B., 118
Edge, Mrs. Selwyn, 130
Edge, Selwyn, 136–37
Edison, Thomas, 21, 70, 73–75, 81,
255n26; phonographic dolls of, 76–77
The Education of Henry Adams (Adams), 18
Effets du tourniquet sur les jupons en crinoline
(Daumier; lithograph), 55
Ehrhart, S. D., 104, 105
Eisenhower, Dwight D., 212

Electra, 258n57
The Electrical Age for Women (magazine),
84, 227
Electrical Association for Women (EAW),
84
Electrical Handbook for Women, 84
Electrical World (journal), 74
electric chair, 75, 258n48
L'Électricité (Robida; print), 71
Électricité: Dix rayogrammes de Man Ray
(Man Ray; photogravure), 86, 87, 88,
231, 258nn50,52
electricity, xviii, 15, 68–88; and artists, 84–
85; and automobiles, 2, 83, 115, 117,
122, 131, 133, 134, 135, 264n44; experiments with, 12, 68–70, 73, 75, 81–82;
fear of, 185; goddesses of, 19, 21, 69–83,
87–88, 254nn11,12; and jewelry, 73–74;
medical benefits of, 62, 73, 254n14; and
woman as spark plug, 257n48; women as
consumers of, 83–85. *See also* appliances,
electrical; digital art
Electricity (journal), 71
"Electricity in the Household" (article),
25
"Electricity Man's Slave" (Edison; article),
73
Electric Lamp (Goncharova; painting), 84
Electric Prisms (Delaunay; painting), 85
electroplating, 78–79
electrotyping, 77, 78, 79
Elliott, Daisy, 112, 113
Elliott-Lynn, Mrs. *See* Heath, Lady Sophie
Mary
Elopement Extraordinary (lithograph), 3, 4
"Engineering and Aviation" (Heath; arti-
cle), 154
Enola Gay (airplane), 174
En Promenade (*Les Dames Goldsmith au bois
de Boulogne*; Stewart; painting), 119
En route to the Lake (Moore; illustration),
97
Érotique voilée (Man Ray; photograph), 22,
23, 244n41
Erskine, Fanny, 109, 111
Essai sur l'électricité des corps (Nollet; illus-
tration), 12
Etruscan Venus, 233, 278n1
Eugénie, Empress, 48, 52
Eve, 18, 168–69; electronic, 231–38; in
fiction, 74–77, 79–80, 82, 87, 88, 231
L'Ève future (Villiers de l'Isle-Adam), 74–
77, 79–80, 82, 87, 88, 231

Eve in Overalls (Wauters; pamphlet), 168–69

Everybody's Autobiography (Stein), 170

L'Expérience sur l'électricité (print), 69, 70

An Experiment on a Bird in the Air Pump (Wright; painting), 11

An Expert Lady Driver (photograph), 119

Exposition Universelle (Paris; 1900), 71, 76

factories, 50, 82, 185; munitions, 183–84, 189–91, 193, 194, 196, 209, 220–21, 272n12; women working in, ix, xi, xiv, xvi, 1, 33, 184, 196–206, 219, 220–21, 243n35, 244n46, 272n12, 273n31, 274n39

Farley, Harriet, 41

Farman, Henri, 152

Farm Security Administration (FSA), 202

farthingales (vertingales), 49, 53. *See also* crinolines

fashion. *See* clothing

Fashion and Eroticism (Steele), 62

Fashion and Fetishism (Kunzle), 59

Faust (Gounod; opera), 80, 81

Feininger, Andreas, 224

Female Bomb (Spero; painting), 179

"The Female Colossus" (Bathrick; essay), 19

The Female Race! or Dandy Chargers Running into Maidenhead (Cruikshank; engraving), 91

The Female Slaves of New York — "Sweaters" and Their Victims (illustration), 34, 35

femininity: and cosmetics, 210–16; and glamor, xii–xiii, 210, 220; and identity, xvi, 239n2; and mechanical imagery, 162, 172, 219; and patriotism, 211; and technology, 161; and war jobs, 164, 166, 169, 185, 191, 215, 218

films, xv, 22, 24, 74, 81–83, 87, 206

First Class: The Meeting . . . and at First Meeting Loved (Solomon; oil), 8

Fitzgerald, F. Scott, 133

Flack, Audrey, 177–78, 179

"The Fleece" (Dyer), 38

The Flight of Fashion (Heath; print), 57

Flügel, J. Carl, 61

Flying Fortress (Celmins; painting), 177

Folk, Thomas C., 172

"Folsom's Motion for Sewing Machines" (illustration), 30

A Fool There Was (film), 22

Ford, Mrs. Henry, 130

A Forge (Delasalle; painting), 11

For King and Country (Skinner; painting), 193

Fortune (magazine), 161, 162, 196, 218

Frankenstein (Shelley), 12, 34, 247n72

Frank Leslie's Illustrated Newspaper, 1, 31, 34, 35

Frantz, Henri, 11

Fraser, Helen, 185

French, Anne Rainsford, 116

French, Daniel Chester, 116

Freud, Sigmund, 46, 250n4

From the Air (Anderson; song), 234

FSA. *See* Farm Security Administration

The Fun of It (Earhart), 155

Galsworthy, Olive Edis, 196

Gamy. *See* Montaut, Marguerite

Garbo, Greta, 120

Garvey, Ellen Gruber, 103

Gascoigne, Mrs., 10

Gears (Rosenquist; painting), xiv

Geldard, James, 40

gender: and advertising, xviii, 1, 23–24; ambiguities of, 10, 166, 182, 183, 197, 210–11; conventional view of, 3, 20–21, 140; and identity, 9, 152, 153, 235, 241n11; and role changes, xv, 3–7, 104, 106, 107, 240nn4,5

Gender at Work (Milkman), 222

gender stereotypes, 10, 79; and automobiles, 115, 127–33, 134, 137; in aviation, 155–57, 167–68, 269n51; in bicycling, 89, 106, 114

General Electric Company, 22, 134, 199; Mazda advertisements of, 21, 84

Genius Rewarded; or The Story of the Sewing Machine (Scott), 27

German Airplane (Celmins; painting), 177

Gérôme, Jean-Léon, 119, 231

Gibson, Charles Dana, 103, 159

Gillespie, Harriet Sisson, 183

Gillray, James, 63

Girls at Work in Aviation (Meyer), 167

"The Girl Who Drives a Car" (Murdock; article), 140–41

"The Girl Who Used to Drive a Nail With Her Hair Brush" (*Life*; cartoon), 186

Girth Control (Lowell; drawing), 64

Glidden, Carlos, 23

Glidden, Mrs. Charles J., 117, 120

Goddard, Paulette, 197
goddesses, xviii, 17; and advertising, 21; of electricity, 19, 21, 69–83, 87–88, 254nn11,12; of industry, 19–20; Venus, 6, 11, 12, 161, 162, 231–33, 278n1
Godey's Lady's Book and Magazine, 20, 28, 63, 108
Golem (Crean; kinetic sculpture), 233–34
Golub, Leon, 178
Goncharova, Natalia, 43, 84, 152
Good Housekeeping (magazine), xii
Gorbachev, Mikhail, 237
Gould, Bruce, 155
Gounod, Charles, 80
Gower, Pauline, 164
Grable, Betty, 24, 197, 198
Grahame-White, Claude, 154, 267n12
Great Exhibition (London; 1851), 60. *See also* amazons
The Great Gatsby (Fitzgerald), 133
Griff, C., 157, 268n24
Grimesthorpe Steel and Ordnance Works, 193
Grosz, George, 154
Grover and Baker Company, 28–29, 32
Grumman Aircraft, 164
Gunship and Victims (Spero; painting), 180

Hammer, Mary, 71
Hammer, William, 70
Handbook for Lady Cyclists (Davidson), 98, 100
Hand-Book on Cotton Manufacture (Geldard), 40
Handbook on Turning (Gascoigne), 10
"Hands on the Job" (*Harper's Bazaar*; photo essay), 214
Hansen, Geneva, 214–15
Haraway, Donna, 88, 237
Harberton, Lady, 107
Harbou, Thea von, 74
Hargreaves, James, 38
Harmsworth, Mrs. Harold, 130
Harper's Bazaar (magazine), 33, 53, 162–63, 206, 213–19
Harper's Weekly (magazine), 44, 50–51, 58, 59, 92–93, 97, 101
Hattoom, Fred L., 155–57
Head (Leaf; sculpture), 236
health, women's: and bicycles, 29, 98–100, 102–3, 260n39; and factory work, 189–90, 247n73; and sewing machines, 28–29, 31

Heartfield, John, 154
Heath, Lady Sophie Mary (Mrs. Elliott-Lynn), 154, 157, 163, 170, 269n40
Heath, William (Paul Pry), 57
Heim, Jacques, 147, 148
Held, Anna, 105, 106
Hewlett, Hilda, 157, 224
High, Wide, and Frightened (Noyes), 149
Hine, Lewis W., 34, 42, 43
Hine, Thomas, 230
Hints on Household Taste (Eastlake), 52
A History of American Manufactures (Bishop), 51
History of the Cotton Manufacture in Great Britain (Baines), 39
Hitchcock, Mrs. A. Sherman, 136
hobby horse. *See* bicycle: accelerator
Höch, Hannah, 86, 153–54
Hoffmann, E.T.A., 46
Hogarth, William, 62, 252n41
Holzer, Jenny, 235
Homer, Winslow, 40, 41
Honey, Maureen, 204
Hood, Thomas, 27
Hoover Company, 26, 31
Horn, Rebecca, 236, 278n11
Hornaday, Mary, 210
Hoyningen-Huene, George, 120
Hughes, Hugh, 5, 9

identity, and femininity, xvi, 239n2
identity, female: ambiguous, xvii, 182; and artists, x, xvii, 6, 115, 170–71, 176–78, 192, 231–38; and automobiles, ix–x, xii, xiii, xvi, xviii, 3, 9, 28, 67, 114, 115, 128–29; and aviation, 152, 153; and changing roles, 183, 186–87; fears of change in, xiv, xv, 88; reconfiguring of, x, xv, xvi, 235; sense of, 135–36, 265n63; and technology, 9, 241n11
images: advertising, xii, xiii, xviii, xix, 18, 83, 84, 145, 240n7; in aviation, 158, 159, 169–72, 176, 207; digital, 86–88, 231–38, 258n58, 279nn12,13,14; eighteenth-century, xvii; erotic, xiv, 20–23, 243n39; government-created, 216; of housewives, 84, 225–26, 228–30, 277n108; of industry and machines, 171, 220–21; mechanical, 162, 172, 219; nineteenth-century, ix–xii, 79, 250n97; stereotypical, 3, 79, 89, 106, 127–33, 134–35; twentieth-century, xii; of women as machines, xiv, 153; of women in auto-

images (*cont.*)
  mobiles, 115, 118–23; of women in
    World War II, 196–210, 219, 221, 225
immaculates, 171. *See also* precisionists
Imperial War Museum, Women's Work
    Subcommittee of, 192, 193, 194
"Industrial Amazons" (Williams; article),
    184–85
industrialization, xiv, xv, xvii, 10, 37
Iribe, Paul, 160
The Iron Forge (Wright; painting), 11

Jacobi, Moritz Hermann von, 78
Jacquet-Droz, Henri-Louis, 77, 78, 236
Jacquet-Droz, Pierre, 77, 78
Jacquette, Yvonne, 270n60
James, Henry, 170
Jarches, James, 83
Jarrell, Randall, 220
Jenik, Adriene, 236
jewelry, electric, 73–74
Johnson, Amy, 151, 164
Johnson, Denys, 89

Kasson, Joy, 82
Kauffmann, Angelica, 13, 242n26
Kennard, Edward, 121
Kennard, Mary, 121, 132, 134
Kilburn, B. W., 7
Kimball, Eunice, 216, 217
King, Mrs. E. M., 107
Kipling, Rudyard, 22
Kiss, August, 19, 20, 60
Knight, Dame Laura, 168, 172, 207
Kors, Michael, xii
Kruger, Barbara, 235
Kulagina, Valentina N., xviii, 250n98
Kunzle, David, 59, 63

The Ladies Accelerator (Cruikshank; engrav-
    ing), 90–91
Ladies' Automobile Club, 116, 122
Ladies' Home Journal (magazine), 121, 135,
    140, 183, 212
Lady Cycling (Erskine), 109
Lady Cyclists' Association, 98
Lancaster, Maud, xv
Lane, Theodore, 16, 17
Lang, Fritz, xv, 74, 81, 82
Lange, Dorothea, 203
Langtry, Lillie, 65
Laver, James, 59
Law, Ruth. *See* Oliver, Ruth Law

Leaf, June, 235–36
Lefebvre, Jules-Joseph, 69
Léger, Fernand, 171
Lehman, Herman F., 228, 229
"Letters from Susan" (Farley; articles), 41
Levitt, Dorothy, 2, 123, 136–40
Lewis, George P., 194–96
Lex-Nerlinger, Alice, 247n71
Life (comic magazine), xvi, 108, 160, 186
Life (magazine), 116, 146, 173, 175, 176,
    196–98, 206, 212, 226
Light of the Home (engraving), 20
Lippard, Lucy, 178
Little Friend, Little Friend (Jarrell; anthol-
    ogy), 220
A Little Tighter (Rolandson; etching), 63
Lloyd George, David, 141
Locomotion (Seymour; etching), 14, 16, 57
Loewy, Raymond, 171
Looking Glass (magazine), x
Loutherbourg, Philip James de, 192
Love, Nancy, 164
Lowell, Francis Cabot, 38
Lowell, Orson, 64
Lowell Offering (magazine), 40–41
Lowell textile mills, 38, 40–41, 248n84
Lowry, Helen Bullitt, 9, 144
Lucid, Shannon, 181

"The Machine Aesthetic" (Léger; essay),
    171
Machine-Age Exposition (New York;
    1927), 171
"Machine Invented by an Enemy of Crino-
    line Jupons" (Daumier; lithographs), 56
machines, 60, 88, 191, 250n98; advertising
    of, xii–xiii, 20, 24, 25, 27–33; design of,
    30, 31, 226–30; and human body, 77,
    255n28; icons of, 22–23; images of, xiv,
    153, 171, 220–21; office, xviii, 1, 23–24;
    repair of, by women, 194–95; and slav-
    ery, 37, 80, 256n39; transformation of
    women by, ix, x, xv–xvii, 27, 89, 120,
    183; women as, xiv, 85, 152, 153. *See also*
    airplanes; automobiles; bicycles; rail-
    roads; sewing machines; steam engines
machine tools, 10, 187, 188, 209, 224
MacPherson, D., 192
"Madge?" (*New Yorker*; cartoon), 182
Manière d'utiliser les jupons (Daumier; litho-
    graph), 56
Mann, Harrington, 141
mannequins. *See* automatons

Man Ray, 22, 23, 86–87, 88, 231, 244n41, 258nn50,52

Marchand, Roland, 135

The March of Intellect (Heath; engraving), 57

"Mariana" (Tennyson), 13

Marilyn Monroe (Warhol; silkscreen), 231

Marilyn (Vanitas) (Flack; painting), 178

Marsh, Charlotte, 141

masks, vii. *See also* identity, female

Matisse, Henri, 170

Mauve Desert (Jenik; CD-ROM), 236

McCall's (magazine), 206

McCartney, Paul, 237

McClintock, Laura Breckinridge, 143, 144

Mechanics, Discovering Perpetual Motion (in a Wife's Tongue) (Lane; etching), 16, 17

mechanics, female, 155, 168, 187, 196, 207, 269n51

Meikle, Jeffrey, 161

Mein Kampf (Hitler), 208

Melville, Herman, xiv, 34–37, 247n77

Meredith, George, 54

Merington, Marguerite, 107, 109

Metropolis (Lang; film), xv, 74, 81, 82–83

Meyer, Dickey. *See* Chapelle, Dickey

"Milady Takes the Air" (Gould; essay), 155

military service, 141, 157, 158, 186, 196, 217. *See also* World War I; World War II

Milkman, Ruth, 222–23

Miller, J. Howard, xi, 207

Miller, Lee, 87, 258n53

Milliet, R. C., 49, 251n13

Milton, John, 81

Mind the children, finish the washing, and have dinner at 12 (stereograph), 107

"The Modern Woman" (Hoyningen-Huene; photograph), 120

ModestWitness@Second_Millennium . . . (Haraway), 237

Moisant, Matilde, 157

Monkhouse, Victoria, 192

Monroe, Harriet, 85, 257n47

Monroe, Marilyn, 178, 231, 237

Monsiau, Nicholas-Andre, 63

The Monster Lady of Crinoline at Turin (engraving), 58, 59

Montaigne, Michel E. de, 54, 252n28

Montaut, Ernest, 152, 158, 159, 267n9

Montaut, Marguerite (Gamy), 152–53, 267n9

Moore, George, 97, 102, 104

Morrison, Enoch Rice, 46

Motor (magazine), 9, 118, 123, 144, 145, 234

"Motor Racing for Women" (Pink; article), 129

Mudge, Genevera Delphine, 116

Muirheid, Mrs. (James Watt's aunt), 14

Munsey's (magazine), 97–100, 101, 105, 108, 118

Murdock, Ann, 140–41

Murray, Stella Wolfe, 172

music, 19, 27, 40–41, 82, 234, 235; sheet-music, 47, 158

Musschenbroek, Pieter van, 68

My Electric Girl (song), 82

National Youth Administration, 223, 224, 225

Naylor, Genevieve, 218–20

Nestle, Charles, 84

A New Machine for Winding Up the Ladies (etching), 64

New Method of Lacing à l'Anglaise, for Slender Waists, by Milord Bricklinghton (print), 63

Newton, Joan. *See* Cuneo, Joan Newton

New Woman, xii, 6, 101, 104, 106, 107, 261n47

The New Woman Takes Her Husband Out for a Ride (Ehrhart; stereograph), 104, 105

New Yorker (magazine), 183

Nice, Hellé (Hélène Delangle), 129–30, 265n49

Nicholls, Horace W., 2, 139, 140, 194–96

Nichols, Ruth, 149, 151

Nochlin, Linda, 66

Nollet, Abbé Jean-Antoine, 12, 68, 69

Noonan, Fred, 152

Noyes, Blanche, 149

Nye, David, 71

O'Brien, Sicéle, 157

Office of War Information (OWI), xvi, 196, 202–4, 206, 216, 217, 219, 224–25

O'Hagen, Anne, 98

Öhne Titel (Höch; photocollage), 154

O'Keeffe, Georgia, 23

Oliver, Ruth Law, 157, 162, 269n36

Oppenheim, Meret, 22, 23

Our Girls (Caine), 188–89

Ovid, xvii, 231

Owen, Janet, 209–10

OWI. *See* Office of War Information

Palmer, Alfred, xvi, 205, 219

Palmer, Minnie, 116

paper industry, 34–35, 36, 37, 247nn73,77

"Paradise Lost" (Milton), 81

The Paradise of Bachelors and The Tartarus of Maids (Melville), xiv, 34–37

"Paris Black" paintings (Spero), 178

Paris Exhibition (1889), 70

Parks, Gordon, 224–25

Parrish, Maxfield, 244n39

patriotism, 210–13, 218

Paul, Lewis, 38

Paull, E. T., 19

Paxton, Joseph, 60

Pegasus (magazine), 166, 167, 215

Penfield, Edward, 103, 104

Perot, Philippe, 59

Phillips, Coles, xii, 168

Phillips, Mrs. John Howell, 116

Philpott, Edward, 52

phonograph, 70, 74, 76, 77

phonographic dolls, 76–77, 78, 255n26

photography: of automobiles, 118–23, 139–40; of bicycles, 105–6, 111–14; fashion, 218–20; machine-age, 173; techniques of, 77, 87, 258n54; wartime, xi, 183, 186–88, 194–207, 214–15, 220; women in, 196, 198–99, 203. *See also* digital art; *and particular photographers*

Picabia, Francis, xiv, 85, 86, 257n48

Picasso (Stein), 170

pilentum. *See* bicycles: accelerator

A Pilentum or Lady's Accelerator, 90

Pink, Lady, 235

Pink, Winifred E., 129

Pittsburgh (Driggs; painting), 171

"Plane Clothes" (Earhart; article), 162–63

Pleasures of Factory Life (Bagley), 41

The Pleasures of the Rail-Road (Hughes; etching), 5, 9

poetry, 85, 169–70, 172, 220

Poetry (magazine), 85

The Poetry of Flight (Murray; anthology), 172

The Politics of Vision (Nochlin), 66

Popova, Liubov, xviii

Portrait d'une jeune fille américaine dans l'état de nudité (Picabia; drawing), 85

Portrait of Myself (Bourke-White), 174–76

Post, Daisy, 116

Post, Robert, 239n2

Powder Puff Derby. *See* Woman's Air Derby

precisionists, 170–72, 271n64

Presley, Elvis, 237

The Progress of Steam (Alken; print), xi

Pry, Paul. *See* Heath, William

Puck (comic magazine), 104, 105

Pugin, A. W. N., 78

Punch (magazine), 8, 13, 25, 47–48, 53, 54, 105, 252n32

Push Button (Rosenquist; painting), 229, 230

Pygmalion and Galatea (Gérôme; painting), 231

Pygmalion myth, xvii, 231

Queen Mary's Army Auxiliary Corps, 196

Quimby, Harriet, 151, 163

Raiche, Bessica, 157, 268n23

railroads, 5, 8–9, 14, 97, 102, 207

Ramelli, Agostino, 18

Ramsey, Alice Huyler, 124, 129

Rational Dress Society, 107

rayogrammes, 86–87, 258n50

A Reform (Seymour; lithograph), ix, x

Relyea, C. M., 100

Renoir, Pierre Auguste, 65

representations: of automobiles, 147–48; of the female, 231–32; role of, xviii, xix

The Return (Solomon; oil), 8

Rhyston mangle, 25

Richardson, Nell, 126–27

Richey, Helen, 154, 157

"Ride a Stearns and be content" (Penfield; poster), 103, 104

Rinehart, Howard, 150, 268n20

Le Rire (Bergson), 46

Ritchie, Andrew, 91

Robida, Albert, 70, 71, 84

robots. *See* automatons

Rochegrosse, Georges, 20

Rockwell, Norman, xi, 199, 208

Rogers, Will, 149, 163

Rolandson, Thomas, 63

Rosener, Ann, 202, 203, 216–17

Rosenquist, James, xiv, 219, 229, 230

Rosie the Riveter, 199, 208–9, 275n64; and Rosie the Housewife, 225–30

Rosie the Riveter (Rockwell; oil), xi, 199, 208

Royal Aero Club, 154

Royal Flying Corps, 157

Royal Ordnance Factory, 221

Ruby Loftus Screwing a Breech-ring (Knight; painting), 207

Rukeyser, Muriel, 169–70
Rupp, Leila, 204
Ruskin, John, 78

SAC. *See* Strategic Air Command
"Safe Clothes for Women War Workers"
 (Rosener; photo essay), 216
Salomon, Colette, 120
The Sandman (Hoffmann), 46
Saprelotte! si les femmes continuent . . .
 (Daumier; lithograph), 55
satires, ix, 3, 5; of men, 12; nineteenth-
 century, xi, 13; of women, 17, 53
Saturday Evening Post (magazine), xi, 131,
 133, 134, 184, 199, 208, 226
Saxon Motor Company, 126, 127
Sayer, Robert, 12
Scharff, Virginia, 123, 134
Schiebinger, Londa, 12
Schiffer, Michael, 122
Das schöne Mädchen (Höch; photocollage),
 86
Scientific American (journal), 25, 26, 31, 39,
 60, 73, 77, 184
Scott, Blanche Stuart, 124–25, 157, 159,
 268n23
Scott, John, 27
Scott, Septimus, 183
Screwing Up the Filter Press (Lewis; photo-
 graph), 195
Scribner's (magazine), 28, 103, 107, 109,
 131
The Seamstress (daguerreotype), 33
Search and Destroy (Spero; painting), 180,
 272n85
Seurat, Georges, 65–66, 67, 231
The Sewing Girl (Conant; engraving), 33
sewing machines, xviii, 19, 25, 26, 189,
 246n57, 247n71; advertisements for, 27–
 33; in art, 30; and crinolines, 251n14;
 and health, 28–29, 31; and putting-out
 system, 33, 246n69; and sweatshop conditions, 33–34, 35
sexuality, 24, 48, 62, 191, 258n48; in advertising, xii, xiii, 17–23, 31, 243n39; stereotypes of female, 17, 22, 79, 80, 83,
 243n39; and technology, 35–37, 73
Seymour, Robert, ix, x, 14, 16, 17, 57
Sguario, Eusebio, 69
SHAEF. *See* Supreme Headquarters, Allied
 Expeditionary Force
Shapiro, Meyer, 66
Shaving by Steam (Seymour; engraving), 17

Sheeler, Charles, 171, 271n64
Shelley, Mary, 12, 34
Shelley, Percy Bysshe, 12
Sherman, Cindy, 236
Shipyard Diary of a Woman Welder (Claw-
 son), 206
The Shocking Miss Pilgrim (film), 24
Sholes, Christopher Latham, 23, 24,
 244n43
Sholes, Lillian, 24, 244n44
Simians, Cyborgs, and Women (Haraway), 88
Singer, Isaac Merrit, 26
Singer Manufacturing Company, 27, 32,
 33, 246n57
Skinner, Edward F., 192–93, 208
Sobchack, Vivian, 237–38
Solomon, Abraham, 8
"The Song of the Shirt" (Hood), 27
"The Song of the Sower" (Bryant), 40, 41
Songs for A.E. (Anderson; musical piece),
 234–35
space exploration, 181
Speiser, Stuart, 178
Spencer, Thomas, 78
Spero, Nancy, 178–81
Sphere (magazine), 28, 141, 183, 184, 192,
 193, 224
spinning, 1, 38, 44, 248n84
spinning jenny, 38, 43, 247n82, 248n83
Spitfire (Flack; painting), 177–78
Standard Flying Schools, 155, 156
Starley, John Kemp, 97
steam engines, 14–17, 38; automobiles
 powered by, 2, 115, 122; and explosions,
 75–76, 82; trains powered by, 14, 102
The Steam King (print), 17
Steele, Valerie, 62
Steel in the War (U.S. Steel), 205
Stein, Gertrude, 142, 170, 266n87,
 270n59
stereographs, 6–7
Stewart, Julius L., 119
Stewart, William G., 9, 128
Stieglitz, Alfred, 23, 257n48
Stinson, Katherine, 142, 151, 157
Stinson, Marjorie, 142, 150, 151, 157, 159,
 224, 268n20
Stowe, Harriet Beecher, 229
Strategic Air Command (SAC), 176
Strauss, Malcolm, 118
Stryker, Roy E., 202
stunt fliers, 124, 142, 150–51, 155, 157,
 159–60

Summersby, Kay, 212
A Sunday on La Grande Jatte — 1884 (Seurat; painting), 65–66, 231
S.U.P.E.R.P.A.C.I.F.I.C.A.T.I.O.N. (Spero; gouache and ink), 180–81
Supreme Headquarters, Allied Expeditionary Force (SHAEF), 212
Surprised (stereographic photograph), 58
Suspended Plane (Celmins; painting), 177
Sutter, Eugene, 84
Sutton, Geoffrey, 68
Swanson, Gloria, 22

Take Off (Knight; oil), 173
Talbot, William Henry Fox, 77, 86
"Teaching Women to Fly" (Hattoom; article), 155–57
technology: anxieties about, xiii, 5, 7–9, 34, 75–76, 81; communications, 181; and the dynamo, 18; fascination with, 75, 81; fashion, xii–xiii, 45, 53, 61; and femininity, 161; and gender identity, 9, 241n11; impact of, xvii, xviii, 1; and representations of the female, 231–32; satires of, ix, x, 5, 9, 14, 16, 17, 57; and sexuality, 35–37; and war contracts, 223. *See also* digital art; machines
telegraph, 8, 19, 97, 240n6, 253n1
telephone, 8, 19, 72, 73, 97, 240n6
Tennyson, Alfred, Lord, 13
Tentamina de causa electricitatis (Windler; illustration), 68–69
textile industry, 38–43, 248nn84,86, 249nn87,88,96, 250n98; and power-loom weaving, 39, 41–42
Thaden, Louise, 149, 150, 156
"Theory of Flight" (Rukeyser), 169–70
"These Women Are Doing Their Bit" (Scott; poster), 183
This is No Courting Night (engraving), 44
Through the Windshield (Matisse; painting), 170
Thulstrup, Thure de, 101
Tibbets, Paul W., Jr., 174, 176
Tiburzi, Bonnie, 158
Time (magazine), 218
Tingle, J., 39
Tired (photograph), 1
Tissot, James, 46, 65
Titian, 6
Tito, Ettore, 133
The Toilet of Venus (Velázquez; oil), 6
Toklas, Alice B., 142

Too Early (Tissot; painting), 46
Towle, Herbert, 28, 131, 132
tractors, 141, 266n91
trade cards, 28, 30, 31, 246n64
tricycles, 94, 96, 97, 130, 259nn14,17. *See also* bicycles
Trouvé, Monsieur, 73
"The Turbine" (Monroe; poem), 85
Turning Bay (Booth; photograph), 221
typewriters, xviii, 1, 23–24, 244n43

Ullman, Marie Russell, 144
The Uncanny (Freud; essay), 46
United Airlines, 163
United States (Anderson; musical piece), 234
United States Army Air Forces, 157, 165, 169, 207, 211
Untitled (Burson; photograph), 87
Urania Coeli motus scratatur, et astra (Kauffmann and Zucchi; engraving), 13

vacuum cleaners, xv, 24, 26–27, 31, 206, 232, 245nn53,54
VAD. *See* Voluntary Aid Detachments
The Vampire (Kipling), 22
Vanderbilt, Mrs. Cornelius, 70, 72
Veblen, Thorstein, 48
Velázquez, Diego, 6
velocipedes, 91–95, 114, 259n5
Venus & Milo (Cox; animation), 231–32
Venus in Time (Cox; silver print), 232–33
Venus of Willendorf (Casidin-Silver; hologram), 231–33
La Vérité (Lefebvre; painting), 69
Vesna, Victoria, 236, 279n13
vibrators, electromechanical, 73. *See also* sexuality
La Vie électrique (Robida), 70, 84
Vietnam War, 178–79, 219, 272nn85,86
Viewing the Transit of Venus (Sayer; engraving), 12
Villemert, Boudier de, 12
Villiers de l'Isle-Adam, Auguste, 74, 76, 79–80, 87, 88, 230, 231, 255n26
Vogue (magazine), 108, 120, 130, 134, 135, 160–61, 201
Voluntary Aid Detachments (VAD), 141
Vultee Aircraft, 223

WAAC. *See* Women's Army Auxiliary Corps
WAAF. *See* Women's Auxiliary Air Force

WAFS. *See* Women's Auxiliary Ferrying Squadron
Walker, Madam C. J., 120, 121
Walpole, Horace, 64, 253n48
Wanted: Women War Workers (Office of War Information; film), 206
War Advertising Council, 204
War Artists Advisory Committee, 207
Ward, Maria, 111–13
Warhol, Andy, 231, 237
War Manpower Commission, 204
Warner, Emily Howell, 157–58
War Production Board, 212, 224
War Production Coordinating Committee, 207
"War Series" paintings (Spero), 179–80
war workers, xi, 28; advertising for, 205–9; clothing of, 216–18; demobilization of, xvi, xvii, 225, 228; ethnic diversity of, 205–6; job training for, 223. *See also* World War I; World War II
WASPs. *See* Women Airforce Service Pilots
Watt, James, 14, 15, 242n27
Watt's First Experiment of Steam (Scott; engraving), 14, 15
Wauters, Arthur, 168–69
The Weaver (Loom + Woman) (Goncharova; painting), 43
The Web (Beaton; photograph), 201
Webster, Patricia K., 121, 263n20
"We Can Do It!" (Miller; poster), xi, 207
Weguelin, Mrs. Bernard, 117, 118, 119
Weimar Republic, 153–54
What a Ridiculous Fashion! (song), 53
A Wheel within a Wheel (Willard), 113–14
When Katy the Waitress (Berlin; song), 158
When Woman Drives (Stewart; illustration), 9, 128
"Why Take a Man Along?" (McClintock; article), 143, 144
Willard, Frances E., 113–14
Williams, Mary Brush, 184–85
Windler, Peter Johann, 68, 69
Winfrey, Oprah, 237
Wolcott, Marion Post, 203
"Woman and the Bicycle" (Merington; article), 109
The Woman and the Car (Levitt), 2, 123, 136, 139
Woman at the Shelton Looms (Hine; photograph), 42, 43

"The Woman at the Wheel" (Towle; article), 28, 131
The Woman Blacksmith (Nicholls; photograph), 196
Woman Engineer (journal), 129, 157
Woman's Air Derby (Powder Puff Derby), 149, 150, 163
Woman's Christian Temperance Union, 113
Woman's Day (magazine), 146
Woman sorting tin plate at Fairfield Works (Aikins; photograph), 200
The Woman's Part (Yates), 191
women: airborne, 56–57; as allegorical figures, 13, 14, 17–23, 69–83, 87–88, 243n38, 254nn11,12; and cleanliness, xi, 101, 163, 211, 239n2; as consumers, xvii, xviii, 3, 83–84, 89, 128, 133; cyborg, 88; dualistic view of, 17, 79, 82–83, 198; electric, 85–88; facsimile, 74–83; and families, 135; farm, 217; gigantic, xiii, 57–61, 105; glamorization of, xii–xiii, 220; grounded, xii, xiii, 3, 132, 158–59, 166–69; hands of, 185, 209–10, 213–14, 230; as hindrance to mechanical invention, 16, 17; as housewives, xvi, xvii, 84, 198, 204; as industrial workers, 3, 32, 33–43; as machines, xiv, 85, 152, 153; and mirrors, xv–xvii, 6, 9, 29, 240nn2,8; power of, 256n39; as protectors, 57, 58, 61; and science, 10–13, 14, 70, 71, 241n20, 242nn21,22, 258n57; as slaves to machines, 37; and sports, 98, 110, 111, 129, 260n21; stereotyping of, 3, 10, 79, 89, 98, 106, 114, 127–35, 138–39; technical competence of, xix, 24, 185, 187, 188, 209, 210, 222–23, 226, 248n85, 270n51; technical incompetence of, 9–10, 127–33, 136, 138; timidity of, 11, 89, 114, 137, 138, 270n60
Women Airforce Service Pilots (WASPs), 157, 165, 211
"Women and Their Cars" (*Vogue*; photo essay), 130
Women and the Trades: Pittsburgh, 1907–1908 (Butler), 34
Women and War Work (Fraser), 185–86
Women at War: Engineering (Booth), 220–21
Women Charging a Gas Retort (MacPherson; drawing), 192
Women's Army Auxiliary Corps (WAAC), 141, 186

Women's Auxiliary Air Force (WAAF), 169, 207

Women's Auxiliary Ferrying Squadron (WAFS), 164, 207, 211

Women's Bureau (Department of Labor), 205, 209

Women's Flying Training Detachment, 164–65, 166, 167

Women's Land Army, 141, 217

Women's Motor Corps, 142

Women's Reserve Ambulance Corps, 141

Women's Rights: The Rehearsal (stereographic image), 6–7

Women's Royal Air Force (WRAF), 186–87

Women's Service Bureau, 193, 224, 228

Woolwich Arsenal, 189

World War I, x, xvi, 128, 140, 223; attitudes toward women after, 190–91; female workers in, 183–96; paintings in, 192–93; women automobilists in, 141–47, 266n85; women substituting for men in, 186–87, 193

World War II, x, xvi–xvii, 28, 162; aviation in, 164–70, 207, 215, 276n80; images of women in, 196–210, 219, 221, 225; job segregation during, 222; manufacturing changes in, 274n37; paintings in, 207; recruitment of women for, 204–5, 219; women instructors in, 223–24

WRAF. *See* Women's Royal Air Force

Wright, Joseph, 10–11

Wright, Orville, 150, 160

Wright, Wilbur, 162

Yates, L. K., 191

"You Are Trapped on the Earth So You Will Explode" (Holzer/Pink; montage), 235

"Your Manias Become Science" (Holzer/Pink; montage), 235

Ziegfeld, Florenz, 106

Zucchi, Joseph, 13